Lecture Notes in Computer Sci

Commenced Publication in 1973
Founding and Former Series Editors:
Gerhard Goos, Juris Hartmanis, and Jan van Leeuwen

M. Ümit Uyar
Ali Y. Duale
Mariusz A. Fecko (Eds.)

Testing
of Communicating
Systems

18th IFIP TC6/WG6.1 International Conference, TestCom 2006
New York, NY, USA, May 16-18, 2006
Proceedings

 Springer

Volume Editors

M. Ümit Uyar
The City College of The City University of New York
Electrical Engineering Department
140th Street at Convent Avenue, New York, NY 10031, USA
E-mail: umit@ccny.cuny.edu

Ali Y. Duale
IBM, Systems Assurance Kernel (SAK)
Poughkeepsie, NY 12601, USA
E-mail: duale@us.ibm.com

Mariusz A. Fecko
Telcordia Technologies, Inc.
Applied Research
One Telcordia Dr RRC-1E326, Piscataway, NJ 08854, USA
E-mail: mfecko@research.telcordia.com

Library of Congress Control Number: 2006925107

CR Subject Classification (1998): D.2.5, D.2, C.2

LNCS Sublibrary: SL 5 – Computer Communication Networks and
Telecommunications

ISSN   0302-9743
ISBN-10   3-540-34184-6 Springer Berlin Heidelberg New York
ISBN-13   978-3-540-34184-0 Springer Berlin Heidelberg New York

Springer is a part of Springer Science+Business Media

springer.com

© IFIP International Federation for Information Processing 2006
Printed in Germany

Typesetting: Camera-ready by author, data conversion by Scientific Publishing Services, Chennai, India
Printed on acid-free paper   SPIN: 11754008   06/3142   5 4 3 2 1 0

Preface

This volume contains the proceedings of the 18th IFIP International Conference on Testing Communicating Systems (TestCom 2006). It was the 18th event in a series of international workshops and conferences which started in 1989 as the International Workshop for Protocol Test Systems (IWPTS); it then became the International Workshop on Testing Communicating Systems (IWTCS) in 1997. The conference has been called TestCom since 2000.

TestCom 2006 was organized under the auspices of IFIP TC 6 WG 6.1 by the City University of New York (CUNY), in cooperation with IBM and Telcordia Technologies, Inc. The conference was held in New York City, USA, May 16–18.

The proceedings contain 23 papers that were selected from 48 submissions in a carefully designed and transparent paper selection process. TestCom 2006 consisted of 8 sessions, whose scope covered:

- Testing theory and foundations
- Testing non-deterministic and probabilistic systems
- Testing the Internet and industrial systems
- TTCN-3
- Compositional and distributed testing
- FSM-based testing and diagnosis
- Testing timed systems
- Testing for security

We would like to thank the numerous people who contributed to the success of TestCom 2006. The reviewing process involved all TPC members and a number of additional reviewers (delegates of the TPC members) who are all listed in this book. We would like to thank the local organizers for their excellent work to make the conference run smoothly: Conference Manager Edward Baurin (CCNY) for his dedicated and hard work, Joseph Driscoll (GC of CUNY) for efficiently handling the local arrangements at the Graduate Center, Connie Shao (CCNY) for flawlessly processing the registration funds, and the CUNY graduate students Samrat Batth, İbrahim Hökelek, Jianping Zou, and Constantinos Djouvas for handling the technical matters.

We would like to thank our families for their patience, especially during our evening conference calls.

May 2006   M. Ümit Uyar
                                                              Ali Y. Duale
                                                           Mariusz A. Fecko

Organization

TestCom 2006 was organized by the City College and the Graduate Center of the City University of New York (CUNY), in cooperation with IBM and Telcordia Technologies, Inc.

Program Co-chairs

M. Ümit Uyar (City University of New York, USA)
Ali Y. Duale (IBM, USA)
Mariusz A. Fecko (Telcordia Technologies, Inc., USA)

Technical Program Committee

A. Bertolino (ISTI-CNR, Italy)
G.v. Bochmann (University of Ottawa, Canada)
T. Brown (City University of New York, USA)
R. Castanet (LABRI, France)
R. Dssouli (Concordia University, Canada)
P. Frankl (Brooklyn Polytechnic, USA)
J. Grabowski (University of Goettingen, Germany)
N. Griffeth (City University of New York, USA)
R. Hierons (Brunel University, UK)
T. Higashino (Osaka University, Japan)
D. Hogrefe (University of Goettingen, Germany)
G. Holzmann (Jet Propulsion Lab, USA)
C. Jard (IRISA Rennes, France)
T. Jéron (IRISA Rennes, France)
F. Khendek (Concordia University, Canada)
M. Kim (ICU, Korea)
H. König (BTU Cottbus, Germany)
D. Lee (Ohio State University, USA)
G. Maggiore (TIM, Italy)
L. Ness (Telcordia, USA)
M. Núñez (UC de Madrid, Spain)
I. Schieferdecker (Fraunhofer Fokus, Germany)
K. Suzuki (University of Electro-Communications, Japan)
J. Tretmans (Radboud University, The Netherlands)
A. Ulrich (Siemens, Germany)
H. Ural (University of Ottawa, Canada)
M. Veanes (Microsoft, USA)
H. Yenigun (Sabanci University, Turkey)
N. Yevtushenko (Tomsk State University, Russia)

Conference Staff

Conference Manager   Edward Baurin (CCNY, USA)
Local Arrangements   Joseph Driscoll (GC of CUNY, USA)
Registration   Connie Shao (CCNY, USA)

Steering Committee

Chairman   J. Derrick (University of Sheffield, UK)
Members   A.R. Cavalli (INT, France)
                            R. Groz (LSR-IMAG, France)
                            A. Petrenko (CRIM, Canada)

Additional Referees

I. Berrada   J.R. Horgan   A. Rollet
A. Cavalli   J. Huo   K. Rowan
D. Chen   G.-V. Jourdan   F. Rubio
C. Chi   F.-C. Kuo   C. Viho
M. Ebner   K. Li   C. Werner
L. Frantzen   C.M. Lott   A. Williams
A. Gotlieb   M.G. Merayo   A. Petrenko
W. Grieskamp   F. Patrick
Y. Gurevich   I. Rodriguez

Sponsoring Institutions

International Federation for Information Processing (IFIP), Laxenburg, Austria
The Graduate Center and the City College of CUNY, New York, USA
IBM Corporation, Armonk, NY, USA
Telcordia Technologies, Inc., Piscataway, NJ, USA

Table of Contents

Session IV: TTCN-3

Session V: Compositional and Distributed Testing

Session VI: FSM-Based Testing and Diagnosis

Session VII: Timed Systems

Session VIII: Testing for Security

Symbolic Execution Techniques
for Test Purpose Definition

Christophe Gaston[1], Pascale Le Gall[2], Nicolas Rapin[1], and Assia Touil[2,*]

[1] CEA/LIST Saclay,
F-91191 Gif sur Yvette, France
{christophe.gaston, nicolas.rapin}@cea.fr
[2] Université d'Évry, IBISC - FRE CNRS 2873,
523 pl. des Terrasses F-91000 Évry, France
{legall, atouil}@lami.univ-evry.fr

Abstract. We propose an approach to test whether a system conforms to its specification given in terms of an Input/Output Symbolic Transition System (IOSTS). IOSTSs use data types to enrich transitions with data-based messages and guards depending on state variables. We use symbolic execution techniques both to extract IOSTS behaviours to be tested in the role of test purposes and to ground an algorithm of test case generation. Thus, contrarily to some already existing approaches, our test purposes are directly expressed as symbolic execution paths of the specification. They are finite symbolic subtrees of its symbolic execution. Finally, we give coverage criteria and demonstrate our approach on a running example.

Keywords: Conformance testing, Input/Output Symbolic Transition Systems, Test Purposes, Symbolic Execution, Coverage Criteria.

1 Introduction

Symbolic Transition Systems (STS) are composed of a data part and of a state-transition graph part. They specify behaviours of reactive systems with some benefits compared to the use of classical labelled transition systems. Models are often smaller and it is even possible to finitely denote systems having an infinite number of states. In this paper, following the works of [5, 11, 3], we are interested in studying conformance testing in the context of Input/Output Symbolic Transition Systems (IOSTS).

Approaches based on symbolic transformations make possible to exploit a particular analysis technique, the so-called *symbolic execution* [2, 6], to define a test selection strategy. This technique has been first defined to compute program executions according to some constraints expressed on input values. The main idea is to use symbols instead of concrete data as input values and to derive a symbolic execution tree in order to describe all possible computations in a symbolic way. In our contribution, *test purposes* are defined as some particular

---

* This work was partially supported by the RNRT French project STACS.

M.Ü. Uyar, A.Y. Duale, and M.A. Fecko (Eds.): TestCom 2006, LNCS 3964, pp. 1–18, 2006.
© IFIP International Federation for Information Processing 2006

subtrees of this symbolic execution tree. They may be chosen by the user but we also propose criteria to automatically compute tests purposes. This is a response to industrial needs where engineers are not always able to define which behaviours they want to test. We introduce two criteria. The first one is called *all symbolic behaviours of length n* criterion. The second one is called *the restriction by inclusion* criterion: the extracted subtree satisfies a coverage criterion which is based on a procedure of redundancy detection. According to these test purposes, test cases are generated. Our algorithm for test case generation is given by a set of inference rules. Each rule is dedicated to handle an observation from the system under test (SUT) or a stimulation sent by the test case to the SUT. This testing process leads to a verdict being either *PASS*, *FAIL*, *INCONC* or *WeakPASS*. *PASS* means that the SUT succeeded in passing a test. *FAIL* means that a non-conformance has been detected. *INCONC* means that conformance is observed but the test purpose is not achieved while *WeakPASS* means that we are not sure to have achieved the test purpose. This last case is essentially due to the fact that the specifications may be non-deterministic.

Our work on symbolic conformance testing is close to the ones of [3, 5]. Our contribution on generation of test cases is inspired by the one of [3]. But as the data part in [3] was only given according to a pure and abstract theoretical description, the implementation counterpart and examples are clearly missing. Associating a verdict to a test case execution requires to perform reachability analysis. Indeed, one must be able to compute as soon as possible whether or not a conformance may still be observed. In [5] over-approximation mechanisms based on abstract interpretation are used to perform reachability analysis. In our approach, we do not use such abstract interpretation techniques which have the drawbacks of both being difficult to use and of sometimes giving only approximated verdicts. We prefer to use symbolic execution based mechanisms which have been already successfully advocated in [10] to validate IOSTS models by exhibiting pertinent scenarios or deadlock situations.

The paper is structured as follow. In Section 2 we present the IOSTS formalism. Symbolic execution and restriction by inclusion are defined in Section 3. In Section 4, we present on-the-fly rules for generating test cases. In Section 5, we discuss the usage of coverage criterion for test purposes definition.

2   Input Output Symbolic Transition Systems

Input/Output Symbolic Transition Systems (IOSTS) extend Input Output Labelled Transition Systems (*IOLTS*) [10] by including data types. IOSTS are used to specify dynamic aspects. This is done by describing modifications of values associated to some variables, called *attribute variables*, in order to denote system state modifications. These modifications may be due to internal operations denoted by attribute variable substitutions or to interactions with the environment under the form of exchanges through communication channels of input/output messages. Those modifications may be conditioned by guards.

2.1 Data Types

Data types are specified with a *typed equational* specification framework.

Syntax. A data type signature is a couple $\Omega = (S, Op)$ where S is a set of type names, Op is a set of operation names, each one provided with a profile $s_1 \cdots s_{n-1} \to s_n$ (for $i \leq n$, $s_i \in S$). Let $V = \bigcup_{s \in S} V_s$ be a set of typed variable names. The set of Ω-*terms* with variables in V is denoted $T_\Omega(V) = \bigcup_{s \in S} T_\Omega(V)_s$ and is inductively defined as usual over Op and V. $T_\Omega(\emptyset)$ is simply denoted T_Ω.

A Ω-*substitution* is a function $\sigma : V \to T_\Omega(V)$ preserving types. In the following, one notes $T_\Omega(V)^V$ the set of all the Ω-substitutions of the variables V. Any substitution σ may be canonically extended to terms.

The set $Sen_\Omega(V)$ of all typed equational Ω-*formulae* contains the truth values *true*, *false* and all formulae built using the equality predicates $t = t'$ for $t, t' \in T_\Omega(V)_s$, and the usual connectives $\neg, \vee, \wedge, \Rightarrow$.

Semantics. A Ω-*model* is a family $M = \{M_s\}_{s \in S}$ with, for each $f : s_1 \cdots s_n \to s \in Op$, a function $f_M : M_{s_1} \times \cdots \times M_{s_n} \to M_s$. We define Ω-*interpretations* as applications ν from V to M preserving types, extended to terms in $T_\Omega(V)$. A model M satisfies a formula φ, denoted by $M \models \varphi$, if and only if, for all interpretations ν, $M \models_\nu \varphi$, where $M \models_\nu t = t'$ is defined by $\nu(t) = \nu(t')$, and where the truth values and the connectives are handled as usual. M^V is the set of all Ω-interpretations of V in M. Given a model M and a formula φ, φ is said *satisfiable* in M, if there exists an interpretation ν such that $M \models_\nu \varphi$.

In the sequel, we suppose that data types of our IOSTS correspond to the generic signature $\Omega = (S, Op)$ and are interpreted in a fixed model M. In the following, elements of M are called *concrete data* and denoted by terms of T_Ω.

2.2 Input/Output Symbolic Transition Systems

Definition 1 (*IOSTS*-signature). *An IOSTS-signature Σ is a triple (Ω, A, C) where Ω is a data type signature, $A = \bigcup_{s \in S} A_s$ is a set of variable names called* attribute variables *and C is a set of* communication channel *names.*

An IOSTS communicates with its environment through communication actions:

Definition 2 (Actions). *The set of communication actions, denoted $Act(\Sigma) = Input(\Sigma) \cup Output(\Sigma)$, is defined as follows, with $c \in C$, $y \in A$ and $t \in T_\Omega(A)$:*

$$Input(\Sigma) = c?y \mid c? \quad and \quad Output(\Sigma) = c!t \mid c!$$

Elements of $Input(\Sigma)$ are stimulations of the system from the environment: $c?x$ (resp. $c?$) means that the system waits on the channel c for a value that will be assigned to the attribute variable x (resp. for a signal, for example, a pressed button). $Output(\Sigma)$ are responses of the system to the environment: $c!t$ (resp. $c!$) is the emission of the value t (resp. of a message without any sensible argument) through the channel c.

Definition 3 (*IOSTS*). *An IOSTS over Σ is a triple $G = (Q, q_0, Trans)$ where Q is a set of state names, $q_0 \in Q$ is the initial state and $Trans \subseteq Q \times Act_\Sigma(A) \times Sen_\Omega(A) \times T_\Omega(A)^A \times Q$. A transition (q, act, ϕ, ρ, q') of $Trans$ is composed of a source state q, an action act, a guard φ, a substitution of variables ρ and a target state q'. For each state $q \in Q$, there is a finite number of transitions of source state q.*

Observations for a communicating system are made of output actions. However, a system cannot always emit an output message from a given state q. It is then said to be *quiescent* [13]. In particular, quiescence from q depends on the current values of the attribute variables and on the guards of all transitions outgoing from q. As in [13], we can complete an IOSTS to explicit quiescent situations. For that, we add a special output communication action $\delta!$, expressing the absence of output, whose guard is complementary to all other guards of output transitions from q. This *enrichment by quiescence* is given by:

Definition 4 (Enrichment by quiescence). *Let $G = (Q, q_0, Trans)$ be an IOSTS over $\Sigma = (\Omega, A, C)$. The enrichment of G by quiescence is the IOSTS over $\Sigma_\delta = (\Omega, A, C \cup \{\delta\})$, defined by $G_\delta = (Q \cup \{q_\delta\}, q_0, Trans \cup Trans_\delta)$ where $(q, act, \varphi, \rho, q') \in Trans_\delta$ iff:*

- *$act = \delta!$, ρ is the identity substitution and $q' = q_\delta$.*
- *Let us note $tr_1, \cdots, tr_n$ all transitions of the form $tr_i = (q, act_i, \varphi_i, \rho_i, q_i)$ with $act_i \in Output(\Sigma)$. Then φ is $\wedge_{i \leq n} \neg \varphi_i$ if $n > 0$ and is true otherwise[1].*

Example 1. Let us consider an ATM system built over the communicating automaton depicted in Figure 1. This IOSTS specifies a system of cash withdrawal, with the initial state q_0. The user asks for some amount ($amount?x$). The ATM system checks if there is enough money in the account user (represented by the variable m) and if this is the first or the second time that the user withdraws money after a deposit. Then the user receives the asked amount by the channel *cash*. If the user account is less than 1000 then the withdrawal operation is not free and costs 1. Else, if there is not enough money in the account, the user receives an error message by the channel *screen*. The user can also deposit some money (t) in his bank account by the channel *deposit*. This is added to the bank account ($m := m + t$). Moreover, the user can ask for the amount of its account by the channel *check*, and receives the answer by the channel *sum*. There is only one transition labelled by $\delta!$ starting from the state q_0. Indeed, the state q_1 and q_2 are such that whatever the values of the attribute variables are, it is always possible to emit at least a message.

---

[1] If $\wedge_{i \leq n} \neg \varphi_i$ is not a satisfiable formula, the $(q, act, \wedge_{i \leq n} \neg \varphi_i, \rho, q')$ transition may clearly be omitted.

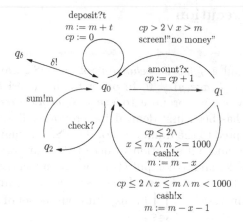

deposit?t
$m := m + t$
$cp := 0$

$cp > 2 \vee x > m$
screen!"no money"

q_δ

$\delta!$

sum!m

amount?x
$cp := cp + 1$

q_0 q_1

check?

q_2

$cp \leq 2 \wedge$
$x \leq m \wedge m >= 1000$
cash!x
$m := m - x$

$cp \leq 2 \wedge x \leq m \wedge m < 1000$
cash!x
$m := m - x - 1$

Fig. 1. Example of ATM system with withdrawal according to some conditions

2.3 Semantics

Definition 5 (Runs of a transition). *Let $tr = (q, act, \varphi, \rho, q') \in Trans$. Let us note $Act(M) = (C \times \{?, !\} \times M) \cup (C \times \{?, !\})$. The set $Run(tr) \subseteq M^A \times Act(M) \times M^A$ of execution runs of tr is such that $(\nu^i, act_M, \nu^f) \in Run(tr)$ iff:*

- *if act is of the form c!t (resp. c!) then $M \models_{\nu^i} \varphi$, $\nu^f = \nu^i \circ \rho$ and $act_M = (c, !, \nu^i(t))$ (resp. $act_M = (c, !)$),*
- *if act is of the form c?y then $M \models_{\nu^i} \varphi$, there exists ν^a such that $\nu^a(z) = \nu^i(z)$ for all $z \neq y$, $\nu^f = \nu^a \circ \rho$ and $act_M = (c, ?, \nu^a(y))$.*
- *if act is of the form c? then $M \models_{\nu^i} \varphi$, $\nu^f = \nu^i \circ \rho$ and $act_M = (c, ?)$.*

We denote $source(tr)$ (resp. $target(tr)$) the source (resp. target) state q (resp. q') and $act(tr)$ stands for act. For a run $r = (\nu^i, act_M, \nu^f)$, we denote $source(r)$, $act(r)$ and $target(r)$ respectively ν^i, act_M and ν^f.

Definition 6 (Finite Paths of an $IOSTS$). *The set of finite paths in G, denoted $FP(G)$ contains all finite sequence $tr_1 \ldots tr_n$ of transitions in $Trans$ such that $source(tr_1) = q_0$ and for all $i < n$, $target(tr_i) = source(tr_{i+1})$.*

The runs of a finite path $tr_1 \ldots tr_n$ in $FP(G)$ are sequences $r_1 \ldots r_n$ such that for all $i \leq n$, r_i is a run of tr_i and for all $i < n$, $target(r_i) = source(r_{i+1})$.

The set of concrete traces of a finite path $p = tr_1 \ldots tr_n$, denoted $Trace(p)$ is the set of finite action sequences $act(r_1) \ldots act(r_n)$ for any run $r_1 \cdots r_n$ of p.

In the following, and as usual, for any $p \in FP(G)$, $length(p)$ denotes the number of occurrences of the transitions in the definition of p. We also note $Ext_{out}(p)$ the set of finite paths of G extending p by a transition introducing an output action. Formally, $Ext_{out}(p) = \{p' \in FP(G) \mid p' = p.tr \wedge act(tr) \in Output(\Sigma)\}$.

Definition 7. *The semantics of an $IOSTS$ G is $Trace(G) = \bigcup_{p \in FP(G)} Trace(p)$.*

3 Symbolic Execution

3.1 Definition

In our context, we call a *symbolic behaviour* of an IOSTS any finite path p of this IOSTS for which $Trace(p) \neq \emptyset$. In order to characterize the set of traces of a symbolic behaviour we propose to use a *symbolic execution* mechanism. Symbolic execution has been first defined for programs [6, 2, 9]. This technique can naturally be adapted to the framework of IOSTS. The main idea is to replace concrete input values and initialization values of attribute variables by symbolic ones with fresh variables and to compute the constraints on these variables: those constraints are called *path conditions*. In the sequel we assume that those fresh variables are chosen in a set $F = \bigcup_{s \in S} F_s$ disjoint from the set of attribute variables A. We now give the intermediate definition of *symbolic extended state* which is a structure allowing to store information about a symbolic behaviour: the IOSTS current state (target state of the last transition of the symbolic behaviour), the path condition and the symbolic values associated to attribute variables.

Definition 8 (Symbolic extended state). *A symbolic extended state over F for an IOSTS $G = (Q, q_0, Trans)$ is a triple $\eta = (q, \pi, \sigma)$ where $q \in Q$, $\pi \in Sen_\Omega(F)$ is called a* path condition *and $\sigma \in T_\Omega(F)^A$. $\eta = (q, \pi, \sigma)$ is said to be satisfiable if π is satisfiable[2]. One notes $\mathcal{S}$ (resp. $\mathcal{S}_{sat}$) the set of all the (resp. satisfiable) symbolic extended states over F.*

We now define the symbolic execution of an IOSTS. Intuitively, the symbolic execution of an IOSTS can be seen as a tree whose edges are symbolic extended states and vertexes are labelled by symbolic communication actions. The root is a symbolic extended state made of the IOSTS initial state, the path condition *true* (there is no constraint to begin the execution) and of an arbitrary initialization σ_0 of variables of A in F. Vertexes are computed by choosing a source symbolic state η already computed and by symbolically executing a transition of the IOSTS whose source is the state introduced in η. The symbolic communication action is computed from the transition communication action and from the symbolic values associated to attribute variables in η. A target symbolic extended state is then computed. It stores the target state of the transition, a new path condition derived from the path condition of η and from the transition guard, and finally the new symbolic values associated to attribute variables.

Definition 9 (Symbolic execution of an IOSTS). *Let $G = (Q, q_0, Trans)$ be an IOSTS over $\Sigma = (\Omega, A, C)$. Let us note $\Sigma_F = (\Omega, F, C)$. A full symbolic execution of G over F is a triple $(\mathcal{S}, init, R)$ with $init = (q_0, true, \sigma_0)$ where σ_0 is an injective substitution in F^A and $R \subseteq \mathcal{S} \times Act(\Sigma_F) \times \mathcal{S}$ such that for any two transitions in R respectively of the form $(\eta^i, c?x, \eta^f)$ and $(\eta'^i, d?y, \eta'^f)$, the variables x and y are distinct and $\forall a \in A, \sigma_0(a) \neq x$. For any $\eta \in \mathcal{S}$ of the form*

---

[2] Let us recall that here, π is *satisfiable* if and only if there exists $\nu \in M^F$ such that $M \models_\nu \pi$ since variables of π are by construction in F.

(q, π, σ), for all $tr \in Trans$ of the form $(q, act, \varphi, \rho, q')$, there exists an unique symbolic transition $st = (\eta, sa, \eta')$ in R such that

- if $act = c!t$ (resp. $c!$), then $sa = c!\sigma(t)$ (resp. $c!$) and $\eta' = (q', \pi \wedge \sigma(\varphi), \sigma \circ \rho)$,
- if $act = c?x$ with x in A then $sa = c?z$ with z in F, and $\eta' = (q', \pi \wedge \sigma(\varphi), \sigma \circ (x \mapsto z) \circ \rho)$,
- if $act = c?$ then $sa = c?$, and $\eta' = (q', \pi \wedge \sigma(\varphi), \sigma \circ \rho)$.

The symbolic execution of G over F is the triple $SE(G) = (\mathcal{S}_{sat}, init, R_{sat})$ where R_{sat} is the restriction of R to $\mathcal{S}_{sat} \times Act(\Sigma_F) \times \mathcal{S}_{sat}$.

The trace semantics for a symbolic execution tree is defined in a natural way. If one solves the path condition of a given path (i.e. the path condition of its last state) one can then evaluate all symbolic actions labelling this path and extract the corresponding trace. Since $SE(G)$ is obtained from the symbolic execution tree of G by removing only un-solvable paths, one can easily prove that $Trace(G) = Trace(SE(G))$. Finally, since an IOSTS and its symbolic execution share the same trace semantics, it is equivalent to study an IOSTS or its symbolic execution in the context of conformance testing.

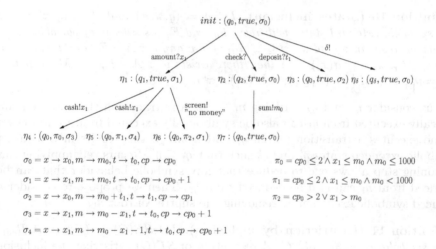

Fig. 2. Symbolic execution tree

Example 2. Figure 2 illustrates the beginning of the symbolic execution of the ATM system presented in Figure 1.

3.2   Inclusion Criterion

A reactive system is supposed to continuously interact with its environment. Thus, behaviours viewed as sequences of interactions are very often arbitrary long. It explains that IOSTS specifications of reactive systems often contain internal loops. This implies that the symbolic execution of the corresponding IOSTS has infinite paths. However, in practice, one can consider an arbitrary long behaviour as a sequence of "basic" behaviours. For example, the ATM system basically offers few basic behaviours. It may: (1) provide the user with money, (2) receive deposit from the user or (3) give the current level of the user account. Any "complex" behaviour of the ATM system can be seen as a sequence of such basic behaviours. Now if one considers the symbolic execution of the ATM system, one would observe a lot (or even an infinite number) of occurrences of those basic behaviours. In other words, information on symbolic behaviours provided by the symbolic execution may be highly redundant in terms of basic behaviours. We propose to cut the symbolic execution of an IOSTS in order to lower this redundancy. Definition 9 of symbolic execution shows that behaviours are indeed determined by states, that is why our procedure to cut the tree is grounded on a relation upon states. From a symbolic state $\eta = (q, \pi, \sigma)$ one can extract constraints on the set A of attribute variables : the set of all possible interpretations $\nu_A : A \to M$ corresponding to η are restrictions[3] to A of all interpretations $\nu : A \cup F \to M$ such that[4] $M \models_\nu \bigwedge_{x \in A}(x = \sigma(x)) \wedge \pi$. If the set of possible interpretations of A for η_1 is included in the one of η_2 one says that $\eta_1 \subseteq \eta_2$.

Definition 10 (States inclusion). *Let $\eta = (q, \pi, \sigma)$ and $\eta' = (q, \pi', \sigma')$ be two symbolic extended states with resp. F_η and $F_{\eta'}$ as subsets of variables in F occurring resp. in π and π'. $\eta \subseteq \eta'$ iff, if for any $\nu : A \cup F_\eta \to M$ such that $M \models_\nu (\bigwedge_{x \in A}(x = \sigma(x)) \wedge \pi)$ then there exists $\nu' : A \cup F_{\eta'} \to M$ such that $\nu_{|A} = \nu'_{|A}$ and $M \models_{\nu'} (\bigwedge_{x \in A}(x = \sigma'(x)) \wedge \pi')$.*

Let us consider η_1 and η_2 verifying $\eta_1 \subseteq \eta_2$. Any transition that can be symbolically executed from η_1 can also be symbolically executed from η_2. Moreover if one executes a transition t from η_1 and from η_2, this results in two target symbolic extended states η'_1 and η'_2 such that $\eta'_1 \subseteq \eta'_2$. Recursively applying this reasoning step allows one to deduce that any symbolic behaviour that can be deduced from η_1 can also be deduced from η_2. Then we propose to consider a reduced symbolic execution by removing the subtree of root η_1.

Definition 11 (Restriction by inclusion). *Let $SE(G) = (\mathcal{S}_{sat}, init, R_{sat})$ be a symbolic execution of G. A restriction of $SE(G)$ satisfying the inclusion criterion is a triple $SE(G)^\subseteq = (\mathcal{S}_{sat}{}^\subseteq, init, R_{sat}{}^\subseteq)$ where:*

- *$\mathcal{S}_{sat}{}^\subseteq \subseteq \mathcal{S}_{sat}$, $init \in \mathcal{S}_{sat}{}^\subseteq$, and $R_{sat}{}^\subseteq \subseteq R_{sat}$.*
- *For any $\eta \in \mathcal{S}_{sat}{}^\subseteq$ if there is no $(\eta, sa, \eta') \in R_{sat}{}^\subseteq$ then either there is no $(\eta, sa, \eta') \in R_{sat}$ or there exists $\eta'' \in \mathcal{S}_{sat}{}^\subseteq$ such that $\eta \subseteq \eta''$.*
- *For any $\eta \in \mathcal{S}_{sat}{}^\subseteq$, if there exists $(\eta, sa, \eta') \in R_{sat}{}^\subseteq$ then for all $(\eta, sa', \eta'') \in R_{sat}$, $(\eta, sa', \eta'') \in R_{sat}{}^\subseteq$.*

---

[3] As usual, the restriction of an application $f : X \to Y$ to a subset Z of X will be denoted by $f_{|Z}$.

[4] When reading $x = \sigma(x)$ for $x \in A$ in the formula, the reader should be aware that $\sigma(x)$ denotes in fact an expression in terms of variables of F.

Definition 11 does not require that the restriction gets a finite number of symbolic extended states: it may happen that symbolic extended states cannot be compared through $\subseteq$. However, in practice, reactive systems generally have the property that they regularly come back to already encountered states, as for example the initial state. For such systems, the restriction by inclusion of their symbolic execution generally gives a finite tree.

Example 3. Figure 2 corresponds in fact to a restriction by inclusion of the symbolic execution of the ATM system. Indeed, $\eta_4 \subseteq init$ since η_4 contains the same state q_0 as $init$ and the constraints in η_4, i.e. $\pi_0 = cp_0 \leq 2 \wedge x_1 \leq m_0 \wedge m_0 \geq 1000$, are stronger that those in $init$ (*true*). The symbolic extended states η_3, η_5, η_6 and η_7 are handled in the same way.

4 Conformance Testing for IOSTS

4.1 Our Approach

Conformance testing supposes that a formal conformance relation is given between the specification G and the system under test SUT. We propose to adapt the *ioco* relation used for example in [12]. As usual for conformance testing, we consider that the SUT is only observable by its input/output sequences. In particular, data handled in these sequences are concrete values which may be denoted by ground terms of T_Ω. By hypothesis, the SUT may be modelled as a labelled transition system for which transitions are simple emissions (output) or receptions (input) carrying concrete values. Moreover, as usual, the SUT is supposed to accept all inputs in all states. The set of traces which can be observed for the SUT, denoted by $Trace(SUT)$, is a subset[5] of $(Act(M) \cup \{(\delta, !)\})^*$. Intuitively a SUT is conform to its specification with respect to *ioco* if the reactions of the SUT are the same than those specified when it is stimulated by inputs deduced from the specification.

Definition 12. *SUT conforms to G if and only if for any $tra \in Trace(G_\delta) \cap Trace(SUT)$, if there exists $act \in Act(M) \cup \{(\delta, !)\}$ of the form $(c, !, t)$ or $(c, !)$ such that $tra.act \in Trace(SUT_\delta)$, then $tra.act \in Trace(G_\delta)$.*

Test purposes are used to select some behaviours to be tested. In our case, test purposes consist of some finite paths of the symbolic execution of the specification. For each of those paths, the last symbolic extended state is the target state of an output action and is labelled by the keyword *accept*. All states belonging to a chosen path (except the last one labelled by *accept*) are labelled by *skip*. So, a *skip* label simply means that it is still possible to reach an *accept* state by emitting or receiving additional messages . So, a test purpose is a finite subtree of the symbolic execution whose leaves are labelled by *accept* and intermediate nodes are labelled by *skip*. All other states, external to the test purpose, are

---

[5] The absence of outputs from SUT can be observed through the emission $\delta!$, and in this case, this cannot be directly followed by another emission.

labelled by $\odot$: they are not meaningful with respect to the selected paths of the test purpose.

Definition 13. *Let G be an $IOSTS$ with $SE(G_\delta) = (\mathcal{S}_{sat}, init, R_{sat})$ its associated symbolic execution. A symbolic test purpose for G is an application $TP : \mathcal{S}_{sat} \to \{skip, accept, \odot\}$ such that:*

- *there exists η verifying $TP(\eta) = accept$,*
- *for any η, η' verifying $TP(\eta) = TP(\eta') = accept$, there is no finite path $st_1 \cdots st_n$ such that for some $i \leq n$, $source(st_i) = \eta$ and $target(st_n) = \eta'$,*
- *for any η' verifying $TP(\eta') = accept$, there exists (η, sa, η') in $SE(G_\delta)$ such that sa is of the form $c!t$ or $c!.$*
- *$TP(\eta) = skip$ iff there exists a finite path $st_1 \cdots st_n$ such that for some $i \leq n$, $source(st_i) = \eta$ and $TP(target(st_n)) = accept$. Otherwise $TP(\eta) = \odot$.*

Unlike [5], our test purposes directly characterize by construction a subset of the specified behaviours since they are extracted from the symbolic execution of the specification. In the following sections, the considered test purposes will refer to an arbitrary test purpose generically denoted by TP.

4.2   Preliminary Definitions and Informal Description

A test execution consists in executing on the SUT a transition system, called a test case and devoted to produce testing verdicts as $PASS$ or $FAIL$. The test case and the SUT share the same set of channels and are synchronized by coupling emissions and receptions on a given communication channel. We focus on the sequence of data exchanged between the test case and the SUT. These data are in fact elements of M (the model of the data part) and will be denoted by ground terms of T_Ω. We use the following notations: $obs(c!t)$ with t in T_Ω to characterize that the SUT emits through the channel c the concrete value denoted t and $stim(c?t)$ to represent stimulations of the SUT, occurring when the data t is sent by the test case to the SUT. We also use the following generic notation $[ev_1, ev_2, \ldots, ev_n | Verdict]$ for a sequence of synchronized transitions between a test case and the SUT leading to the verdict $Verdict$, each action ev_i being issued either from an observation $obs(ev_i)$ or a stimulation $stim(ev_i)$.

Testing a SUT with respect to a given symbolic test purpose amounts to look for stimulating and observing the SUT in such a way that when conformity is not violated, the sequence of stimulations and observations corresponds to a trace (belonging to semantics) of at least one path of the test purpose.

To reach this goal, the testing process achieves two tasks. The first one consists in computing, each time it is required, a stimulation compatible with reaching an *accept* state. The second one consists in computing all the symbolic states which may have been reached taking into account the whole sequence of observations/ stimulations already encountered.

We firstly define *contexts* composed of a symbolic state and of a formula expressing constraints induced by the sequence of previously encountered inputs/outputs.

Definition 14 (Context). *A context is a couple (s, f) where $s \in S_{sat}$ and f is a formula whose variables are in F.*

As previously pointed out, there may be more than one single context compatible with a sequence of observations/stimulations. This is taken into account by using a set of contexts, generically noted SC (for Set of Contexts), representing the set of all potential appropriate contexts for a given sequence of stimulations/observations. We introduce some auxiliary functions useful to reason about sets of contexts, in particular in order to be able to compute the sequence of sets of contexts resulting from the successive application of elementary actions.

Definition 15 (Function $Next(ev, SC)$). *Let SC be a finite set of contexts and $ev \in Act(\Sigma_F)$. If ev is of the form $c \triangle t$ (resp. $c \triangle$) with $\triangle \in \{?, !\}$ then $(s', f') \in Next(ev, SC)$ with $s' = (q', \pi', \sigma')$ iff:*

- *there exists $(s, f) \in SC$ such that $(s, c \triangle u, s') \in R$ (resp. $(s, c \triangle, s') \in R$)*
- *f' is $f \wedge (t = u)$ (resp. f) and $f' \wedge \pi'$ is satisfiable.*

Thus, $Next(ev, SC)$ computes the set of all contexts following directly the context SC with the event ev. When stimulating the SUT, it matters to check whether the computation of a stimulation is compatible with the goal of finally reaching an *accept* state. For that, for any context ct, the $targetCond(ct)$ predicate allows us to confront constraints inherited from the first observations or stimulations to the target states, those labelled by *accept* by the test purpose.

Definition 16 ($targetCond(ct)$). *Let $ct = (s, f)$ be a context such that $TP(s) = skip$ and[6] $E = \{s' \in S_{sat} \mid \exists m \in (Act(\Sigma_F))^*, s \xrightarrow{m} s' \text{ and } TP(s') = accept\}$, then $targetCond(ct)$ is the formula :*

$$\bigvee_{(q, \pi, \sigma) \in E} \pi.$$

Given a set of contexts SC, we distinguish among all contexts in $Next(ev, SC)$ those which are pertinent with respect to the considered test purpose:

Definition 17 (Functions $NextSkip(ev, SC)$ and $NextPass(ev, SC)$). *Let SC be a finite set of contexts and $ev \in Act(\Sigma_F)$. If ev is of the form $c \triangle t$ (resp. $c \triangle$) with $\triangle \in \{?, !\}$ then $(s', f') \in NextSkip(ev, SC)$ iff:*

- *there exists $(s, f) \in SC$ such that $(s, c \triangle u, s') \in R$ (resp. $(s, c \triangle, s') \in R$) with $TP(s') = skip$*
- *f' is $f \wedge (t = u)$ (resp. f) and $f' \wedge targetCond(s')$ is satisfiable.*

$NextPass(ev, SC)$ is defined in the same way with the difference that $TP(s')$ is required to be accept instead of skip.

Let us remark that for a given symbolic state $s' = (q', \pi', \sigma')$, the predicate $targetCond(s')$ is necessarily stronger[7] than π' since by definition of symbolic

---

[6] For a labelled graph G and a word $m = a_1. \cdots . a_n$, the notation $s_0 \xrightarrow{m} s_n$ stands for any path $s_0 \xrightarrow{a_1} s_1 \cdots s_{n-1} \xrightarrow{a_n} s_n$ where each $s_i \xrightarrow{a_i} s_{i+1}$ is a transition of G.

[7] π' is said to be stronger than π iff for any interpretation ν, if $M \models_\nu \pi'$, then $M \models_\nu \pi$.

execution, the set of constraints is increasing at each new transition. Thus, we get $NextSkip(ev, SC) \subseteq Next(ev, SC)$ and $NextPass(ev, SC) \subseteq Next(ev, SC)$ for all contexts SC and events ev. Emptiness of $NextSkip(ev, SC)$ means that no more $accept$ is now reachable while non emptiness of $NextPass(ev, SC)$ means that at least an $accept$ has been reached.

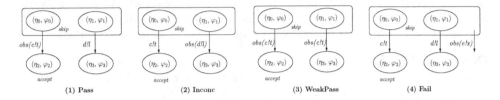

Fig. 3. Algorithm's explanations

Let us illustrate our algorithm with Figure 3 and describe an execution step based on an emission ev from the SUT and starting from $SC=\{(\eta_0, \varphi_0), (\eta_1, \varphi_1)\}$. If $Next(ev, SC)$ is empty, that is the case for $ev = e!x$, this means that the emission is not specified and so we conclude $FAIL$ (see Figure 3 (4)). If an $accept$ is reached ($NextPass(ev, SC)$ non empty) we conclude $PASS$ when no other context is reached, see for example Figure 3 (1) with $NextPass(c!t, SC) = \{(\eta_2, \varphi_2)\}$, or $WeakPASS$ when others contexts are also reached, see for example Figure 3 (3) with $Next(c!t, SC) = \{(\eta_2, \varphi_2), (\eta_3, \varphi_3)\}$. In this last case, we cannot distinguish whether the inner state of the SUT is represented by the reached $accept$ state (η_2, φ_2) or by the state (η_3, φ_3) outside of the test purpose. At last, if $NextSkip(ev, SC)$ is empty while $Next(ev, SC)$ is not, see Figure 3 (2) for $ev = d!l$, this means that the emission was specified but was not aimed by the test purpose. Then, we conclude by an inconclusive verdict $INCONC$.

4.3   Inference Rules

Let us recall that our goal is to compute sequences $[ev_1, \ldots, ev_n | Verdict]$ representing synchronized transitions between a test case and the SUT leading to the verdict $Verdict$, each action ev_i being derived either from an observation $obs(ev_i)$ or a stimulation $stim(ev_i)$, and $Verdict$ belonging to this set of keywords : $\{PASS, WeakPASS, INCONC, FAIL\}$. For that, we will take into account the knowledge of the associated contexts. Each step of the construction of such a sequence will be described by means of inference rules. Those rules are structured as follows[8] $\frac{SC}{Result}$ $cond(ev)$ where SC is a set of contexts, $Result$ is either a set of contexts or a verdict, $cond(ev)$ is a set of conditions including the observation $obs(ev)$ or the stimulation $stim(ev)$. One should read

---

[8] The initialisation rule will not respect this generic structure since it will simply consist in introducing the starting context.

a rule as follows: *Given the current set of contexts SC, if cond(ev) is verified then the algorithm may achieve a step of execution, with ev as elementary action.* As long as *Result* is a set of contexts, a new rule may be applied to pursue the computation of the sequence. And of course, reaching a verdict stops the algorithm.

Rule 0: Initialisation rule

$$\overline{\{(init, true)\}}$$

Rule 1: The emission is compatible with the purpose but no *accept* is reached.

$$\frac{SC}{Next(ev, SC)} \quad obs(ev), NextSkip\ (ev, SC) \neq \emptyset, NextPass(ev, SC) = \emptyset$$

Rule 2: The emission is not expected with regards to the specification.

$$\frac{SC}{FAIL} \quad obs(ev), Next(ev, SC) = \emptyset$$

Rule 3: The emission is specified but not compatible with the test purpose.

$$\frac{SC}{INCONC} \quad obs(ev), Next(ev, SC) \neq \emptyset, NextSkip(ev, SC) = \emptyset, NextPass(ev, SC) = \emptyset$$

Rule 4: All next contexts are *accept* ones.

$$\frac{SC}{PASS} \quad obs(ev), Next(ev, SC) = NextPass(ev, SC), Next(ev, SC) \neq \emptyset$$

Rule 5: Some of the next contexts are labelled by *accept*, but not all of them.

$$\frac{SC}{WeakPASS} \quad obs(ev), NextPass(ev, SC) \neq \emptyset, NextPass(ev, SC) \subsetneq Next(ev, SC)$$

Rule 6: Stimulation of the *SUT*

$$\frac{SC}{Next(ev, SC)} \quad stim(ev), NextSkip\ (ev, SC) \neq \emptyset$$

Rules from 1 to 5 concern observations while only Rule 6 concerns stimulations. Rule 5 calls for some comments: a verdict $WeakPASS$ means both that the test purpose is reached and that the sequence of observations/stimulations may correspond to another behaviour of the symbolic execution. This verdict is thus a kind of *warning*. One should pursue the test execution sequence to distinguish which states really correspond to the performed execution sequence.

We can consider a transition system, denoted $TS(TP)$, from a test purpose TP and the set of inference rules. The states are the sets of contexts appearing in the rules and four special states labelled by the verdicts. Two states are related by a transition labelled by an emission $ev = c!t$ or $ev = c!$ (resp. a receipt $ev = c?t$ or $ev = c?$) if they can be relied by the application of the unique rule conditioned by $stim(ev)$ (resp. of one of the rules conditioned by $obs(ev)$). Such a transition system is a simple labelled one. If such a transition system is synchronized with the system under test in such a way that emissions and receptions are synchronized by sharing the same communication channel and the same data, then any licit sequence of synchronized transitions is necessarily finite and leads to one of the four verdicts [4]. In fact, this transition system may be viewed as a test case in the sense of [4], except that our transition system may be non-deterministic. Indeed, for a given set of contexts, several rules may be applied. In particular, depending of the form of the specification, one can choose to send to the system a message, to wait for an emission or to observe quiescence. Even worse, for a given rule, several choices are often possible for the data carried by the associated observation or stimulation.

We note $st(TP, SUT)$ the set of $[ev_1, \ldots, ev_n | Verdict]$ such that $ev_1 \ldots ev_n$ is a sequence of synchronized transitions between $TS(TP)$ and SUT leading to the final state labelled by $Verdict$ in $TS(TP)$. Finally, we introduce the notation: $vdt(TP, SUT) = \{Verdict \mid \exists ev_1, \ldots ev_n, [ev1, \ldots, evn | Verdict] \in st(TP, SUT)\}$

Using these notations, we can now state the correctness and the completeness of our algorithm:

Theorem 1. *For any IOSTS G and any SUT:*

Correctness: *If SUT conforms to G, for any symbolic test purpose TP, $FAIL \notin vdt(TP, SUT)$.*
Completeness: *If SUT does not conform to G, there exists a symbolic test purpose TP such that $FAIL \in vdt(TP, SUT)$.*

The completeness property holds up to all the non-deterministic choices induced by our set of rules and captured in the set $vdt(TP, SUT)$.

5   Criterion-Based Test Purposes

Most of the times, the set of all finite symbolic behaviours associated to a specification is lucky enough to be infinite. In such a case, one generally uses coverage criteria to define test purposes.

The first idea is to simply cut the (infinite) symbolic execution of a specification according to a parameter n indicating the length of the paths to be tested. The corresponding test purpose will contain all the paths of length n derived from the symbolic execution, provided that they are terminated by an output action.

Definition 18 ("all paths of length n"). *Let G be an IOSTS on the signature Σ and let us consider $SE(G_\delta) = (\mathcal{S}_{sat}, init, R_{sat})$ its associated symbolic execution. Let $n \geq 0$. The test purpose "all paths of length n" for G is the test purpose $TG_n : \mathcal{S}_{sat} \to \{skip, accept, \odot\}$ such that the only symbolic states labelled by accept by TG_n are given by the following property. For any path $p = t_1 \cdots t_n$ of $SE(G_\delta)$ starting from init and verifying $length(p) = n$:*

– either $act(t_n) \in Output(\Sigma_\delta)$ and $TG_n(target(t_n)) = accept$,
– or for any[9] $p.t \in Ext_{out}(p)$, $label(target(t)) = accept$.

The criterion "all paths of length n" allows one to characterize a countable family of test purposes, approaching more and more the whole symbolic execution of the specification. Then, the tester can make an trade-off between the size of the test purpose (in relation with the parameter n) and the testing cost. Moreover, such a test purpose may be decomposed in as many test purposes as *accept* states: indeed, for each *accept* state, we can build a dedicated test purpose with this state as unique *accept* state. Such a decomposition allows the tester to systematically try to reach each *accept* state, thus, to reach each path of length n (up to the fact that they are not necessarily terminated by an output action).

The criterion "all paths of length n" allows us to build test purposes. However, the pertinent length n to be chosen is up to the tester. In order to help the tester to chose this parameter n, we propose to use the restriction by inclusion defined in Definition 11. This characterizes a subpart of a symbolic execution with no redundant behaviours. The inclusion criterion gives some clear indications about the size of basic behaviours of the specification. Intuitively, one can choose for the value of the parameter n the length p_{max} of the longest path of a restriction by inclusion of a symbolic execution. More generally, one can compose basic behaviours by juxtaposing them. It suffices to take for the parameter n, p_{max}, $2 \times p_{max}, \ldots$ or $k \times p_{max}$ if we want to consider all the combinations of k basic behaviours.

Definition 19 ("k-inclusion" criterion). *Let G be an IOSTS and $SE(G_\delta)^\subseteq = (S_{sat}^\subseteq, init, R^\subseteq)$ be a restriction of $SE(G_\delta)$ satisfying the inclusion criterion. Let us note $p_{max} \in FP(SE(G_\delta)^\subseteq)$ such that for all $p \in FP(SE(G_\delta)^\subseteq)$, $length(p_{max}) \geq length(p)$. Let $k > 0$. The test purpose "k-inclusion criterion" associated to $SE(G_\delta)^\subseteq$ is the test purpose "all paths of length $k \times length(p_{max})$" for G.*

Example 4. Figure 4 illustrates the construction of test purposes. The leaves of the symbolic tree constructed in Figure 2, *i.e.* the restriction by inclusion of the ATM system, are represented by a circle $\bigcirc$. This tree is completed from the symbolic state η_3 with symbolic states (that are represented by a triangle $\triangle$) to have the same length for all paths of the tree (the dotted line marks the length 2). For each leaf above the dotted line which does not result from an output, some additional outputs are considered to ensure that paths to be tested are observed by outputs. This last step introduces the states η_{11}, η_{12}, η_{13}, η_{14}, η_{15}. Finally, the states that are in a square $\square$ are those labelled with *accept*. It corresponds to the 1-inclusion criterion (for lack of space, we cannot unfold the symbolic tree until the 2-inclusion criterion but it would be the same construction). Now, we can apply the rules of our algorithm over the paths of the symbolic tree of Figure 4. We explain the computation of the final verdict by making explicit the

---

[9] Let us recall that by extension, symbolic executions of IOSTS inherit from notions associated to IOSTS: here, we have translated the notion Ext_{out} defined for IOSTS.

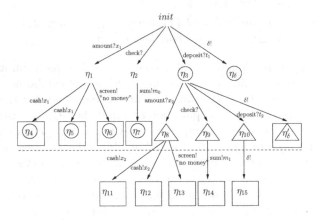

Fig. 4. Construction of test purposes

intermediate applications of rules over the current set of contexts (SC) using the following notation : $SC \xrightarrow[rule]{action} SC'$ where *action* denotes the current element of the considered trace (either of the form $c!t$, $c!$, $c?t$ or $c?$), *rule* indicates which rule is applied to get the next set of context SC'.

Let us consider the trace $[deposit?250\ amount?50\ cash!50 \mid WeakPASS]$.

$$SC_0 = \{(init, true)\} \xrightarrow[rule6]{deposit?250} SC_1 = \{(\eta_3, t_1 = 250)\} \xrightarrow[rule6]{amount?50} SC_2$$

$$SC_2 = \{(\eta_8, (t_1 = 250 \wedge x_2 = 50))\} \xrightarrow[rule5]{cash!50} WeakPASS$$

The $WeakPASS$ verdict is due to the 2 equalities:

- $Next(cash!50, SC_2) = \{(\eta_{11}, (t_1 = 250 \wedge x_2 = 50)), (\eta_{12}, (t_1 = 250 \wedge x_2 = 50))\}$
- $NextPass(cash!50, SC_2) = \{(\eta_{11}, (t_1 = 250 \wedge x_2 = 50)\}$.

One cannot decide whether the test purpose has been achieved (the real state corresponds to η_{11}) or missed (the real state corresponds to η_{12}).

5.1   Implementation Issues

The work presented here is implemented as an extension of the $AGATHA$ tool set [7, 10] which uses symbolic execution techniques to debug and validate specifications. The $AGATHA$ tool allows to unfold IOSTS specifications in the form of trees provided with path conditions for all paths of trees. Trees are computed according to coverage criteria including those grounding test purpose definitions discussed in Section 5. Those test purposes are thus obtained for free. All rules defined in Section 4.3 are implemented. However applying those rules does not necessarily lead to a deterministic process. Implementing deterministic strategies for rules appliance is still an open issue. Presburger arithmetics [8] constitutes the data part of IOSTS treated by AGATHA. The algorithm requires some decision procedures (for inclusion criterion) and constraint solving (to compute stimulations). This is done thanks to the Omega Library [1].

6 Conclusion

We have proposed an approach to test reactive systems specified as Input/ Output Symbolic Transition Systems (IOSTS). Symbolic execution allows us to re-express the specification in the form of a tree whose set of paths denotes the set of all behaviours of the specification. We propose to define test purposes by selecting a finite set of behaviours (*i.e.* paths) in the symbolic execution tree. We define an algorithm to test SUT with regard to a test purpose. This algorithm is given by a set of rules, both to compute stimulations of SUT which are adequate to achieve the test purpose and to assign a verdict to a test execution. There may be four verdicts: $PASS, FAIL, INCONC$ and $WeakPASS$. $WeakPASS$ is a verdict which expresses that conformance is observed but we are not able to ensure that the test purpose is really achieved. Indeed, it may happen that an input/output sequence observed during a test execution can be related to several behaviours, not all being accepting paths of the test purpose.

Test purposes may be defined manually but we also propose to use some coverage criteria to automatically extract them. The first one is the *all symbolic behaviours of length n* criterion which requires to cover all paths of length n in the symbolic execution. The other one, so-called *restriction by inclusion* criterion, extracts a subset of all paths of the symbolic execution tree by avoiding redundancies. Concerning coverage criteria, we are currently investigating other kinds of criteria. Our aim is to help the tester in defining test purposes. Indeed on the one hand test purposes are difficult to define manually as soon as the specification has a realistic size, but on the other hand the intervention of a human is often necessary to characterize "clever" test purposes (*i.e.* allowing to discover subtle non conformance).

References

1. Omega 1.2. The Omega Project: Algorithms and Frameworks for Analyzing and Transforming Scientific Programs. 1994.
2. L.-A. Clarke. A system to generate test data and symbolically execute programs. *IEEE Transactions on software engineering*, 2(3):215–222, September 1976.
3. L. Frantzen, J. Tretmans, and T. A.C. Willemse. Test generation based on symbolic specifications. In J. Grabowski and B. Nielsen, editors, *FATES 2004*, number 3395 in LNCS, pages 1–15. Springer-Verlag, 2005.
4. C. Jard and T. Jéron. TGV: theory, principles and algorithms, a tool for the automatic synthesis of conformance test cases for non-deterministic reactive systems. *Software Tools for Technology Transfer (STTT)*, 6, October 2004.
5. B. Jeannet, T. Jéron, V. Rusu, and E. Zinovieva. Symbolic test selection based on approximate analysis. In *11th Int. Conference on Tools and Algorithms for the Construction and Analysis of Systems (TACAS)*, volume 3440, Edinburgh, April 2005.

6. J.-C. King. A new approach to program testing. *Proceedings of the international conference on Reliable software, Los Angeles, California*, 21-23:228–233, April 1975.
7. D. Lugato, N. Rapin, and J.-P. Gallois. Verification and tests generation for SDL industrial specifications with the AGATHA toolset. In P. Petterson and S. Yovine, editors, *Proceedings of the Workshop on Real-Time Tools affiliated to CONCUR01*, Department of Information Technology UPPSALA UNIVERSITY Box 337, SE-751 05 Sweden, August 2001. ISSN 1404-3203.
8. M. Presburger. Über die Vollständigkeit eines gewissen Systems der Arithmetic. *Comptes rendus du premier Congres des Math. des Pays Slaves*, pages 92–101,395, 1929.
9. C.-V. Ramamoorthy, S.-F. Ho, and W.-T. Chen. On the automated generation of program test data. *IEEE Transactions on software engineering*, 2(4):293–300, September 1976.
10. N. Rapin, C. Gaston, A. Lapitre, and J.-P. Gallois. Behavioural unfolding of formal specifications based on communicating automata. In *Proceedings of first Workshop on Automated technology for verification and analysis*, Taiwan, 2003.
11. V. Rusu, L. du Bousquet, and T. Jéron. An approach to symbolic test generation. In *IFM '00: Proceedings of the Second International Conference on Integrated Formal Methods*, pages 338–357, London, UK, 2000. Springer-Verlag.
12. J. Tretmans. Conformance Testing with Labelled Transition Systems: Implementation Relations and Test Generation. *Computer Networks and ISDN Systems*, 29:49–79, 1996.
13. J. Tretmans. Test generation with inputs, outputs and repetitive quiescence. *Software—Concepts and Tools*, 17(3):103–120, 1996.

Controllable Combinatorial Coverage in Grammar-Based Testing

Ralf Lämmel[1] and Wolfram Schulte[2]

[1] Microsoft Corp., Webdata/XML, Redmond, USA
[2] Microsoft Research, FSE Lab, Redmond, USA

Abstract. Given a grammar (or other sorts of meta-data), one can trivially derive combinatorially exhaustive test-data sets up to a specified depth. Without further efforts, such test-data sets would be huge at the least and explosive most of the time. Fortunately, scenarios of grammar-based testing tend to admit non-explosive approximations of naive combinatorial coverage.

In this paper, we describe the notion of controllable combinatorial coverage and a corresponding algorithm for test-data generation. The approach is based on a suite of control mechanisms to be used for the characterization of test-data sets as well-defined and understandable approximations of full combinatorial coverage.

The approach has been implemented in the C#-based test-data generator *Geno*, which has been successfully used in projects that required differential testing, stress testing and conformance testing of grammar-driven functionality.

1 Introduction

This paper is about *grammar-based testing* of software. We use the term 'grammar' as a placeholder for context-free grammars, algebraic signatures, XML schemas, or other sorts of meta-data. The system under test may be a virtual machine, a language implementation, a serialization framework for objects, or a Web Service protocol with its schema-defined requests and responses. It is generally agreed that manual testing of grammar-driven functionality is quite limited. *Grammar-based test-data generation* allows one to explore the productions of the grammar and grammatical patterns more systematically. The test-oracle problem has to be addressed in one of two ways: either multiple implementations are subjected to differential testing (e.g., [20]), or the intended meaning of each test case is computed by an extra model (e.g., [23]).

Prior art in grammar-based testing uses *stochastic test-data generation* (e.g., [19, 20, 23]). The canonical approach is to annotate a grammar with probabilistic weights on the productions and other hints. A test-data set is then generated using probabilistic production selection and potentially further heuristics. Stochastic approaches have been successfully applied to practical problems. We note that this approach requires that coverage claims are based on stochastic arguments. In our experience, the actual understanding of coverage may be

M.Ü. Uyar, A.Y. Duale, and M.A. Fecko (Eds.): TestCom 2006, LNCS 3964, pp. 19–38, 2006.

challenging due to intricacies of weights and other forms of control that 'feature-interact' with the basic stochastic model.

The work reported in this paper adopts an alternative approach to test-data generation. The point of departure is full *combinatorial coverage* of the grammar at hand, up to a given depth. Without further efforts, such test-data sets would be huge at the least and explosive most of the time. Hence, approximations of combinatorial coverage are needed. To this end, our approach provides *control mechanisms* which can be used in modeling the test problem. For instance, one may explicitly limit the recursive applications for a given sort ('nonterminal')[1], and thereby scale down the 'productivity' of that sort. The control mechanisms are designed in such a way that the approximations of combinatorial coverage are intelligible. In particular, the effect of each use of a mechanism can be perceived as a local restriction on the operation for term construction.

The approach has been implemented in the C#-based test-data generator *Geno*.[2] The input language of *Geno* is a hybrid between EBNF and algebraic signatures, where constructors and sorts can be annotated with control parameters. *Geno* has been successfully used in development projects over the last 2+ years at Microsoft. These projects required differential testing, stress testing and conformance testing of grammar-driven functionality.

The paper is structured as follows. The overall approach is motivated and illustrated in Sec. 2. The basics of combinatorial test-data generation are laid out in Sec. 3 – Sec. 5. The control mechanisms are defined in Sec. 6. A grammar-based testing project is discussed in Sec. 7. Related work is reviewed in Sec. 8. The paper is concluded in Sec. 9.

2   Controllable Combinatorial Coverage in a Nutshell

The following illustrations will use a trivial expression language as the running example, and it is assumed that we want to generate test-data for testing a code generator or an interpreter. We further assume that we have access to a test-oracle; so we only care about test-data generation at this point. Using the grammar notation of *Geno*, the expression language is defined as follows:

```
Exp = BinExp ( Exp , BOp, Exp )   // Binary expressions
    | UnaExp ( UOp , Exp )        // Unary expressions
    | LitExp ( Int ) ;            // Literals as expressions

BOp = "+" ;  // A binary operator
UOp = "-" ;  // A unary operator
Int = "1" ;  // An integer literal
```

---

[1] We use grammar- vs. signature-biased terminology interchangeably. That is, we may say nonterminal vs. sort, production vs. constructor, and word vs. term.

[2] *Geno* reads as "*Gen*erate *o*bjects" hinting at the architectural property that test data is materialized as objects that can be serialized in different ways by extra functionality.

Depth	G_a	G_b	G_c	G_d
1	0	0	0	1
2	1	3	6	29
3	2	42	156	9.367
4	10	8.148	105.144	883.148.861
5	170	268.509.192	–	–
6	33.490	–	–	–
7	–	–	–	–

Fig. 1. Number of terms with the given depth for different grammars ('–' means outside the long integer range 2.147.483.647); G_a is the initial grammar from the beginning of this section; G_b comprises 3 integer literals (0, 1, 2), 2 unary operators ('+', '−'), and 4 binary operators ('+', '−', '*', '/'); G_c further adds variables as expression form along with three variable names (x, y, z); G_d further adds typical expression forms of object-oriented languages such as C#

We can execute this grammar with *Geno* to generate all terms over the grammar in the order of increasing depth. The following C# code applies *Geno* programmatically to the above grammar (stored in a file `"Expression.geno"`) complete with a depth limit for the terms (cf. 4). The `foreach` loop iterates over the generated test-data set such that the terms are simply printed.

```
using Microsoft.AsmL.Tools.Geno;

public class ExpressionGenerator {
  public static void Main (string[] args) {
    foreach(Term t in new Geno(Geno.Read("Expression.geno"), 4))
      Console.WriteLine(t);
  }
}
```

Let us review the combinatorial complexity of the grammar. We note that:

- there is no term of sort `Exp` with depth 1 (we start counting depth at 1);
- there is 1 term of sort `Exp` with depth 2: `LitExp("1")`;
- ... 2 terms ... with depth 3:
  - `UnaExp("-",LitExp("1"))`,
  - `BinExp(LitExp("1"),"+",LitExp("1"))`;
- ... 10 terms ... with depth 4;
- hence, there are 13 terms of sort `Exp` up to depth 4;
- the number of terms explodes for depth 6 — 7.

In Fig. 1, the number of terms with increasing depth is shown. We also show the varying numbers for slightly extended grammars. We note that all these numbers are about expression terms *alone*, neglecting the context in which such expressions may occur in a non-trivial language. Now suppose that we consider a

grammar which has nonterminals for programs, declarations and statements —
in addition to expressions that are used in statement contexts. With full combinatorial exploration, we cannot expect to reach expression contexts and to
explore them for some non-trivial depth.

Combinatorial coverage can be *approximated* in a number of ways. One option
is to give up on combinatorial completeness for the argument domains when
constructing terms. In particular, one could choose to exhaust the argument
domains *independently* of each other. Such an approximation is justified when
the grammar-driven functionality under test indeed processes the corresponding
arguments independently, or when the test scenario is not concerned with the
dependencies between the arguments.

In reference to pairwise testing [18] (or two-way testing), we use the term *one-
way testing* for testing argument domains independently of each other. Combinatorially exhaustive testing of arguments domains is then called *all-way testing*.
In the running example, we want to require one-way testing for the the constructor of binary expressions. Here we assume that the system under test is a simple
code generator that performs *independent* traversal on the operands of BinExp.

A *Geno* grammar can be directly annotated with control parameters:

```
Exp = [Oneway] BinExp ( Exp , BOp, Exp )
    | UnaExp ( UOp , Exp )
    | LitExp ( Int ) ;
```

Alternatively, one may also collect control parameters in a separate test specification that refers to an existing grammar. The above example is then encoded
as follows:

```
[Oneway] Exp/BinExp ;
```

Let us consider another opportunity for approximation. We may also restrict
the normal or recursive *depth* of terms on a specific argument position of a
specific constructor. By the latter we mean the number of nested applications
of recursive constructors. Such an approximation is justified when the grammar-
driven functionality under test performs only straightforward induction on the
argument position in question, or when the specific test scenario is not concerned
with that position. In the running example, we want to limit the recursive depth
of expressions used in the construction of unary expressions:

```
[MaxRecDepth = 1] Exp/UnaExp/2 ;
```

Here "2" refers to the 2nd parameter position of UnaExp. The helpful effect of the
Oneway and MaxRecDepth approximations is calculated in Fig. 2. We showcase
yet another form of control, for which purpose we need to slightly extend the
grammar for expressions. That is, we add a nonterminal, Args, for sequences of
arguments, just as in a method call.

```
Args = (Exp*) ;
```

Grammar	Depth = 1	Depth = 2	Depth = 3	Depth = 4	Depth = 5	Depth = 6
Full	0	1	2	10	170	33490
Oneway	0	1	2	5	15	45
MaxRecDepth	0	1	1	3	21	651

Fig. 2. Impact of control mechanisms on size of test-data sets; 1st row: unconstrained combinatorial coverage (same as G_a in Fig. 1); 2nd row: one-way testing for binary expressions — the resulting numbers reflect that we have eliminated the only source of explosion for this trivial grammar; 3rd row: the recursive depth for operands of unary operators is limited to 1 — explosion is slightly postponed

Now suppose that we face a test scenario such that we need to consider argument sequences of different length, say of length 0, 1 and 2. We may want to further constrain the sequences in an attempt to obtain a smaller data set or simply because we want to honor certain invariants of the grammar-driven functionality under test. Suppose that the order of arguments is irrelevant, and that duplicate arguments are to be avoided. The following annotations express these different approximation intents:

```
[MinLength = 0, MaxLength = 2, NoDuplicates, Unordered] Args ;
```

To enforce a finite data-set, we may impose a depth constraint on expressions:

```
[MaxDepth = 5] Exp ;
```

To summarize, we have illustrated several control mechanisms for combinatorial coverage. These mechanisms require that the test engineer associates approximation intents with sorts, constructors or constructor arguments. The control parameters can be injected into the actual grammar productions, and they can also be given separately.

3   Definition of Combinatorial Coverage

For clarity, we will briefly define combinatorial coverage. (We will use folklore term-algebraic terminology for some formal bits that follow.) Given is a signature Σ and a distinguished root sort, *root* (the latter in the sense of a start symbol of a context-free grammar). As common, we use $T_\Sigma(\sigma)$ to denote *the set of all ground terms* of sort σ. A *test-data set* for Σ is a subset of $T_\Sigma(root)$. (We may also consider test-data sets for sorts other than *root*, but a *complete* test datum is of sort *root*.)

We say that $T \subseteq T_\Sigma(\sigma)$ achieves combinatorial coverage up to depth d for σ if:

$$T \supseteq \{t \mid t \in T_\Sigma(\sigma), depth(t) \le d\}$$

Depth of terms is defined as follows; each term is of depth 1 at the least:

$$depth(c) = 1$$
$$depth(c(t_1, \ldots, t_n)) = max(\{depth(t_1), \ldots, depth(t_n)\}) + 1$$

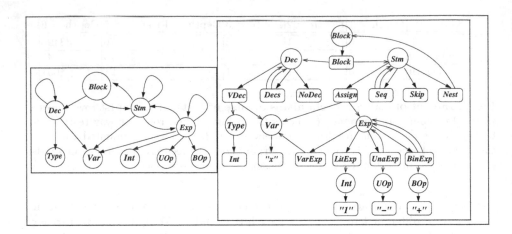

Fig. 3. Sort and constructor graphs for an imperative programming language; there are sorts for program blocks, declarations, statements, expressions, etc. with all the usual constructors; the sort graph is clearly an abstraction of the constructor graph.

It is clear that terms over a signature can be enumerated in increasing depth — the basic algorithm, given below, does just that, in a certain way. *More notation:* We use $\mathcal{T}_\Sigma^d(\sigma)$ to denote the set of all terms of sort σ *at* a given depth d, and we use $\mathcal{T}_\Sigma^{\leq d}(\sigma)$ to denote the set of all terms of sort σ *up to* a given depth d — the latter being the union over all $\mathcal{T}_\Sigma^i(\sigma)$ for $i = 1, \ldots, d$. By definition, $\mathcal{T}_\Sigma^{\leq d}(\sigma)$ is the smallest set that achieves combinatorial coverage up to depth d for sort σ.

4   Grammar Properties Related to Combinatorial Coverage

We will discuss several grammar properties in this section. They are meant to be useful for two purposes: (i) for the *implementation* of test-data generation; (ii) as a *feedback mechanism* for the test engineer who needs to understand the combinatorial complexity of a grammar in the process of modeling the test scenario.

Example. The grammar of expressions in Sec. 2 did not admit any expression terms of depth 1 since all constructors of sort Exp have one or more arguments; the minimum depth for expression terms is 2. We call this the *threshold* of a sort. Clearly, one should not attempt to specify a depth limit below the threshold.

All the properties of this section are conveniently described in terms of sort and constructor graphs, which can be derived from any grammar; cf. Fig. 3 for an illustration. The nodes in the *sort graph* are the sorts, while an edge from σ to σ' means that σ' occurs as argument sort of some constructor of sort σ. The *constructor graph* provides a more detailed view with two kinds of nodes, namely constructors and sorts. There are edges from each sort to all of its constructors. There is one edge for each argument of a constructor — from the constructor node to the node of the argument's sort. Our implementation, *Geno*, compiles the input grammar into an internal representation containing the sort and constructor graph. It uses this graph to direct the generation of objects.

Reachability of sorts from other sorts is extractable from the sort graph. Sort σ' is reachable from σ, denoted by $\rho_\Sigma(\sigma, \sigma')$, if there is a path from σ to σ'. In similarity to terminated context-free grammars, we require that all sorts are reachable from *root*, except perhaps *root* itself. Based on reachability, we can define *recursiveness* of sorts. A sort σ is recursive, denoted by $\mu(\sigma)$, if $\rho_\Sigma(\sigma, \sigma)$. (For a mutually recursive sort, there is a path through other sorts, thereby uncovering a recursive clique. For a directly recursive sort, there is a self-edge.) A stricter form of reachability is *dominance*. Sort σ dominates σ', denoted as $\delta_\Sigma(\sigma, \sigma')$, if all paths from *root* to σ' go through σ. (*root* trivially dominates every sort.) If σ' is reachable from σ, then there is a *distance* between the sorts, denoted as $\varepsilon_\Sigma(\sigma, \sigma')$, which is defined as the shortest path from σ to σ'.

Example. Suppose that the test engineer aims at combinatorial coverage of a specific sort σ up to a given depth d_σ. This implies that the root depth must be at least $d_\sigma + \varepsilon_\Sigma(root, \sigma)$. In case of explosion, the test engineer may review all dominators of σ and limit the recursive depth for them so that the sort of interest, σ, is reached more cheaply.

Using the constructor graph, we can extract the *threshold* of a sort σ, denoted as $\theta_\Sigma(\sigma)$; it is the smallest i such that $T_\Sigma^i(\sigma) \neq \emptyset$. A more specific threshold can be inquired for each constructor c as denoted by $\theta_\Sigma(c)$. The constructor graph also facilitates shortest term completions both top-down and bottom-up.

5 The Basic Algorithm for Test-Data Generation

There are two overall options for test-data generation: top-down vs. bottom-up. The top-down approach would lend itself to a recursive formulation as follows. Given a depth and a sort, the recursive function for test-data generation constructs all terms of the given sort by applying all possible constructors to all possible combinations of subterms of smaller depths; the latter are obtained through recursive calls.

In Fig. 4, we define an algorithm that adopts the *bottom-up* approach instead. This formulation is only slightly more complex than a top-down recursive formulation, while it offers one benefit. That is, an implementation (using reference semantics for terms) can *immediately* use sharing for the constructed terms; each term construction will be effectively a constant-time operation then (given that the arities of constructors are bounded). It is true that the top-down approach could employ some sort of memoization so that sharing is achieved, too. The bottom-up approach also happens to provide useful feedback to the test engineer. That is, the various sorts are inhabited in increasing depth; so "one can observe explosion", right when it happens.

Arguments
- Signature Σ with root sort, *root*
- Depth $d \geq 1$ for combinatorial coverage

Result Test-data set T that covers Σ up to depth d

Variables
- at_σ^i — terms of sort σ at depth i (i.e., $\mathcal{T}_\Sigma^i(\sigma)$)
- *kids* — an array of sets of terms for building new terms
- *len* — current length of the *kids* array

Notation
- Σ_σ — the set of constructors from Σ that are of sort σ
- $args(c)$ — the sequence of argument sorts for the constructor c
- $kids[1], kids[2], \ldots$ — array subscripting
- $\mathsf{combine}(c, kids, len)$ — build terms with constructor c and subterms from *kids*

Algorithm

```
for i = 1, . . . , d do begin              // Term construction in increasing depth
   for each σ in Σ do begin                      // Iterate over all sorts
      at_σ^i := ∅;
      if d − ε_Σ(root, σ) ≥ i then begin   // Skip depth depending on distance from root
         if i ≥ θ_Σ(σ) then begin // Skip depth if threshold has not been reached yet
            for each c in Σ_σ do begin         // Iterate over all constructors of sort
               len :=  0;
               for each a in args(c) do begin      // Iterate over all arguments of c
                  len := len + 1;
                  kids[len] :=  at_a^1 ∪ · · · ∪ at_a^{i−1};    // Determine argument terms
               end;
               at_σ^i := at_σ^i ∪ combine(c, kids, len);       // Build and store terms
            end;
         end;
      end;
   end;
end;
T := at_root^1 ∪ · · · ∪ at_root^d;                          // Compose result
```

Fig. 4. Basic algorithm for bottom-up test-data generation

We denote the combinatorial construction of terms by $\mathsf{combine}(c, kids, len)$; cf. Fig. 4. Initially, this operation calculates the *Cartesian product* over the term sets for the argument sorts of a constructor (i.e., over *kids*) — modulo a slight detail. That is, a legal combination must involve at least one term of depth $i - 1$ (as opposed to $1, \ldots, i - 2$); otherwise we were not constructing a term of depth i. Controlled combinatorial coverage caters for options other than the Cartesian product. Hence, $\mathsf{combine}(c, kids, len)$ is subject to redefinition by dependence control; cf. Sec. 6.4.

6   Control Mechanisms for Combinatorial Coverage

We will now define the mechanisms for controlling combinatorial coverage. The basic algorithm, as presented above, will only need simple and local amendments for each mechanism. The following mechanisms will be described:

- Depth control — limit depth of terms; not just for the root sort.
- Recursion control — limit nested applications of recursive constructors.
- Balance control — limit depth variation for argument terms.
- Dependence control — limit combinatorial exhaustion of argument domains.
- Construction control — constrain and enrich term construction.

Several of these mechanisms were illustrated in Sec. 2 complete with additional mechanisms for lists (cf. `MinLength`, `MaxLength`, `Unordered`, `NoDuplicates`). The latter mechanisms will not be formalized here because they are just list-specific instantiations of depth control and dependence control.

6.1   Depth Control

With d as the limit for the depth of terms of the root sort, the *depth limits* for all the other sorts are *implied*. For any given σ, the implied depth limit is $d - \varepsilon_\Sigma(root, \sigma)$, and the actual depth may actually vary per occurrence of the sort. This fact suggests a parameterization of the basic algorithm such that a depth limit, d_σ, can be supplied explicitly for each sort σ. The algorithm evolves as follows:

Before refinement

       if $d - \varepsilon_\Sigma(root, \sigma) \geq i$ then begin   // *Skip depth depending on distance from* root

After refinement

       if $d_\sigma \geq i$ then begin   // *Skip depth depending on sort-specific limit*

All but the depth limit for the root sort are considered optional. (Per notation, d becomes d_{root}.) One should notice that per-sort limits can only *lower* the actual depth limit beyond the limit that is already implied by the *root* limit. More generally, the depth limit for any sort is also constrained by its dominators. Hence, we assume that the explicit depth limits respect the following sanity check:

$$\forall \sigma, \sigma' \in \Sigma. \ \delta_\Sigma(\sigma, \sigma') \Rightarrow d_{\sigma'} \leq d_\sigma - \varepsilon_\Sigma(\sigma, \sigma')$$

Any control mechanism that works per sort, works *per argument position* of constructors, too. We can view the control-parameter value for a sort as the default for the control parameters for all argument positions of the same sort. Let us generalize control depth in this manner. Hence, we parameterize the algorithm by depth limits, $d_{c,j}$, where c is a constructor and $j = 1, \ldots, arity(c)$. The algorithm evolves as follows:

Before refinement

$$kids[len] := at_a^1 \cup \cdots \cup at_a^{i-1}; \qquad // \textit{ Determine argument terms}$$

After refinement

$$kids[len] := at_a^1 \cup \cdots \cup at_a^{min(i-1,d_{c,len})}; // \textit{ Determine argument terms}$$

We note that some argument position of a given sort may exercise a given depth, whereas others do not. This is the reason that the above refinement needs to be precise about indexing sets of terms.

6.2 Recursion Control

Depth control allows us to assign *low priority to sorts* in a way that full combinatorial coverage is consistently relaxed for subtrees of these sorts. Recursion control allows us to assign *low property to intermediary sorts* only until combinatorial exploration hits *sorts of interests*. To this end, the *recursive depth* of terms of intermediary sorts can be limited. (For simplicity, we ignore the issues of recursive cliques in the following definition.) The recursive depth of a term t for a given sort σ is denoted as $rdepth_{\Sigma,\sigma}(t)$ and defined as follows:

$$
\begin{aligned}
rdepth_{\Sigma,\sigma}(c) &= \text{if } c \in \Sigma_\sigma \text{ then } 1 \text{ else } 0 \\
rdepth_{\Sigma,\sigma}(c(t_1,\ldots,t_n)) &= \text{if } c \in \Sigma_\sigma \text{ then } 1 + ts \text{ else } ts \\
\text{where } ts &= max(\{rdepth_{\Sigma,\sigma}(t_1),\ldots,rdepth_{\Sigma,\sigma}(t_n)\})
\end{aligned}
$$

Recursion control is enabled by further parameterization of the algorithm. The limits for recursive depth amount to parameters $r_{c,j}$, where c is a constructor and $j = 1,\ldots,arity(c)$. An unspecified limit is seen as ∞. The algorithm evolves as follows:

Before refinement

$$kids[len] := at_a^1 \cup \cdots \cup at_a^{min(i-1,d_{c,len})}; // \textit{ Determine argument terms}$$

After refinement

$$kids[len] := \left\{ t \in at_a^1 \cup \cdots \cup at_a^{min(i-1,d_{c,len})} \mid rdepth_{\Sigma,a}(t) \leq r_{c,len} \right\};$$

The actual term traversals for the calculation of (recursive) depth can be avoided in an efficient implementation by readily maintaining recursive depth and normal depth as term properties along with term construction.

6.3 Balance Control

Depth and recursion control cannot be used in cases where terms of *ever-increasing depth* are needed (without causing explosion). This scenario is enabled by balance control, which allows us to limit the variation on the depth of argument terms. Normally, when we build terms of depth i, we consider argument terms of depth $1,\ldots,i-1$. An extreme limitation would be to only consider terms of depth $i - 1$. In this case, the constructed terms were balanced — hence, the name: balance control. In this case, it is also easy to see that the number of terms would only grow by a constant factor. Balance control covers the full spectrum of options — with $i - 1$ being one extreme and $1, \ldots, i - 1$ the other. We further parameterize the algorithm by limits, $b_{c,j} > 1$, where c is a constructor and $j = 1, \ldots, arity(c)$. Again, this parameter is trivially put to work in the algorithm by adapting the step for argument terms (details omitted). An unspecified limit is seen as ∞.

6.4 Dependence Control

We will now explore options for controlling combinations of arguments for forming new terms; recall the discussion of all-way vs. one-way coverage in Sec. 2. The main idea is to specify whether arguments should be varied dependently or independently.

One-Way Coverage. The completely independent exhaustion of argument domains is facilitated by a dedicated coverage criterion, which requires that each argument term appears at least once in a datum for the constructor in question; we say that $T \subseteq \mathcal{T}_\Sigma(\sigma)$ achieves *one-way coverage* of $c : \sigma_1 \times \cdots \times \sigma_n \to \sigma \in \Sigma$ relative to $T_1 \subseteq \mathcal{T}_\Sigma(\sigma_1), \ldots, T_n \subseteq \mathcal{T}_\Sigma(\sigma_n)$ if:

$$\forall i = 1, \ldots, n. \ \forall \ t \in T_i. \ \exists c(t_1, \ldots, t_n) \in T. \ t_i = t$$

We recall that one-way coverage is justified if dependencies between argument positions are not present in the system under test, or they are negligible in the specific scenario. If necessary, we can even further relax one-way coverage such that exhaustion of candidate sets is not required for specific argument positions.

Multi-way Coverage. In between all-way and one-way coverage, there is multi-way coverage reminiscent of multi-way testing (see, e.g., [9]). Classic multi-way testing is intended for testing functionality that involves several arguments. For example, *two-way* testing (or pair-wise testing) assumes that only pair-wise combinations of arguments are to be explored as opposed to all combinations. The justification for limiting combinations in this manner is that functionality tends to branch on the basis of binary conditions that refer to two arguments. In grammar-based testing, we can adopt this justification by relating to the functionality that handles a given constructor by pattern matching or otherwise. For example, some functionality on terms of the form $f(t_1, t_2, t_3)$ might perform parallel traversal on t_1 and t_2 without depending on t_3. Then, it is mandatory to exhaust combinations for t_1 and t_2, while it is acceptable to exhaust t_3 independently. Hence, we face two-way coverage for t_1, t_2 and one-way coverage for t_3.

We further parameterize the algorithm by o_c for each constructor c. The parameters affect the workings of combine($c, kids, len$). In turns out that there is a fundamental way of *specifying* combinations. Suppose, c is of arity n. A valid specification o_c must be a subset of $\mathcal{P}(\{1, \ldots, n\})$. (Here, $\mathcal{P}(\cdot)$ is the power-set constructor.) Each element in o_c enumerates indexes of arguments for which combinations need to be considered. For instance, the aforementioned example of $f(t_1, t_2, t_3)$ with two-way coverage for t_1 and t_2 vs. one-way coverage for t_3 would be specified as $\{\{1, 2\}, \{3\}\}$. Here are representative specifications for the general case with n arguments, complete with their intended meanings:

1. $\{\{1, \ldots, n\}\}$: all-way coverage.
2. $\{\{1\}, \ldots, \{n\}\}$: one-way coverage with exhaustion of all components.
3. $\emptyset$: no exhaustion of any argument required.
4. $\{\{1, 2\}, \ldots, \{1, n\}, \{2, 3\}, \ldots, \{2, n\}, \ldots, \{n-1, n\}\}$: two-way coverage.

This scheme makes sure that all forms of multi-way coverage can be specified. Also, by leaving out certain components in o_c, they will be ignored for the combinatorial exploration. The default for an unspecified parameter o_c is the full Cartesian product. We require minimality of the specifications o_c such that $\forall x, y \in o_c. \; x \not\subseteq y$. (We can remove x because y provides a stronger requirement for combination.) Options (1.)–(3.) are readily implemented. Computing minimum sets for pair-wise coverage (i.e., option (4.)), or more generally — multi-way coverage — is expensive, but one can employ efficient strategies for near-to-minimum test sets (see, e.g., [26]).

6.5   Construction Control

A general control mechanism is obtained by allowing the test engineer to customize term construction through *conditions and computations*. This mechanism provides expressiveness that is reminiscent of attribute grammars [15]. Thereby, we are able to semantically constrain test-data generation and to complete test data into *test cases* such that additional data is computed by a test oracle and attached to the constructed terms.

We require that conditions and computations are evaluated during bottom-up data generation as opposed to an extra phase. Hence, 'invalid' terms are eliminated early on — before they engage in new combinations and thereby cause explosion. The early evaluation of computations allows conditions to take advantage of the extra attributes. As an aside, we mention that some of the previously described control mechanisms can be *encoded* through construction control. For instance, we could use computations to actually compute depths as attributes to be attached to terms, while term construction would be guarded by a condition that enforced the depth limit for all sorts and constructor arguments. A *native* implementation of depth control is simply more efficient.

We associate conditions and computations to constructors. Given a condition (say, a predicate) p_c for a constructor c, both of arity n, term construction $c(x_1, \ldots, x_n)$ is guarded by $p_c(x_1, \ldots, x_n)$. Given a computation (say, a function) f_c for a constructor $c : \sigma_1 \times \cdots \times \sigma_n \to \sigma_0$ is of the following type:

$$f_c : (\sigma_1 \times A_{\sigma_1}) \times \cdots \times (\sigma_n \times A_{\sigma_n}) \to A_{\sigma_0}$$

Here, A_σ is a domain that describes the attribute type for terms of sort σ. The function observes argument terms and attributes, and computes an attribute value for the newly constructed term. This means that we assume purely synthesized attribute grammars because immediate completion of computations and conditions is thereby enabled. Hence, no expensive closures are constructed, and both conditions and computation may effectively narrow down the combinatorial search space. For brevity, we do not illustrate attributes, but here are some typical examples:

- Expression types in the sense of static typing.
- The evaluation results with regard to some dynamic semantics.
- Complexity measures that are taken into account for combination.

There is one more refinement that increases generality without causing overhead. That is, we may want to customize term construction such that the proposed candidate is replaced by a different term, or by several terms, or it is rejected altogether. This provides us with the following generalized type of a conditional computation which returns a set of attributed terms:

$$f_c : (\sigma_1 \times A_{\sigma_1}) \times \cdots \times (\sigma_n \times A_{\sigma_n}) \to \mathcal{P}(\sigma_0 \times A_{\sigma_0})$$

Geno — our implementation of controllable combinatorial coverage — also provides another form of computations: one may code extra passes over the generated object structures to be part of the serialization process of the in-memory test data to actual test data. Both kinds of computations (attribute grammar-like and serialization-time) are expressed as functions in a .NET language.

7   Testing an Object Serialization Framework

The described grammar-based testing approach has been applied in the mean time to a number of problems, in particular, to differential testing, stress testing and conformance testing of language implementations and virtual processors (virtual machines). *Geno* has been used to generate test-data from problem-specific grammars for Tosca [25], XPath [28], XML Schema [29], the Microsoft Windows Card runtime environment [11], the Web Service Policy Framework [22], and others. Measurements for some applications of *Geno* are shown in Fig. 5.

We will now discuss one grammar-based testing project in more detail. The project is concerned with testing a framework for object serialization, i.e., a framework that supports conversion of in-memory object instances into a form that can be readily transported over the network or stored persistently so that these instances can be de-serialized at some different location in a distributed system, or at some later point in time. The specific technology under test is 'data contracts' as part of Microsoft's WCF. This framework allows one to map classes (CLR types) to XML schemas and to serialize object instances as XML. Data contracts also support some sort of loose coupling.

The overall *testing problem* is to validate the proper declaration of data contracts by CLR types, the proper mapping of CLR types (with valid data contracts) to XML schemas, the proper serialization and de-serialization of object

Status	Grammar	Depth	Time	Terms	Memory
Uncontrolled	WindowsCard	5..	0.05	7.657	1.489.572
	WS Policy	5	1.57	313.041	41.121.608
	Tosca	4	0.08	27.909	2.737.204
	XPath	2	0.09	22.986	2.218.004
Controlled	Tosca	8	0.14	42.210	5.669.616
	Data Contract	6	22.33	2.576.177	365.881.216

Fig. 5. Measuring some applications of _Geno_. Runtime is in seconds, generation time on a Compaq OPTIPLEX GX280, Pentium 4, 3.2 Ghz, 2 Gigabyte of memory. Memory consumption is in bytes. Column 'Terms' lists the number of different terms for the root sort. The _'uncontrolled' measurements_ combinatorially exhaust the grammar, except that the length of lists must be in the range 0,1,2. The maximum depth before proper explosion ('out of memory') is shown. In the WindowsCard case, the test set is actual finite; so we write "5.." to mean that test-data generation has converged for depth 5. The depth for Tosca is insufficient to explore expression forms in all possible contexts. The depth for XPath indicates that control is indispensable for generating non-trivial selector expressions. The _'controlled' measurements_ take advantage of problem-specific grammar annotations. In the case of Tosca, the corresponding test-data set achieves branch-coverage of a reference implementation. In the case of Data Contract, all essential variation points of the serialization framework are exercised for up to three classes with up to three fields each, complete with the necessary attributes and interface implementations.

instances including round-tripping scenarios. (There are also numerous requirements regarding the expected behavior in case of invalid schemas or CLR types.) Essentially, _Geno_ is used in this project to generate classes like the following:

```
[DataContract]
public class Car : IUnknownSerializationData {

  [DataMember]
  public string Model;

  [DataMember]
  public string Maker;

  [DataMember(Name="HP", VersionAdded=2, IsOptional=true)]
  public int Horsepower;

  public virtual UnknownSerializationData UnknownData {
      get { ... } set { ... } // omitted
  }
}
```

In these classes, specific custom attributes are used to inform the serialization framework. The `DataContract` attribute expresses that the class can be serialized. Likewise, fields and properties are tagged for serialization using the

DataMember attribute. There is a default mapping from CLR names to XML names, but the name mapping can be customized; see the attribute Name="HP". There are several other attributes and features related to versioning and loose coupling; cf. the implementation of IUnknownSerializationData which supports round-tripping of XML data that is not understood by a given CLR type.

The project delivered 7 *Geno* grammars for different validation aspects and different feature sets. The baseline grammar, from which all other grammars are derived by slight extensions, factoring and annotation has 21 nonterminals and 34 productions ("alternatives"). Eventually, these grammars generated about 200.000 well justified test cases. As shown in Fig. 5, *Geno* scales well for grammars of this size. We have also tried to use state-of-the-art test-data generation techniques such as Korat [5], AsmL-Test tool [10] or Unit Meister [27]. However these techniques were not able to cope with the complexity of the serialization problem. (We continue this discussion in the related work section.) The combinatorial search space is due to class hierarchies with multiple classes, classes with multiple fields, various options for custom attributes, different primitive types, potentially relevant interface implementations, etc. The *Geno*-generated test cases uncovered around 25% of all filed bugs for the technology.

8   Related Work

Coverage Criteria for Grammars. Controlled combinatorial coverage is a coverage criterion for grammars, which generalizes on other such criteria. Purdom devised a by-now folklore algorithm to generate a small set of short words from a context-free grammar where each production of the grammar is used in the derivation of at least one word [21], giving rise to *rule coverage* as a coverage criterion. The first author (Lämmel) generalized rule coverage such that all the different occurrences of a nonterminal are distinguished [16] — denoted as *context-dependent rule coverage* (and *context-dependent branch coverage* for EBNF-like expressiveness). Harm and Lämmel defined a simple, formal framework based on *regular path expressions on derivation trees* that can express various grammar-based coverage criteria including rule coverage and context-dependent rule coverage [17]. The same authors also designed a coverage notion for attribute-grammar-like specifications, *two-dimensional approximation coverage*, using recursive depth in the definition of coverage [12]. Controlled combinatorial coverage properly generalizes the aforementioned coverage criteria by integrating depth of derivation, recursive depth of derivation, the dichotomy one-way, two-way, multi-way, all-way as well as the point-wise specification of these controls per sort, per constructor or even per constructor argument.

Grammar-Based Testing. Maurer designed a general grammar-based test-data generator: DGL [19]. The grammar notation is presumably the most advanced in the literature. Productions are associated with weights, but there also features for actions, systematic enumeration, ordered selection of alternatives, and others. McKeeman described differential testing for compilers and potentially other grammar-driven functionality [20], while test-data generation is accomplished by a 'stochastic grammar'. (Differential testing presumes the availability of multiple implementations whose behavior on a test datum can be compared such that a discrepancy reveals a problem with at least one of the implementations.) Slutz used a similar stochastic approach for differential testing of SQL implementations and databases, even though the grammar knowledge is concealed in the actual generator component [24]. Sirer and Bershad tested Java Virtual machines [23] using 'production grammars' that involve weights as well as guards and actions in order to control the generation process. The weights are actually separated from the grammar so that the grammar may be used in different configurations. This project did not use differential testing but more of a model-based approach. That is, an executable specification of the meaning of the generated JVM byte-code sequences served as an oracle for testing JVM implementations. Claessen and Hughes have delivered a somewhat grammar-based testing approach for Haskell [8], where programmers are encouraged to annotate their functions with properties which are then checked by randomized test data. The approach comprises techniques for the provision of test-data generators for Haskell types including algebraic data types ('signatures'). Again, constructors are associated with probabilistic weights.

Testing Hypotheses. The seminal work on testing hypotheses by Gaudel et al. [4, 2] enables rigorous reasoning about the completeness of test-data sets. Our control mechanisms are well in line with this work. Most importantly, depth control corresponds to a form of a *regularity hypothesis*, which concerns the complexity of data sets. That is, suppose we have a model m and an implementation i, both given as functions of the same type over the same signature, be it $m, i : \mathcal{T}_\Sigma(root) \to r$, where the result type r admits intensional equality. We say that i correctly implements m under the regularity hypothesis for sort *root* and depth d if we assume that the following implication holds:

$$(\forall t \in \mathcal{T}^1_\Sigma(root) \cup \cdots \mathcal{T}^d_\Sigma(root).\ m(t) = i(t)) \implies (\forall t \in \mathcal{T}_\Sigma(root).\ m(t) = i(t))$$

Hence, *any use of a control mechanism for depth or recursive depth for either sorts or constructors or constructors arguments can be viewed as an expression of a regularity hypothesis.* However, our approach does not presume that the complexity measure for regularity is a property of the grammar of even the sort; thereby we are able to express very fine-grained regularity hypotheses, as necessary for practical problems. Dependence control does not map to regularity hypotheses; instead it maps to *independence hypotheses* as common in classic multi-way testing. So our approach integrates common kinds of hypotheses for use in automated grammar-based testing.

Symbolic and Monitored Execution. Our approach does not leverage any sort of existing model or reference implementation for test-data generation. By contrast, approaches based on symbolic execution or execution monitoring support the derivation of test data from models or implementations. For instance, the Korat framework for Java [5] is capable of generating systematically all non-isomorphic test cases for the arguments of the method under test — given a bound on the size of the input. To this end, Korat uses an advanced backtracking algorithm that monitors the execution of predicates for class invariants, and makes various efforts to prune large portions of the search space. This technique is also embodied in the AsmL-Test tool [10]. Even more advanced approaches use symbolic execution and constraint solving for the purpose of test-data generation [14, 27]. Approaches for execution monitoring and symbolic execution can be very efficient when small intricate data structures need to be generated. An archetypal example of a system under test is a library for AVL trees. These approaches commit to a 'small scope hypothesis' [1, 13], assuming that a high portion of bugs can be found by testing the program for all test inputs within a small scope. (In an OO language, a small scope corresponds to a small number of conglomerating objects.) Hence, these techniques do not scale for the 'large or huge scopes' needed for testing grammar-driven functionality, as discussed in Sec. 7.

9 Concluding Remarks

Summary and Results. Testing language implementations, virtual machines, and other grammar-driven functionality is a complexity challenge. For instance, highly optimized implementations of XPath (the selector language for XML) execute different branches of code depending on selector patterns, the degree of recursion, the use of the reverse axis and the state of the cache. In this context, it is important to automate testing and to enable the exploration of test data along non-trivial complexity metrics such as deep grammar patterns and locally exhaustive combinations. We have described an approach to test-data generation for grammar-based testing of grammar-driven functionality. This approach has been implemented in a tool, *Geno*, and validated in software development projects. The distinguished characteristics of the approach is that test data is generated in a combinatorially exhaustive manner modulo approximations defined by the test engineer. It is indispensable that approximations can be expressed: test engineers can generate test cases that focus on particular problematic areas in language implementations like capacity tests, or the interplay between loading, security permissions and accessibility. We contend that the approach is very powerful, and we have found that test-data generation is unprecedentedly efficient because of the possibility of a backtracking-free bottom-up algorithm that cheaply allows for maximum sharing and semantic constraint checking. Of course, test-data generation is only one ingredient of a reasonable test strategy (others are: grammar development, test oracle, test-run automation), but doing test-data generation systematically and efficiently is beneficial.

Whether or Not to Randomize. Randomized test data generation is well established [3, 7] in testing, in general, and in grammar-based testing, in particular.

The underlying assumption is that the resulting test sets — if large enough — will include all 'interesting cases'. In grammar-based testing, randomized test data generation is indeed prevalent, but we fail to see that this approach would be as clear as ours when it comes to reasoning about testing hypotheses [4, 2], which are crucial in determining appropriateness of test sets. We contend that the weights, which are typically associated with grammar productions, end up fulfilling two *blurred* roles: (i) they specify the relative frequency of an alternative *and* (ii) they control termination of recursive deepening. Instead, controlled combinatorial coverage appeals to hypotheses for regularity and subtree independence by providing designated control mechanisms. Users of *Geno* have expressed that they would like to leverage weights as an *additional* control mechanism, very much in the sense of (i), and we plan to provide this mechanism in the next version. In fact, it is a trivial extension as opposed to the dual marriage: adding systematic test-data generation to a randomized setup is complicated [19, p.54] implementation-wise, and its meaning is not clear either. In our case, weights essentially define filters on subtree combinations.

Future Work. *Geno* and other tools for grammar-based testing are batch-oriented: the test engineer devises a grammar and test-data generation is initiated in the hope that it terminates (normally). The actual generation may provide little feedback to the test engineer; refinement of generator grammars requires skills and is tedious. We envisage that the expression of testing hypotheses could be done more interactively. To help in this process, a generator should provide feedback such that it reports (say, visualizes) thresholds, distances and explosive cliques in the grammar. (Some ideas have been explored in an experimental extension of *Geno* [30].) A testing framework could also help in taking apart initial test scenarios and then managing the identified smaller scenarios.

Another important area for improvement is the handling of problem-specific identifiers in test-data generation. (Think of variable names in an imperative language.) In fact, this issue very much challenges the generation of statically correct test data. There exist pragmatic techniques in the literature on compiler testing and grammar-based testing; see, e.g., [6, 19, 12]. *Geno* users are currently advised to handle identifiers during test-data serialization in ad-hoc manner. That is, a generator grammar only uses placeholders. Actual identifier generation and the establishment of use-def relationships must be coded in extra strategies that are part of the serialization process. We contend that the overall topic of *general, declarative and efficient identifier handling* deserves further research. For some time in the past, we were hoping that symbolic execution of attribute grammars, as in [12], potentially involving constraint solving, would be a solution to that problem, but its scalability is not acceptable as far as we know of.

Acknowledgments. We are grateful for contributions by Vadim Zaytsev and Joe Zhou. We also acknowledge discussions with Ed Brinksma at an earlier stage of this research. The TestCom 2006 referees have made several helpful proposals.

References

1. A. Andoni, D. Daniliuc, S. Khurshid, , and D. Marinov. Evaluating the "Small Scope Hypothesis". Unpublished; Available at http://sdg.lcs.mit.edu/ publications.html, Sept. 2002.

2. G. Bernot, M. C. Gaudel, and B. Marre. Software testing based on formal specifications: a theory and a tool. *Software Engineering Journal*, 6(6):387–405, 1991.

3. D. L. Bird and C. U. Munoz. Automatic generation of random self-checking test cases. *IBM Systems Journal*, 22(3):229–245, 1983.

4. L. Bouge, N. Choquet, L. Fribourg, and M.-C. Gaudel. Test sets generation from algebraic specifications using logic programming. *Journal of Systems and Software*, 6(4):343–360, 1986.

5. C. Boyapati, S. Khurshid, and D. Marinov. Korat: automated testing based on java predicates. In *Proc. International Symposium on Software testing and analysis*, pages 123–133. ACM Press, 2002.

6. C. Burgess. The Automated Generation of Test Cases for Compilers. *Software Testing, Verification and Reliability*, 4(2):81–99, June 1994.

7. C. J. Burgess and M. Saidi. The automatic generation of test cases for optimizing Fortran compilers. *Information and Software Technology*, 38(2):111–119, Feb. 1996.

8. K. Claessen and J. Hughes. QuickCheck: a lightweight tool for random testing of Haskell programs. In *ICFP '00: Proceedings of the fifth ACM SIGPLAN international conference on Functional programming*, pages 268–279, New York, NY, USA, 2000. ACM Press.

9. D. Cohen, S. Dalal, M. Fredman, and G. Patton. The AETG system: An approach to testing based on combinatorial design. *IEEE Transactions on Software Engineering*, 23(7):437–443, July 1997.

10. Foundations of Software Engineering, Microsoft Research. AsmL — Abstract State Machine Language, 2005. http://research.microsoft.com/fse/AsmL/.

11. Y. Gurevich and C. Wallace. Specification and Verification of the Windows Card Runtime Environment Using Abstract State Machines. Technical report, Microsoft Research, Feb. 1999. MSR-TR-99-07.

12. J. Harm and R. Lämmel. Two-dimensional Approximation Coverage. *Informatica*, 24(3):355–369, 2000.

13. D. Jackson and C. A. Damon. Elements of Style: Analyzing a Software Design Feature with a Counterexample Detector. *IEEE Transactions on Software Engineering*, 22(7):484–495, 1996.

14. J. C. King. Symbolic execution and program testing. *Communications of the ACM*, 19(7):385–394, July 1976.

15. D. Knuth. Semantics of context-free languages. *Mathematical Systems Theory*, 2:127–145, 1968. Corrections in 5:95-96, 1971.

16. R. Lämmel. Grammar Testing. In H. Hussmann, editor, *Proc. of Fundamental Approaches to Software Engineering (FASE) 2001*, volume 2029 of *LNCS*, pages 201–216. Springer-Verlag, 2001.

17. R. Lämmel and J. Harm. Test case characterisation by regular path expressions. In E. Brinksma and J. Tretmans, editors, *Proc. Formal Approaches to Testing of Software (FATES'01)*, Notes Series NS-01-4, pages 109–124. BRICS, Aug. 2001.

18. Y. Lei and K.-C. Tai. In-parameter-order: A test generation strategy for pairwise testing. In *HASE*, pages 254–261. IEEE Computer Society, 1998.

19. P. Maurer. Generating test data with enhanced context-free grammars. *IEEE Software*, 7(4):50–56, 1990.

20. W. McKeeman. Differential testing for software. *Digital Technical Journal of Digital Equipment Corporation*, 10(1):100–107, 1998.
21. P. Purdom. A sentence generator for testing parsers. *BIT*, 12(3):366–375, 1972.
22. J. Schlimmer et al. Web Services Policy Framework, Sept. 2004. Available at http://www-128.ibm.com/developerworks/library/specification/ws-polfram/.
23. E. G. Sirer and B. N. Bershad. Using production grammars in software testing. In USENIX, editor, *Proceedings of the 2nd Conference on Domain-Specific Languages (DSL '99), October 3–5, 1999, Austin, Texas, USA*, pages 1–13, Berkeley, CA, USA, 1999. USENIX.
24. D. Slutz. Massive Stochastic Testing for SQL. Technical Report MSR-TR-98-21, Microsoft Research, Redmond, 1998. A shorter form of the paper appeared in the Proc. of the 24th VLDB Conference, New York, USA, 1998.
25. S. Stepney. *High Integrity Compilation: A Case Study*. Prentice Hall, 1993.
26. K. Tai and Y. Lei. A Test Generation Strategy for Pairwise Testing. *IEEE Transactions on Software Engineering*, 28(1):109–111, 2002.
27. N. Tillmann, W. Schulte, and W. Grieskamp. Parameterized Unit Tests. Technical report, Microsoft Research, 2005. MSR-TR-2005-64; also appeared in FSE/ESEC 2005.
28. W3C. XML Path Language (XPath) Version 1.0, Nov. 1999. http://www.w3.org/TR/xpath.
29. W3C. XML Schema, 2000–2003. http://www.w3.org/XML/Schema.
30. V. V. Zaytsev. Combinatorial test set generation: Concepts, implementation, case study. Master's thesis, Universiteit Twente, Enschede, The Netherlands, June 2004.

A Logic for Assessing Sets of Heterogeneous Testing Hypotheses*

Ismael Rodríguez, Mercedes G. Merayo, and Manuel Núñez

Dept. Sistemas Informáticos y Programación,
Universidad Complutense de Madrid, 28040 Madrid, Spain
isrodrig@sip.ucm.es, mgmerayo@fdi.ucm.es, mn@sip.ucm.es

Abstract. To ensure the conformance of an *implementation under test* (IUT) with respect to a specification requires, in general, the application of an infinite number of tests. In order to use finite test suites, most testing methodologies add some feasible hypotheses about the behavior of the IUT. Since these methodologies are designed for considering a *fix* set of hypotheses, they usually do not have the capability of dealing with other scenarios where the set of assumed hypotheses varies. We propose a logic to infer whether a set of *observations* (i.e., results of test applications) allows to claim that the IUT conforms to the specification *if* a specific set of hypotheses (taken from a repertory) is assumed.

1 Introduction

The time a tester can spend testing an IUT with respect to a specification is finite, whereas IUTs define, in general, arbitrarily long behaviors. Hence, it takes infinite time to assess the validity of all these behaviors with respect to a specification. In order to overcome this problem, testers add some reasonable assumptions about the implementation regarding the knowledge about its construction. For example, the tester can suppose that the implementation can be represented by means of a deterministic finite state machine, that it has at most n states, etc. A lot of testing methodologies have been proposed which, for a specific set of initial hypotheses, guarantee that a test suite extracted from the specification is correct and complete to check the conformance of the IUT with respect to the specification (e.g. [2, 8, 15, 12]).

However, a framework of hypotheses established in advance is very strict and limits the applicability of a specific testing methodology. For example, in a concrete environment, the tester could assume that the behavior in four specific states of the implementation is deterministic and that two of them represent equivalent states of the implementation. The tester could also make more complex assumptions such as *"non-deterministic states of the IUT cannot show outputs that the machine did not show once the state has been tested 100 times."* In

---

* Research partially supported by the Spanish MCYT project TIC2003-07848-C02-01, the Junta de Castilla-La Mancha project PAC-03-001, and the Marie Curie project MRTN-CT-2003-505121/TAROT.

M.Ü. Uyar, A.Y. Duale, and M.A. Fecko (Eds.): TestCom 2006, LNCS 3964, pp. 39–54, 2006.

a different scenario the tester could not believe this but think that "*if she observes two sequences of length 200 and all their inputs and outputs coincide then they actually traverse the same IUT states.*" If the tester assumes the validity of a set of hypotheses to test a given IUT, then a specific test suite would be appropriate, while by using other hypotheses, the test suite could not be so.

It would be desirable to provide the tester with a tool to let her analyze the impact of considering a given set of hypotheses in the testing process, as well as the consequences of adding/eliminating hypotheses from the set. The goal of this methodology would be to ascertain if a given finite set of observations extracted by a test suite is *complete* in the case that the considered hypotheses hold, that is, we assess whether obtaining these observations from the IUT implies that the IUT conforms to the specification *if* the hypotheses hold. In this paper we propose a *logic* called $\mathcal{HOTL}$ (*Hypotheses and Observations Testing Logic*). Its aim is to assess whether a given set of observations implies the correctness of the IUT under the assumption of a given set of hypotheses. In order to allow the tester to compose sets of hypotheses, the logic provides a repertoire of hypotheses, including some of the ones appearing in known testing methodologies.

Our logic allows to perform at least *three* different tasks. First, a tester can use it to customize the testing process to her specific environment. By using the logic, she can infer not only the consequences of adding a new test, but also the consequences of adding a new hypothesis. In this way, the tester has control over a wide range of testing variables. In particular, the construction of test suites to extract observations and the definition of hypotheses can influence each other. This provides a dynamic testing scenario where, depending on the specification and the tester's knowledge of the IUT, different sets of tests and hypotheses can be considered. Second, such logic allows the tester to evaluate the *quality* of a test suite to discover errors in an implementation: If the observations that could be extracted by the test suite require (for their completeness) a set of hypotheses that is *harder* to be accepted than those required by another suite, then the latter suite should be preferred. This is because this suite could allow the tester to reach diagnostics in a less restrictive environment. Finally, our logic provides a conceptual bridge between different testing approaches. In particular, we may use it to represent the (fix) sets of hypotheses considered by different approaches. Then, by considering the observations each test suite could obtain, a test suite that is complete in an approach could be turned into a complete suite in another. Similarly, we can analyze how the size of test suites is affected by hypotheses. Moreover, we can use the logic to create intermediate approaches where sets of hypotheses are appropriately mixed.

Let us concentrate on how our logic is applied to perform the first of the previous tasks, that is, serving as core of a (dynamic) testing methodology. The methodology is applied in two phases. The first phase consists in the classical application of tests to the IUT. By using any of the available methods in the literature, a test suite will be derived from the specification. If the application of this test suite finds an unexpected result then the testing process stops: The IUT is not conforming. However, if such a wrong behavior is not detected then the tester cannot be sure that the IUT is correct. In this case, the second phase begins, that is, the tester applies the logic described in this paper to infer whether passing these tests *implies* that the IUT is correct if a given set of hypotheses is assumed. If it does then the IUT is assumed to be correct; otherwise, the tester may be interested in either applying more tests or in assuming more hypotheses (in the latter case, on the cost of *feasibility*) and then applying the logic again until the correctness of the IUT is effectively granted. In order to appropriately apply the logic, the behavior of the IUT observed during the application of tests must be properly represented. For each application of a test to the IUT, we construct an *observation*, that is, a sequence of inputs and outputs denoting the test and the response produced by the IUT, respectively. Both observations and the assumed hypotheses will be represented by appropriate *predicates* of the logic. Then, the deduction rules of the logic will allow to infer whether we can claim that the IUT conforms to the specification (actually, the logic will check whether *all* the implementations that could produce these observations and fulfill the requirements of the hypotheses conform to the specification).

We distinguish two kinds of hypotheses in the predefined repertory: Hypotheses concerning specific parts (states) of the IUT and hypotheses concerning the whole IUT. In order to unambiguously denote the states regarded by the former, they will be attached to the corresponding observations that reached these states. For example, if the IUT was showing the sequence of outputs $o_1, o_2, \ldots, o_n$ as answer to the sequence of inputs $i_1, i_2, \ldots, i_n$ provided by the tester, the tester may think that the state reached after performing i_1/o_1 is deterministic or that the state reached after performing the sequence $i_1/o_1, i_2/o_2$ is the same as the one reached after performing the whole sequence $i_1/o_1, i_2/o_2, \ldots, i_n/o_n$. Let us remark that these are *hypotheses* that the tester is assuming. Thus, she might be wrong and reach a wrong conclusion. However, this is similar to the case when the tester assumes that the implementation is deterministic or that it has at most n states and, in reality, this is not the case. In addition to using hypotheses associated to observations, the tester can also consider global hypotheses that concern the whole IUT. These are assumptions such as the ones that we mentioned before: Assuming that the IUT is deterministic, that is has at most n states, that is has a unique initial state, etc. In order to denote the assumption of this kind of hypotheses, specific logic predicates will be used.

Regarding related work, there are several papers where testing hypotheses are used to perform the testing process. For example, we may consider that the implementation is deterministic (e.g. [13]), that we are testing the coupling of several components by assuming that all of them are correct or that at most one of them is incorrect (e.g. [9]), etc. Our methodology provides a *generalization* of these frameworks because it allows to decide the specific hypotheses we will consider. In this line, we can compare the suitability of different test suites or test criteria in terms of the hypotheses that are considered (e.g. [10]); some formal relations to compare them have been defined [6]. Since our logic provides a mechanism to effectively compare sets of hypotheses, it may help to compute relations defined in these terms. Even though we work with rules and properties, our work is not related to *model checking* [4] since we do not *check* the validity of properties: We assume that they hold and we infer results about the conformity of the IUT by using this assumption. In the same way, this work is not related to some recent work on passive testing where the validity of a set of properties (expressed by means of *invariants*) is checked by passively observing the execution of the system (e.g. [7, 3, 1]).

The rest of the paper is organized as follows. In Section 2 we present some basic concepts related to the formalisms that we will use. In Section 3 we introduce the predicates of $\mathcal{HOTL}$, while in Section 4 we present the deduction rules. Finally, in Section 5 we present our conclusions and some directions for further research. Due to the lack of space, some auxiliary definitions and rules have not been included in this paper. All of them can be found in [14].

2   Formal Model

In this section we introduce some basic concepts that will be used along the paper to formally present our methodology. Specifically, we introduce the notion of finite state machine and a conformance relation.

Definition 1. A *finite state machine*, in short FSM, is a tuple of five elements $M = (\mathcal{S}, \text{inputs}, \text{outputs}, \mathcal{I}, \mathcal{T})$ where $\mathcal{S}$ is the set of *states*, inputs is the set of *input actions*, outputs is the set of *output actions*, $\mathcal{I} \subseteq \mathcal{S}$ is the set of *initial states*, and $\mathcal{T}$ is the set of *transitions*. A transition is a tuple $(s, i, o, s') \in \mathcal{T}$ where $s, s' \in \mathcal{S}$ are the *initial* and *final* states, respectively, $i \in$ inputs is the *input* that activates the transition, and $o \in$ outputs is the *output* produced in response. A transition $(s, i, o, s') \in \mathcal{T}$ is also denoted by $s \xrightarrow{i/o} s'$.

We say that $(i_1/o_1, \ldots, i_n/o_n)$ is a *trace* of M if there exists $s_1 \in \mathcal{I}$ and $s_2, \ldots, s_{n+1} \in \mathcal{S}$ such that $s_1 \xrightarrow{i_1/o_1} s_2, s_2 \xrightarrow{i_2/o_2} s_3, \ldots, s_n \xrightarrow{i_n/o_n} s_{n+1}$ are transitions of $\mathcal{T}$. The set of all traces of M is denoted by $\text{traces}(M)$. Let us consider $s, s' \in \mathcal{S}$. We say that s' is *reachable* from s, denoted by $\text{isReachable}(M, s, s')$, if either there exist u, i, o such that $s \xrightarrow{i/o} u \in \mathcal{T}$ and $\text{isReachable}(M, u, s')$ holds, or $s = s'$. The set $\text{reachableStates}(M, s)$ contains all $s' \in \mathcal{S}$ such that $\text{isReachable}(M, s, s')$.

Let $s \in \mathcal{S}$ and $i \in$ inputs. $\text{outs}(M, s, i)$ denotes the set of outputs that can be produced in s in response to i, that is, the set $\{o \mid \exists s' : s \xrightarrow{i/o} s' \in \mathcal{T}\}$.

We say that $s \in \mathcal{S}$ is *deterministic*, denoted by $\text{isDet}(M, s)$, if there do not exist $s \xrightarrow{i/o'} s', s \xrightarrow{i/o''} s'' \in \mathcal{T}$ such that $o' \neq o''$ or $s' \neq s''$.   □

In the previous definition, let us note that machines are allowed to be nondeterministic. In order to fix the kind of formalisms our logic will deal with, the following hypothesis will be imposed: Both implementations and specifications can be represented by appropriate FSMs. As a consequence, we have that when an input is offered to an IUT it always produces an observable response (that is, *quiescent* states not producing any output are not considered). Next we present the basic conformance relation that will be considered in our framework. This relation is similar to ioco [15] but in the framework of FSMs. This relation has been used in [11] as a preliminary step to define timed conformance relations. Intuitively, an IUT is conforming if it does not *invent* behaviors for those traces that can be executed by the specification. We will assume that the IUT is *input-enabled*, that is, for all state s and input i there exist o, s' such that $s \xrightarrow{i/o} s'$ belongs to the set of transitions of the IUT. During the rest of the paper, and when no confusion arises, we will assume that the FSM representing a generic specification is given by $spec = (\mathcal{S}_{spec}, \text{inputs}_{spec}, \text{outputs}_{spec}, \mathcal{I}_{spec}, \mathcal{T}_{spec})$.

Definition 2. Let S and I be two FSMs. We say that I *conforms to* S, denoted by I conf S, if for all $\rho_1 = (i_1/o_1, \ldots, i_{n-1}/o_{n-1}, i_n/o_n) \in \text{traces}(S)$, with $n \geq 1$, we have $\rho_2 = (i_1/o_1, \ldots, i_{n-1}/o_{n-1}, i_n/o'_n) \in \text{traces}(I)$ implies $\rho_2 \in \text{traces}(S)$. □

Example 1. A simple example, adapted from [5], will be used along the paper to illustrate our framework. A medical ray beaming system is controlled by using three buttons: A button for charging the machine (a single button press increases the voltage by 10 mV), another one for the beam activation, and the last one for resetting the machine at any time. The system will only charge the machine twice (increasing the voltage up to 20 mV) and it only lets to beam twice. Any further attempt to either increase the charge of the machine or to activate the beaming will be rejected because there is a danger of seriously injuring the patient. The FSM specifying this behavior is depicted in Figure 1 (left) and it is defined as $spec\_ray = (\mathcal{S}_{spec\_ray}, \text{inputs}_{spec\_ray}, \text{outputs}_{spec\_ray}, \mathcal{I}_{spec\_ray}, \mathcal{T}_{spec\_ray})$. We have $\mathcal{S}_{spec\_ray} = \{r, c1, c2, b1, \bar{b}2\}$, where r denotes the *ready* state, $c1/c2$ denote the states where the beamer has been charged one/two times, and $b1/b2$

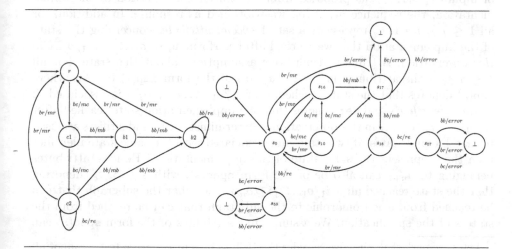

Fig. 1. Finite State Machines $spec\_ray$ (left) and $worst_{spec\_ray}$ (right)

denote the states where the first/second beaming is performed. $\text{inputs}_{spec\_ray} = \{br, bc, bb\}$, where $br/bc/bb$ respectively denote that the reset/charging/beaming button has been pressed. $\text{outputs}_{spec\_ray} = \{mr, mc, mb, re\}$, where $mr/mc/mb$ respectively denote that the machine is ready/charging/beaming while re denotes that the command has been rejected. Finally, $\mathcal{I}_{spec\_ray} = \{r\}$, that is, the initial state is *ready*.   □

3   Predicates of the Logic

In this section we present the predicates that will be part of $\mathcal{HOTL}$. These predicates allow to represent our knowledge and assumptions about the IUT. In particular, they will allow us to represent the *observations* that we have obtained from the IUT during the preliminary *classical* testing phase. *Observations* denote that, in response to a given sequence of inputs, the IUT produced a given sequence of outputs. Let us remark that if one of the sequences shows a behavior that is forbidden by the specification, then the IUT does not conform to the specification and no further analysis is required, that is, there is no need to apply our logic. As we said before, our notion of observation will be able to include some assumptions about the IUT as well as the observed behavior.

3.1   Manipulating Observations

During the rest of the paper, Obs denotes the set of all the observations collected during the preliminary interaction with the IUT, while Hyp denotes the set of *hypotheses* the tester has assumed. In this latter set, we will not consider the hypotheses that are implicitly introduced by means of observations.

Observations follow the form $ob = (a_1, i_1/o_1, a_2, \ldots, a_n, i_n/o_n, a_{n+1}) \in \text{Obs}$, where ob is a unique identification name. It denotes that when the sequence of inputs $i_1, \ldots, i_n$ was proposed from the initial configuration of the implementation, the sequence $o_1, \ldots, o_n$ was obtained as response. In addition, for all $1 \leq j \leq n+1$, a_j represents a set of *special attributes* concerning the state of the implementation that we reached after performing $i_1/o_1, \ldots, i_{j-1}/o_{j-1}$ in *this* observation. Attributes denote our assumptions about this state. For all $1 \leq j \leq n$ the attributes in the set a_j are of the form $\text{imp}(s)$ or det, where $\text{imp}(s)$ denotes that the state reached after $i_1/o_1, \ldots, i_{j-1}/o_{j-1}$ is associated to a *state identifier* s and det denotes that the implementation state reached after $i_1/o_1, \ldots, i_{j-1}/o_{j-1}$ in this observation is deterministic. State identifiers are used to match equal states: If two states are associated with the same state identifier then they represent the *same* state of the implementation.[1] Besides, attributes belonging to a_{n+1} can also be of the form $\text{spec}(s)$, with $s \in \mathcal{S}_{spec}$, denoting that the state reached after $i_1/o_1, \ldots, i_n/o_n$ is such that the subgraph that can be reached from it is isomorphic to the subgraph that can be reached from the state s of the specification. We assume that attributes of the form $\text{spec}(s)$ can

---

[1] Let us remark that, since we consider the IUT to be a black-box, a tester cannot always be sure of the state where the IUT is placed. However, she may still *hypothesize* that the reached states after performing two subsequences are in fact the same.

appear only at the end of the observation, meaning that the behavior of the implementation from that point on is known and there is no need to check its correctness.

Example 2. For our case study we will consider that the set of observations $\text{Obs}=\{ob_i|1 \leq i \leq 11\}$ was obtained. As illustration, we show the following ones (the full set is given in [14]).

$$ob_1 = (\{det\}, bc/mc, \{imp(q_1)\}, bb/mb, \{imp(q_2)\}, bb/mb, \emptyset)$$
$$ob_6 = (\emptyset, bc/mc, \{imp(q_1)\}, bb/mb, \{imp(q_2)\}, bb/mb, \{imp(q_3)\}, br/mr, \emptyset, bc/mc,$$
$$\{imp(q_1)\}, bb/mb, \{imp(q_2)\})$$
$$ob_{10} = (\emptyset, bc/mc, \{imp(q_1)\}, bb/mb, \{imp(q_2), det\}, bb/mb, \{imp(q_2)\}, bb/re, \emptyset)$$

For example, ob_{10} denotes that we initially do not make assumptions about the state the IUT selected as *initial* for this observation (its set of attributes is $\emptyset$). After pressing the charging button, the beaming system is charged and the state reached is identified by a specific state identifier (denoted by q_1). Next, by pressing the beaming button, the beaming is performed and we reach a certain state, in principle different to previous ones, which is assumed to be deterministic and is denoted by the identifier q_2. After pressing again the beaming button, the beaming action is performed and we assume that we reach again the same state as before (note that it is denoted by the same identifier q_2). We press once again the beaming button but this time the action is rejected, and we make no assumptions about the state reached afterwards.   □

3.2 Model Predicates

Observations will allow to create *model predicates*. A model predicate denotes our knowledge about the implementation. Models will be constructed according to the observations and hypotheses we consider. In particular, they induce a graph consistent with the observations and hypotheses considered so far. As more information is retrieved, models will be refined and particularized. We denote model predicates by $\texttt{model}(m)$, where $m = (\mathcal{S}, \mathcal{T}, \mathcal{I}, \mathcal{A}, \mathcal{E}, \mathcal{D}, \mathcal{O})$. The meaning of the different components of the tuple are the following. $\mathcal{S}$ (*states*) is the set of states that appear in the graph of the model. Despite the fact that this graph attempts to represent (a part of) the behavior of the implementation, any name belonging to $\mathcal{S}$ is fresh and by no means is related to the corresponding state of the implementation. Let us note that after more information is considered, it could turn out that some states belonging to $\mathcal{S}$ coincide. Next, $\mathcal{T}$ (*transitions*) is the set of transitions appearing in the graph of the model. $\mathcal{I}$ (*initial states*) is the set of states that are initial in the model. $\mathcal{A}$ (*accounting*) is the set of *accounting registers*. A register is a tuple $(s, i, outs, n)$ denoting that in the state $s \in \mathcal{S}$ the input i has been offered n times and we have obtained the outputs belonging to the set *outs*. This information allows to handle some hypotheses about nondeterminism. If, due to the hypotheses that we consider, we infer that the number of times we observed an input is enough to believe that the implementation cannot react to that input in a way that has not happened before (that is, either with an output that was not produced before or leading to a state that was not taken before), then the value n is set to $\top$. In this case, we say that the behavior of the state s for the input i is *closed*. Next, $\mathcal{E}$ (*equality relations*) is the set of equalities relating states in $\mathcal{S}$. Equalities have the form s is s'. For example if s_1 is $s \in \mathcal{E}$ and s_2 is $s \in \mathcal{E}$ then we infer that $s_1 = s_2$ and that one of the names could be eliminated afterwards. $\mathcal{D}$ (*deterministic states*) is the set of states that are deterministic (according to the hypotheses considered so far). Finally, $\mathcal{O}$ (*used observations*) is the set of observations we have used so far for the construction of this model. The aim of recording them is to avoid considering the same observation several times, which could ruin the information codified, for instance, in $\mathcal{A}$.

In $\mathcal{HOTL}$, conclusions about the conformance of a model (that is, of the possible IUTs it represents) with respect to a specification will be established only after the full set of observations Obs has been considered. Besides, we will require that no other rule concerning hypotheses in Hyp can be applied. In Section 4 we introduce some of the hypotheses a tester might consider in this set. These hypotheses include usual ones such as to assume an upper bound on the number of states of the IUT, the uniqueness of the initial state, the determinism of the IUT, etcetera.

3.3   Other Predicates

We will also consider other predicates related to the correctness of models. The correct(m) predicate denotes that m is a correct model, that is, it denotes a behavior that has to be conforming to the specification. The allModelsCorrect predicate represents a set of correct models. This predicate is the *goal* of the logic: If it holds then all the IUTs that could produce the observations in Obs and meet all the requirements in Hyp conform to the specification. The consistent(m) predicate means that the model m does not include any *inconsistency*. Note that the requirements imposed by Obs and Hyp could lead to inconsistent models. For example, let us consider a model where a state s is assumed to be deterministic, s is equal to another state s', and s' produces either o_1 or o_2 when i is offered, with $o_1 \neq o_2$. There is no FSM that meets the requirements of this model. Since a user of the logic can create a set of observations and hypotheses leading to that model, inconsistent models may indeed appear. As we will see, the rules of the logic will eliminate inconsistent models by deducing an *empty* set of models from them. In addition, we will be provided with rules that allow to *guarantee* the consistency of a model.

In general, several models can be constructed from a set of observations and hypotheses. Hence, our logic will deal with *sets* of models. If $\mathcal{M}$ is a set of models then the predicate models ($\mathcal{M}$) denotes that, according to the observations and hypotheses considered, $\mathcal{M}$ contains all the models that are valid candidates to properly describe the implementation. Besides, modelsSubset ($\mathcal{M}'$) denotes that for some set $\mathcal{M}$ we have models ($\mathcal{M}$) and $\mathcal{M}' \subseteq \mathcal{M}$.

The formal semantics of predicates, which is defined in terms of the set of FSMs that fulfill each predicate, is introduced in [14]. These concepts are considered there to prove the soundness and completeness of the logic.

4   Deduction Rules of the Logic $\mathcal{HOTL}$

The rules will be presented in the form $\frac{premises}{conclusion}$. If B can be deduced from A then we write $A \vdash B$. If we want to be more specific, we write $A \vdash_r B$ to denote that B is deduced from A by applying the rule r. The ultimate goal is to deduce the *conformance* of a set of observations Obs and hypotheses Hyp, that is, whether *all* the FSMs that meet these conditions conform to the specification. Since inconsistent models may appear, conformance will be granted only if there exists at least one *consistent* model that meets these conditions. Some formal definitions and rules could not be included in this version of the paper and can be found in [14]. In these cases, brief informal explanations will be provided.

$\mathcal{HOTL}$ will consider observations and hypotheses in two phases. First, observations, as well as the hypotheses they can implicitly express, will be collected. Once all of them have been considered (i.e., we have a model predicate with $\mathcal{O} = \mathsf{Obs}$) a second phase, to add the rest of hypotheses, will start.

First, we present a rule to construct a model from an observation. Given a predicate denoting that an observation was collected, the rule deduces some details about the behavior of the implementation. These details are codified by means of a *model* that shows this behavior. Basically, *new* states and transitions will be created in the model so that it can produce the observation. Even though some model states could actually coincide, we will not consider it yet. Thus, we take fresh states to name all of them. Besides, the hypotheses denoted by the attributes of the observation will affect the information associated to the corresponding model states. In particular, if the tester assumes that the last state of the observation is isomorphic to a state of the specification (i.e., $\mathsf{spec}(s)$, for some $s \in \mathcal{S}_{spec}$) then the sets of states, transitions, accounting registers, and deterministic states will be extended with some extra elements taken from the specification and denoted by $\mathcal{S}'$, $\mathcal{T}'$, $\mathcal{A}'$, and $\mathcal{D}'$, respectively. The new states and transitions $\mathcal{S}'$ and $\mathcal{T}'$, respectively, will copy the structure existing among the states that can be reached from s in the specification. The new accounting, $\mathcal{A}'$, will denote that the knowledge concerning the new states is *closed* for all inputs, that is, the only transitions departing from these states are those we copy from the specification and no other transitions will be added in the future. Finally, those model states that are images of deterministic specification states will be included in the set $\mathcal{D}'$ of deterministic states of the model.

$$(obser) \frac{ob = (a_1, i_1/o_1, a_2, \ldots, a_n, i_n/o_n, a_{n+1}) \in \mathsf{Obs} \wedge s_1, \ldots, s_{n+1} \text{ are fresh states}}{\mathtt{model} \begin{pmatrix} \{s_1, \ldots, s_{n+1}\} \cup \mathcal{S}', \\ \{s_1 \xrightarrow{i_1/o_1} s_2, \ldots, s_n \xrightarrow{i_n/o_n} s_{n+1}\} \cup \mathcal{T}', \{s_1\}, \\ \{(s_j, i_j, \{o_j\}, 1) \mid 1 \leq j \leq n\} \cup \mathcal{A}', \\ \{s_j \text{ is } s_j' \mid 1 \leq j \leq n+1 \wedge \mathtt{imp}(s_j') \in a_j\}, \\ \{s_j \mid 1 \leq j \leq n+1 \wedge \mathtt{det} \in a_j\} \cup \mathcal{D}', \{ob\} \end{pmatrix}}$$

The formal definition of $\mathcal{S}'$, $\mathcal{T}'$, $\mathcal{A}'$, and $\mathcal{D}'$ follows. If there does not exist s' such that $\mathsf{spec}(s') \in a_{n+1}$ then $(\mathcal{S}', \mathcal{T}', \mathcal{A}', \mathcal{D}') = (\emptyset, \emptyset, \emptyset, \emptyset)$. Otherwise, that is, if $\mathtt{spec}(s) \in a_{n+1}$ for some $s \in \mathcal{S}_{spec}$, let us consider the following set of states: $U = \{u_j \mid u_j \text{ is a fresh state } \wedge \ 1 \leq j < |\mathtt{reachableStates}(spec, s)|\}$ and a bijective function $f : \mathtt{reachableStates}(spec, s) \longrightarrow U \cup \{s_{n+1}\}$ such that $f(s) = s_{n+1}$. Then, $(\mathcal{S}', \mathcal{T}', \mathcal{A}', \mathcal{D}')$ is equal to

$$
\left(
\begin{array}{c}
U, \quad \{f(s') \xrightarrow{i/o} f(s'') \mid s' \xrightarrow{i/o} s'' \in \mathcal{T}_{spec} \ \wedge \ \mathtt{isReachable}(spec, s, s')\}, \\
\{(u, i, \mathtt{outs}(spec, s, i), \top) \mid u \in U \cup \{s_{n+1}\} \ \wedge \ i \in \mathtt{inputs}_{spec}\}, \\
\{f(s') \mid \mathtt{isReachable}(spec, s, s') \ \wedge \ \mathtt{isDet}(spec, s')\}
\end{array}
\right)
$$

Example 3. If we apply the *obser* deduction rule to the observation ob_6 given in Example 2 then we obtain a model $m_6 = (\mathcal{S}_6, \mathcal{T}_6, \mathcal{I}_6, \mathcal{A}_6, \mathcal{E}_6, \mathcal{D}_6, \mathcal{O}_6)$, where

$\mathcal{S}_6 = \{s_{26}, s_{27}, s_{28}, s_{29}, s_{30}, s_{31}, s_{32}\}$ and $\mathcal{I}_6 = \{s_{26}\}$

$$
\mathcal{T}_6 = \left\{
\begin{array}{l}
s_{26} \xrightarrow{bc/mc} s_{27}, s_{27} \xrightarrow{bb/mb} s_{28}, s_{28} \xrightarrow{bb/mb} s_{29}, s_{29} \xrightarrow{br/mr} s_{30}, \\
s_{30} \xrightarrow{bc/mc} s_{31}, s_{31} \xrightarrow{bb/mb} s_{32}
\end{array}
\right\}
$$

$$
\mathcal{A}_6 = \left\{
\begin{array}{l}
(s_{26}, bc, \{mc\}, 1), (s_{27}, bb, \{mb\}, 1), (s_{28}, bb, \{mb\}, 1), (s_{29}, br, \{mr\}, 1), \\
(s_{30}, bc, \{mc\}, 1), (s_{21}, bb, \{mb\}, 1)
\end{array}
\right\}
$$

$\mathcal{E}_6 = \{s_{27} \text{ is } q_1, s_{28} \text{ is } q_2, s_{29} \text{ is } q_3, s_{31} \text{ is } q_1, s_{32} \text{ is } q_2\}$, $\mathcal{D}_6 = \emptyset$, and $\mathcal{O}_6 = \{ob_6\}$

Similarly, for all $1 \leq i \leq 11$ we can obtain a model m_i by applying the deduction rule *obser* to ob_i. □

We will be able to join different models created from different observations into a single model. The components of the new model will be the union of the components of each model.

$$
(fusion) \frac{
\begin{array}{c}
\mathtt{model}\,(\mathcal{S}_1, \mathcal{T}_1, \mathcal{I}_1, \mathcal{A}_1, \mathcal{E}_1, \mathcal{D}_1, \mathcal{O}_1) \ \wedge \\
\mathtt{model}\,(\mathcal{S}_2, \mathcal{T}_2, \mathcal{I}_2, \mathcal{A}_2, \mathcal{E}_2, \mathcal{D}_2, \mathcal{O}_2) \ \wedge \ \mathcal{O}_1 \cap \mathcal{O}_2 = \emptyset
\end{array}
}{
\mathtt{model}\,\big(\mathcal{S}_1 \cup \mathcal{S}_2, \mathcal{T}_1 \cup \mathcal{T}_2, \mathcal{I}_1 \cup \mathcal{I}_2, \mathcal{A}_1 \cup \mathcal{A}_2, \mathcal{E}_1 \cup \mathcal{E}_2, \mathcal{D}_1 \cup \mathcal{D}_2, \mathcal{O}_1 \cup \mathcal{O}_2\big)
}
$$

The condition $\mathcal{O}_1 \cap \mathcal{O}_2 = \emptyset$ appearing in the previous rule avoids to include the same observation in a model more than once, which would be inefficient. Besides, since models in the second phase must fulfill $\mathcal{O} = \mathtt{Obs}$, we avoid to use the previous rule in the second phase.

By iteratively applying these two first rules, we will finally obtain a model where $\mathcal{O}$ includes all the observations belonging to the set $\mathtt{Obs}$.

Example 4. The deduction rule *fusion* allows to join all the models obtained after applying the deduction rule *obser* to the set of observations given in Example 2. After it, we have a new model m_T defined as follows:

$$
m_T = \mathtt{model}\left(\bigcup_{j=1}^{11} \mathcal{S}_j, \bigcup_{j=1}^{11} \mathcal{T}_j, \bigcup_{j=1}^{11} \mathcal{I}_j, \bigcup_{j=1}^{11} \mathcal{A}_j, \bigcup_{j=1}^{11} \mathcal{E}_j, \bigcup_{j=1}^{11} \mathcal{D}_j, \mathtt{Obs}\right) \qquad □
$$

At this point, the inclusion of those hypotheses that are covered by observations will begin. During this new phase, in general, we will need several models to represent all the FSMs that are compatible with a set of observations and hypotheses. The next simple rule allows to represent a single model by means of a set containing a single element. Since the forthcoming rules will concern only the second phase, in all cases we will have $\mathcal{O} = \texttt{Obs}$.

$$(set)\frac{\texttt{model}\,(\mathcal{S},\mathcal{T},\mathcal{I},\mathcal{A},\mathcal{E},\mathcal{D},\texttt{Obs})}{\texttt{models}\,(\{(\mathcal{S},\mathcal{T},\mathcal{I},\mathcal{A},\mathcal{E},\mathcal{D},\texttt{Obs})\})}$$

In order to reflect how a rule that applies to a single model affects the set including this model, we provide the following rule. Let φ denote a logical predicate (in particular, $\texttt{true}$) and $m = \texttt{model}\,(\mathcal{S},\mathcal{T},\mathcal{I},\mathcal{A},\mathcal{E},\mathcal{D},\texttt{Obs})$. Then,

$$(propagation)\frac{\texttt{models}\,(\mathcal{M}\cup\{m\})\ \wedge\ \varphi\ \wedge\ ((\texttt{model}\,(m)\ \wedge\ \varphi)\vdash\texttt{modelsSubset}\,(\mathcal{M}'))}{\texttt{models}\,(\mathcal{M}\cup\mathcal{M}')}$$

By using the previous rule, we will be able to use other rules that apply to a *single* model and then propagate its change to the set where the model is included as expected: As the previous rule states, this model changes while other models belonging to the set remain unchanged. Most of the forthcoming rules will apply to single models. After each of them is used, the rule *propagation* will be applied to propagate its effect to the corresponding set of models.

Our logic will allow to discover that a state of the model coincides with another one. In this case, we will eliminate one of the states and will allocate all of its constraints to the other one. This will modify all the components that define the model. This functionality is provided by the $\texttt{modelElim}$ function. Specifically, $\texttt{modelElim}(m, s_1, s_2)$ denotes the elimination of the state s_2 and the transference of all its responsibilities to state s_1 in the model m. This function returns a *set* of models. If the transference of responsibilities creates an *inconsistency* in the rest of the model, an empty set of models is returned. Sometimes we will use a *generalized* version of this function to perform the elimination of several states instead of a single state: $\texttt{modelElim}(m, s, \{s_1, \dots, s_n\})$ represents the substitution of s_1 by s, followed by the substitution of s_2 by s, and so on up to s_n. The formal definitions of both forms of the $\texttt{modelElim}$ function are given in [14].

Next we present some rules that use this function. In the first one, we join two states if the set of equalities allows to deduce that both coincide.

$$(equality)\frac{\texttt{model}\,(\mathcal{S},\mathcal{T},\mathcal{I},\mathcal{A},\mathcal{E},\mathcal{D},\texttt{Obs})\ \wedge\ s_1, s_2 \in \mathcal{S}\ \wedge\ \{s_1\ \texttt{is}\ s, s_2\ \texttt{is}\ s\}\subseteq\mathcal{E}}{\texttt{modelsSubset}\,(\texttt{modelElim}((\mathcal{S},\mathcal{T},\mathcal{I},\mathcal{A},\mathcal{E},\mathcal{D},\texttt{Obs}), s_1, s_2))}$$

Another situation where two states can be fused appears when a deterministic state shows two transitions labelled by the same input. Since the state is deterministic, they must also be labelled by the same output. The determinism of the state implies that both destinations are actually the same state. Hence, these two reached states can be fused. Note that if both outputs are different then the model is inconsistent, because the determinism of the state is not preserved. In this case, an empty set of models is produced.

$$(determ)\frac{\texttt{model}\,(\mathcal{S},\mathcal{T},\mathcal{I},\mathcal{A},\mathcal{E},\mathcal{D},\texttt{Obs})\ \wedge}{s, s_1, s_2 \in \mathcal{S}\ \wedge\ s \in \mathcal{D}\ \wedge\ \{s\xrightarrow{i/o_1}s_1, s\xrightarrow{i/o_2}s_2\}\subseteq\mathcal{T}}{\texttt{modelsSubset}\,(\mathcal{M}')\ [\text{if } o_1 = o_2 \text{ then } \mathcal{M}' = \texttt{modelElim}(m, s_1, s_2) \text{ else } \mathcal{M}' = \emptyset]}$$

Next we present the first rule dealing with an hypothesis that is not implicitly given by an observation. This hypothesis allows to assume that the initial state of the implementation is *unique*.

$$(singleInit)\frac{\texttt{model}\,(\mathcal{S},\mathcal{T},\mathcal{I},\mathcal{A},\mathcal{E},\mathcal{D},\texttt{Obs})\,\wedge\,\mathcal{I}=\{s_1,\ldots,s_n\}\,\wedge\,\texttt{singleInitial}\in\texttt{Hyp}}{\texttt{modelsSubset}\,\big(\texttt{modelElim}((\mathcal{S},\mathcal{T},\mathcal{I},\mathcal{A},\mathcal{E},\mathcal{D},\texttt{Obs}),s_1,\{s_2,\ldots,s_n\})\big)}$$

If the tester adds the hypothesis that all the states are deterministic, then the complete set of states $\mathcal{S}$ coincides with the set of deterministic states $\mathcal{D}$.

$$(allDet)\frac{\texttt{model}\,(\mathcal{S},\mathcal{T},\mathcal{I},\mathcal{A},\mathcal{E},\mathcal{D},\texttt{Obs})\,\wedge\,\texttt{allDet}\in\texttt{Hyp}}{\texttt{modelsSubset}\,(\{(\mathcal{S},\mathcal{T},\mathcal{I},\mathcal{A},\mathcal{E},\mathcal{S},\texttt{Obs})\})}$$

The logic $\mathcal{HOTL}$ allows to consider other hypotheses about the IUT. For example, the predicate $\texttt{allTranHappenWith}(n)$ assumes that for all state s and input i such that the IUT behavior has been observed at least n times, *all* the outgoing transitions from s having as input i, have been observed at least once. This means that the IUT state s cannot react to i with an output that has not produced so far or moving to a state it has not moved before. If the hypothesis is assumed then some accounting registers of the model will be set to the value $\top$, denoting that our knowledge about this state and input is *closed*. Depending on the compatibility of the hypothesis with the current model, several models can be produced by this rule. If no model is returned then we infer that the resulting model is inconsistent with the current model requirements. The $\texttt{upperBoundOfStates}(n)$ hypothesis allows to assume that the IUT uses at most n states. The reduction of states, based on the identification of several states with the same state identifiers, will be performed by means of new equalities s is $s' \in \mathcal{E}$. The $\texttt{longSequencesSamePath}(n)$ hypothesis assumes that if two sequences of n transitions produce the same inputs and outputs, then they actually go through the same states. The set $\mathcal{E}$, containing the assumed equalities between states, will be also used in this case. The formal definition of the rules that allow to consider the $\texttt{allTranHappenWith}(n)$, $\texttt{upperBoundOfStates}(n)$, and $\texttt{longSequencesSamePath}(n)$ hypotheses can be found in [14].

We have seen some rules that may lead to inconsistent models. In some of these cases, an empty set of models is produced, that is, the inconsistent model is eliminated. Before granting conformance, we need to be sure that at least one model belonging to the set is consistent. Next we provide a rule that labels a model as consistent. Let us note that the inconsistences created by a rule can be detected by the forthcoming applications of rules. For instance, the *determ* rule can detect that a previous rule matched a deterministic state with another state in such a way that both react to the input i with a different output. Actually, all inconsistencies can be detected by applying suitable rules. Thus, a model is free of inconsistencies if for any other rule either it is not applicable to the model or the application does not modify the model (i.e., it deduces the same model). Next we introduce this concept. In the following definition, $\mathcal{R}$ denotes the set of all rules in $\mathcal{HOTL}$ that follow the form required to apply the *propagation* rule. In particular, it consists of all previous rules from *equality* up to the forthcoming *correct* rule.

Definition 3. We denote the set of all rules in $\mathcal{HOTL}$ that follow the form $(\texttt{model}\,(m) \,\wedge\, \varphi) \vdash \texttt{modelsSubset}\,(\mathcal{M}')$ by $\mathcal{R}$.

Let $r = (\texttt{model}\,(m') \,\wedge\, \varphi) \vdash \texttt{modelsSubset}\,(\mathcal{M}') \in \mathcal{R}$ and m be a model. The *unable predicate* for m and r, denoted by $\texttt{unable}(m, r)$, is defined by the expression $\texttt{unable}(m, r) = \neg\varphi \vee ((\texttt{model}\,(m) \,\wedge\, \varphi) \vdash_r \texttt{modelsSubset}\,(\{m\}))$. We extend this predicate to deal with sets of rules as follows: $\texttt{unable}(m, Q) = \bigwedge\{\texttt{unable}(m, r) | r \in Q\}$. □

The next rule detects that a model is consistent. It requires that no other rule that manages hypotheses can modify the model. These rules consist of all the rules in $\mathcal{R}$ we have seen so far.

$$(consistent)\,\dfrac{m = (\mathcal{S}, \mathcal{T}, \mathcal{I}, \mathcal{A}, \mathcal{E}, \mathcal{D}, \texttt{Obs}) \,\wedge\, \texttt{model}\,(m)\,\wedge}{\texttt{modelsSubset}\,(\{\texttt{consistent}(\mathcal{S}, \mathcal{T}, \mathcal{I}, \mathcal{A}, \mathcal{E}, \mathcal{D}, \texttt{Obs})\})}$$

Since a model is a (probably incomplete) representation of the IUT, in order to check whether a model conforms to the specification, two aspects must be taken into account. First, only the conformance of consistent models will be considered. Second, given a consistent model, we will check its conformance with respect to the specification by considering the *worst* instance of the model, that is, if this instance conforms to the specification then any other instance extracted from the model does so. This worst instance is constructed as follows: For each state s and input i such that the behavior of s for i is not closed *and* either s is not deterministic or no transition with input i exists in the model, a new *malicious* transition is created. The new transition is labelled with a special output *error*, that does not belong to $\texttt{outputs}_{spec}$. This transition leads to a new state $\bot$ having no outgoing transitions. Since the specification cannot produce the output *error*, this worst instance will conform to the specification only if the unspecified parts of the model are not relevant for the correctness of the IUT it represents.

Definition 4. Let $m = (\mathcal{S}, \mathcal{T}, \mathcal{I}, \mathcal{A}, \mathcal{E}, \mathcal{D}, \texttt{Obs})$ be a model. We define the *worst instance* of the model m with respect to the considered specification *spec*, denoted by $\texttt{worstCase}(m)$, as the FSM

$$\left(\begin{array}{c} \mathcal{S} \cup \{\bot\}, \texttt{inputs}_{spec}, \texttt{outputs}_{spec} \cup \{error\}, \\[2mm] \mathcal{T} \cup \left\{ s \xrightarrow{i/error} \bot \,\middle|\, \begin{array}{l} s \in \mathcal{S} \,\wedge\, i \in \texttt{inputs}_{spec} \,\wedge \\ \nexists\, outs : (s, i, outs, \top) \in \mathcal{A} \,\wedge \\ (s \notin \mathcal{D} \vee \nexists\, s', o : s \xrightarrow{i/o} s') \end{array} \right\}, \mathcal{I} \end{array} \right)$$
□

Thus, the rule for indicating the correctness of a model is

$$(correct)\,\dfrac{m = (\mathcal{S}, \mathcal{T}, \mathcal{I}, \mathcal{A}, \mathcal{E}, \mathcal{D}, \texttt{Obs}) \,\wedge\, \texttt{consistent}(m) \,\wedge\, \texttt{worstCase}(m)\; \texttt{conf}\; spec}{\texttt{modelsSubset}\,(\{\texttt{correct}(m)\})}$$

Now we can consider the conformance of a set of models. A set conforms to the specification if all the elements do so and the set contains at least one element. Note that an empty set of models denotes that all the models were inconsistent. Hence, granting the conformance of an empty set would imply accepting models that do not represent any implementation. In fact, although *false implies anything*, accepting inconsistent models is useless for a tester.

$$(allCorrect)\frac{\texttt{models}\,(\mathcal{M})\ \wedge\ \mathcal{M} \neq \emptyset\ \wedge\ \mathcal{M} = \{\texttt{correct}(m_1),\dots,\texttt{correct}(m_n)\}}{\texttt{allModelsCorrect}}$$

Example 5. We consider the model m_R obtained after applying the *determ*, *equality*, *long* (see [14]), and *singleInit* deduction rules. The *long* rule is applied to introduce the hypothesis `longSequencesSamePath`(1). The *singleInit* and *long* rules are applied once, while *determ* and *equality* are applied as long as we can. Let us recall that, after each of these rules is used, the *propagation* rule must be applied as well. When the *determ* and *equality* rules cannot be applied anymore, our model cannot be further manipulated to produce new inconsistencies. Then, we can use the *consistent* and *propagation* rules to deduce `models`($\{\texttt{consistent}(m_R)\}$).

We build an FSM by applying the function `worstCase` to m_R and we verify its conformance with respect to the specification. The obtained FSM, denoted by $worst_{spec\_ray}$, is graphically depicted in Figure 1 (right). For the sake of clarity, we have included four states $\perp$, even though they correspond to only one state.

We have $worst_{spec\_ray}$ `conf` *spec_ray* and, by successively applying the *correct* and *propagation* rules, we obtain `models`($\{\texttt{correct}(m_R)\}$) and deduce, by means of the *allCorrect* deduction rule, `allModelsCorrect`. A more detailed description of the application of rules to this example can be found in [14].   □

Now that we have presented the set of deduction rules, we introduce a correctness criterion. In the next definition, in order to uniquely denote observations, fresh names are assigned to them. Besides, let us note that all hypothesis predicates follow the form $h \in \texttt{Hyp}$ for some h belonging to `Hyp`.

Definition 5. Let *spec* be an FSM, `Obs` be a set of observations, and `Hyp` be a set of hypotheses. Let $A = \{ob = o \mid ob\ \text{is a fresh name}\ \wedge\ o \in \texttt{Obs}\}$ and $B = \{h_1 \in \texttt{Hyp},\dots,h_n \in \texttt{Hyp}\}$, where $\texttt{Hyp} = \{h_1,\dots,h_n\}$. If the deduction rules allow to infer `allModelsCorrect` from the set of predicates $C = A \cup B$, then we say that C *logically conforms to spec* and we denote it by $C\,\texttt{logicConf}\,spec$.   □

In order to prove the validity of our method, we have to relate the deductions that we make by using our logic with the notion of conformance introduced in Definition 2. The *semantics* of a logic predicate is described in terms of the set of FSMs that fulfill the requirements given by the predicate; given a predicate p, we denote this set by $\nu(p)$. As illustration, the semantics of some predicates is formally defined in [14] by means of the function ν. Let us consider that P is the conjunction of all the considered observation and hypothesis predicates. Then, the set $\nu(P)$ denotes all the FSMs that can produce these observations and fulfill these hypotheses, that is, it denotes all the FSMs that, according to our knowledge, can *define* the IUT. So, if our logic deduces that all of these FSMs conform to the specification (i.e., `allModelsCorrect` is obtained), then the IUT actually conforms to the specification.

Theorem 1. Let *spec* be an FSM and C be a set of predicates including at least one observation predicate. Then, C `logicConf` *spec* iff for all FSM $f \in \nu(\bigwedge_{p \in C})$ we have f `conf` *spec* and $\nu(\bigwedge_{p \in C}) \neq \emptyset$. $\square$

Corollary 1. Let IUT and *spec* be FSMs and C be a set of predicates including at least one observation predicate. If $IUT \in \nu(\bigwedge_{p \in C})$ then C `logicConf` *spec* implies IUT `conf` *spec*. If there exists $f \in \nu(\bigwedge_{p \in C})$ such that f `conf` *spec* does not hold then C `logicConf` *spec* does not hold.

5 Conclusions and Future Work

In this paper we have presented a logic to infer whether a collection of observations obtained by testing an IUT together with a set of hypotheses allow to deduce that the IUT conforms to the specification. A repertory of heterogeneous hypotheses providing a tester with expressivity to denote a wide range of testing scenarios has been presented. By considering those observations and hypotheses that better fit into her necessities, the tester can obtain diagnosis results about the conformance of an IUT in a flexible range of situations. Besides, our logic allows her to iteratively add observations (i.e., the results of the application of tests) and/or hypotheses until the complete set of predicates guarantees the conformance. In this sense, our logic can be used to dynamically *guide* the steps of a testing methodology.

As future work, we will study some ways to improve our logic. We plan to include an *incorrectness* rule, that is, a rule that detects whether a model is *necessarily incorrect*. If an incorrect model is detected then the calculus can be early terminated, which improves the efficiency. Moreover, the rule could be used to detect which observations/hypotheses made the model incorrect. Besides, we want to develop a more complex application example in the context of Internet protocols. We would also like to introduce a *feasibility* score for each of the logic rules. For example, for a given framework, we can consider that assuming that all the states are deterministic is harder than assuming that the implementation has less that 50 states. In this case, a lower feasibility score will be assigned to the first hypothesis. By accounting the feasibility of all the hypotheses that we have to add before ensuring conformance, we will obtain a measure of the suitability of the considered observations and, indirectly, of the tests that we used to obtain them. Hence, our logic can help a tester to choose her tests so that more *trustable* diagnosis results are obtained. We also consider to extend the repertory of hypotheses. Finally, we want to extend the logic so that it can deal with *extended finite state machines*. In this case, different formalisms to work with models and different sets of hypotheses will be considered.

Acknowledgements. We would like to thank the anonymous referees for their valuable comments. Though they proposed very interesting ideas to improve our paper (some are commented above), we could not apply all of them due to the lack of space. Certainly, these ideas will be considered in the future.

References

1. E. Bayse, A. Cavalli, M. Núñez, and F. Zaïdi. A passive testing approach based on invariants: Application to the WAP. *Computer Networks*, 48(2):247–266, 2005.
2. B.S. Bosik and M.U. Uyar. Finite state machine based formal methods in protocol conformance testing. *Computer Networks & ISDN Systems*, 22:7–33, 1991.
3. A. Cavalli, C. Gervy, and S. Prokopenko. New approaches for passive testing using an extended finite state machine specification. *Journal of Information and Software Technology*, 45:837–852, 2003.
4. E.M. Clarke, O. Grumberg, and D. Peled. *Model Checking*. MIT Press, 1999.
5. G. Eleftherakis and P. Kefalas. Towards model checking of finite state machines extended with memory through refinement. In *Advances in Signal Processing and Computer Technologies*, pages 321–326. World Scientific and Engineering Society Press, 2001.
6. R. Hierons. Comparing test sets and criteria in the presence of test hypotheses and fault domains. *ACM Transactions on Software Engineering and Methodology*, 11(4):427–448, 2002.
7. D. Lee, D. Chen, R. Hao, R. Miller, J. Wu, and X. Yin. A formal approach for passive testing of protocol data portions. In *10th IEEE Int. Conf. on Network Protocols, ICNP'02*, pages 122–131. IEEE Computer Society Press, 2002.
8. D. Lee and M. Yannakakis. Principles and methods of testing finite state machines: A survey. *Proceedings of the IEEE*, 84(8):1090–1123, 1996.
9. L.P. Lima and A. Cavalli. A pragmatic approach to generating tests sequences for embedded systems. In *10th Workshop on Testing of Communicating Systems*, pages 288–307. Chapman & Hall, 1997.
10. S.C. Ntafos. A comparison of some structural testing strategies. *IEEE Transactions on Software Engineering*, 14:868–874, 1988.
11. M. Núñez and I. Rodríguez. Encoding PAMR into (timed) EFSMs. In *FORTE 2002, LNCS 2529*, pages 1–16. Springer, 2002.
12. A. Petrenko. Fault model-driven test derivation from finite state models: Annotated bibliography. In *4th Summer School, MOVEP 2000, LNCS 2067*, pages 196–205. Springer, 2001.
13. A. Petrenko, N. Yevtushenko, and G. von Bochmann. Testing deterministic implementations from their nondeterministic specifications. In *9th Workshop on Testing of Communicating Systems*, pages 125–140. Chapman & Hall, 1996.
14. I. Rodríguez, M.G. Merayo, and M. Núñez. A logic for assessing sets of heterogeneous testing hypotheses: Extended version. Available at: http://dalila.sip.ucm.es/~manolo/papers/logic-extended.pdf, 2006.
15. J. Tretmans. Test generation with inputs, outputs and repetitive quiescence. *Software – Concepts and Tools*, 17(3):103–120, 1996.

Bounded Sequence Testing from Non-deterministic Finite State Machines

Florentin Ipate

Department of Computer Science and Mathematics,
University of Pitesti, Romania
fipate@ifsoft.ro

Abstract. The widespread use of *finite state machines* (*FSMs*) in modeling of communication protocols has lead to much interest in testing from (deterministic and non-deterministic) FSMs. Most approaches for selecting a test suite from a non-deterministic FSM are based on *state counting*. Generally, the existing methods of testing from FSMs check that the implementation under test behaves as specified for *all* input sequences. On the other hand, in many applications, only input sequences of limited length are used. In such cases, the test suite needs only to establish that the IUT produces the specified results in response to input sequences whose length does not exceed an upper bound *l*. A recent paper devises methods for *bounded sequence testing* from *deterministic* FSM specifications. This paper considers the, more general, situation where the specification may be a *non-deterministic FSM* and extends state counting to the case of bounded sequences. The extension is not trivial and has practical value since the test suite produced may contain only a small fraction of all sequences of length less than or equal to the upper bound.

1 Introduction

Finite state machines (*FSMs*) are widely used in modeling of communication protocols. As testing is a vital part of system development, this has lead to much interest in testing from FSMs [13], [9]. Given a FSM specification, for which we have its transition diagram, and an implementation, which is a "black box" for which we can only observe its input/output behavior, we want to test whether the implementation under test (IUT) conforms to the specification. This is called *conformance testing* or *fault detection* and a set of sequences that solves this problem is called a *test suite*.

Many test selection methods have been developed for the case where the specification is a *deterministic* FSM. The best known methods are: Transition Tour [13], Unique Input Output (UIO) [13], Distinguishing Sequence [13], the *W* method [2], [13] and its variant, the "partial *W*" (*Wp*) method [3]. The *W* and *Wp* methods will find all the faults in the IUT provided that the number of states of the IUT remain below a known upper bound.

M.Ü. Uyar, A.Y. Duale, and M.A. Fecko (Eds.): TestCom 2006, LNCS 3964, pp. 55–70, 2006.

When the specification is deterministic, equivalence is the natural notion of correctness. On the other hand, when the specification is non-deterministic, equivalence may often be too restrictive. Usually, a non-deterministic FSM specification provides a set of alternative output sequences that are valid responses to some input sequence and the IUT may choose from these (when the IUT is deterministic only one choice is allowed, otherwise multiple choices can be made). Consequently, the IUT is correct if and only if every input/output sequence that is possible in the IUT is also present in the specification; we say that the IUT is a *reduction* of the specification. Obviously, equivalence is a particular case of reduction, where all specified choices are implemented. Most approaches for selecting a test suite from a non-deterministic FSM are based on *state counting* [11], [12], [14].

Generally, the existing methods of testing from FSMs check that the IUT behaves as specified for *all* input sequences. On the other hand, in many applications, only input sequences of limited length are used. In such cases, the test suite needs only to establish that the IUT produces the specified results in response to input sequences whose length does not exceed an upper bound l. A recent paper extends the W and Wp methods to the case of *bounded sequences* [8].

This paper considers the, more general, situation where the specification may be a *non-deterministic FSMs* and extends the state counting based test selection method to the case of bounded sequences. The extension is not straightforward since it is not sufficient to extract the prefixes of length at most l from the test suite produced in the unbounded case. Furthermore, the test suite produced may contain only a small fraction of all sequences of length less than or equal to the upper bound.

The paper is structured as follows. Section 2 introduces FSM related concepts and results that are used later in the paper, while section 3 reviews the use of state counting in testing from non-deterministic FSMs. Section 4 presents the testing method for bounded sequences, while the following two sections provide its theoretical basis: the l-bounded product FSM is defined in section 5, while in section 6, state counting is used to validate the test suite given earlier. Conclusions are drawn in section 7.

2   Finite State Machines

This section introduces the finite state machine and related concepts and results that will be used later in the paper.

First, the notation used is introduced. For a finite set A, we use A^* to denote the set of finite sequences with members in A; ϵ denotes the empty sequence. For $a, b \in A^*$, ab denotes the concatenation of sequences a and b. a^n is defined by $a^0 = \epsilon$ and $a^n = a^{n-1}a$ for $n \geq 1$. For $U, V \subseteq A^*$, $UV = \{ab \mid a \in U, b \in V\}$; U^n is defined by $U^0 = \{\epsilon\}$ and $U^n = U^{n-1}U$ for $n \geq 1$. Also, $U[n] = \bigcup_{0 \leq k \leq n} U^k$. For a sequence $a \in A^*$, $\|a\|$ denotes the length (number of elements) of a; in particular $\|\epsilon\| = 0$. For a sequence $a \in A^*$, $b \in A^*$ is said to be a *prefix* of a if there exists a sequence $c \in A^*$ such that $a = bc$. The set of all prefixes of a is denoted by $pref(a)$. For $U \subseteq A^*$, $pref(U)$ denotes the set of all prefixes of the elements in U. For a finite set A, $|A|$ denotes the number of elements of A.

A *finite state machine (FSM)* M is a tuple $(\Sigma, \Gamma, Q, h, q_0)$, where Σ is the finite *input alphabet*, Γ is the finite *output alphabet*, Q is the finite *set of states*, $h : Q \times \Sigma \longrightarrow 2^{Q \ \Gamma}$ is the *transition function* and $q_0 \in Q$ is the *initial state*. A FSM is usually described by a state-transition diagram. Given $q, q' \in Q$, $\sigma \in \Sigma$ and $\gamma \in \Gamma$, the application of input σ when M is in state q may result in M moving to state q' and outputting γ if and only if $(q', \gamma) \in h(q, \sigma)$.

M is said to be *completely specified* if for all $q \in Q$ and $\sigma \in \Sigma$, $|(h(q, \sigma)| \geq 1$. If M is not completely specified, it may be transformed to form a completely specified FSM by assuming that the "refused" inputs produce a designated error output, which is not in the output alphabet of M; this behavior can be represented as self-looping transitions or transitions to an extra (error) state. M is said to be *deterministic* if for all $q \in Q$ and $\sigma \in \Sigma$, $|(h(q, \sigma)| \leq 1$.

The function h may be extended to take input sequences and produce output sequences, i.e. $h : Q \times \Sigma^* \longrightarrow 2^{Q \ \Gamma^*}$. The projections $h_1 : Q \times \Sigma^* \longrightarrow 2^Q$ and $h_2 : Q \times \Sigma^* \longrightarrow 2^{\Gamma^*}$ of h give the states reached (h_1) and the output sequences produced (h_2) from a state, given an input.

A FSM M is said to be *initially connected* if every state q is reachable from the initial state of M, i.e. there exists $s \in \Sigma^*$ such that $q \in h_1(q_0, s)$. If M is not initially connected it may be transformed into an initially connected FSM by removing the unreachable states.

Given a state q, the associated language of q, $L_M(q)$, contains the input/output sequences allowed by M from q. More formally, $L_M(q) = \{(s, g) \mid s \in \Sigma^*, g \in h_2(q, s)\}$. The input/output sequences allowed by M from q_0 make up the associated language of M, denoted by $L(M)$.

States q of M and q' of M' are said to be *equivalent* if $L_M(q) = L_{M'}(q')$. FSMs M and M' are said to be *equivalent* if their initial states are equivalent, i.e. $L(M) = L(M')$. The equivalence relation can be restricted to a set of input sequences $Y \subseteq \Sigma^*$; this is called Y-equivalence.

M is said to be *observable* if for every state q, input σ and output γ, M has at most one transition leaving q with input σ and output γ, i.e. $|\{q' \mid (q', \gamma) \in h(q, \sigma)\}| \leq 1$. In such a FSM, given $q \in Q$, $s \in \Sigma^*$ and $g \in \Gamma^*$, $h_g(q, s)$ is used to denote the state (if exists) where input sequence s takes M from state q while outputting sequence g. Every FSM is equivalent to an observable FSM [10]. It will thus be assumed that any FSM considered is observable.

Suppose M and M' are two completely specified FSMs. Given states q of M and q' of M', q is said to be a reduction of q', written $q \leq q'$, if $L_M(q) \subseteq L_{M'}(q')$. Obviously, q and q' are equivalent if and only if $q \leq q'$ and $q' \leq q$. On the class of deterministic FSMs, the two relations coincide. The FSM M is said to be a reduction of the FSM M', written $M \leq M'$, if $q_0 \leq q_0'$. Given a set of input sequences $Y \subseteq \Sigma^*$, weaker reduction relations, denoted by $\leq_Y$, can be obtained by restricting the above definitions to Y.

3   Testing from Non-deterministic FSMs

This section briefly reviews the use of state counting in testing from (possibly) non-deterministic FSMs [5]. One important case is where the IUT is known to be deterministic [12]. However, the general case where the IUT may also be non-deterministic, is considered. All FSMs referred to are assumed to be initially connected, completely specified and observable.

3.1   Prerequisites

When testing from a formal specification, it is usual to assume that the IUT behaves like some unknown element from a *fault domain*. In the case of a FSM specification $M = (\Sigma, \Gamma, Q, h, q_0)$, the fault domain consists of all initially connected, completely specified and observable FSMs $M' = (\Sigma, \Gamma, Q', h', q_0')$ with the same input and output alphabets as M and at most m' states, where m' is a predetermined integer greater than or equal to the number m of states of M. Furthermore, it will be assumed that the IUT has a reliable reset. A FSM has a reset operation if there is some input r that takes every state to the initial state. A reliable reset is a reset that is known to have been implemented correctly and might be implemented through the system being switched off and then on again. The reset will not be included in the input alphabet.

A *test suite* is a finite set of input sequences that, for every M' in the fault model that is not a reduction of M, shows that M' is erroneous. More formally, $Y \subseteq \Sigma^*$ is a test suite if and only if for every M' in the fault model, $M' \leq M$ if and only if $M' \leq_Y M$. Naturally, when the specification M is deterministic, testing for $M' \leq M$ reduces to testing for the equivalence of the two FSMs

When testing a non-deterministic implementation, it is normal to make a fairness assumption, called the *complete testing assumption* [10], that there is some known N such that if an input sequence is applied N times then every possible response is observed at least once. Naturally, this assumption automatically holds when the implementation is deterministic. This paper will assume that the complete testing assumption can be made.

When testing from a FSM M, sequences that *reach* and *distinguish* the states of M are normally selected. These issues are now discussed.

3.2   Reaching States

Input sequence $s \in \Sigma^*$ is said to *deterministically-reach* (*d-reach*) state q if $h_1(q_0, s) = \{q\}$. That is, q is the only state reached by s. q is said to be *d-reachable* [12]. The initial state is always d-reachable since it is d-reached by the empty sequence ϵ. Naturally, all reachable states of a deterministic FSM are also d-reachable.

A set $S \subseteq \Sigma^*$ of input sequences is called a *state cover* of M if $\epsilon \in S$ and S is a minimal set such that every d-reachable state of M is d-reached by some sequence from S.

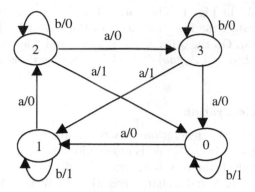

Fig. 1. The state-transition diagram of M

Consider, for example, M as represented in Fig. 1. States 0, 1 and 2 are d-reached by ϵ, a and aa, respectively. On the other hand, state 3 is not d-reachable. Thus $S = \{\epsilon, a, aa\}$ is a state cover of M.

3.3   Distinguishing States

In order for an input sequence s to distinguish two states q and q' of M, it is sufficient that the corresponding sets of output sequences do not intersect, i.e. $h_2(q, s) \cap h_2'(q', s) = \emptyset$. Two states for which there exists an input sequence with this property are said to be *separable*.

In the case of a non-deterministic FSM, however, it may be possible that there is no single input sequence that distinguishes between two states, rather these can be distinguished by a set of sequences. This idea leads to the, more general, concept of r-distinguishable states, formally defined in an inductive manner as follows [12]. States q and q' are said to be *r(1)-distinguishable* if there exists $\sigma \in \Sigma$ such that $h_2(q, \sigma) \cap h_2(q', \sigma) = \emptyset$. States q and q' are said to be *r(k)-distinguishable*, $k > 1$, if either q and q' are r(j)-distinguishable for some j, $1 \leq j < k$, or there is some input $\sigma \in \Sigma$ such that for all $\gamma \in h_2(q, \sigma) \cap h_2(q', \sigma)$, the states $h_\gamma(q, \sigma)$ and $h_\gamma(q', \sigma)$ are r(j)-distinguishable for some j, $1 \leq j < k$. States q and q' are said to be *r-distinguishable* if there exists some $k \geq 1$ such that q and q' are r(k)-distinguishable. Clearly, any two separable states are r-distinguishable, but not vice versa. Naturally, the two notions coincide when the FSM is deterministic.

The definition of r-distinguishable (r(k)-distinguishable) states naturally leads to the concept of *r-distinguishing set* (*r(k)-distinguishing set*) of two states q and q'; this can also be defined inductively [12]. A set of input sequences that contains an r-distinguishing (r(k)-distinguishing) set of q and q' is said to *r-distinguish* (*r(k)-distinguish*) q and q'.

A set $W \subseteq \Sigma^*$ of input sequences is a called a *characterization set* of M if it r-distinguishes each pair of r-distinguishable states of M.

Consider again M in Fig. 1. The pairs of states $(0, 2)$, $(0, 3)$, $(1, 2)$, $(1, 3)$ and $(0, 1)$ are separable; the first four are r-distinguished by $\{b\}$, the last is r-distinguished by $\{ab\}$. On the other hand, states 2 and 3 are not separable, but they are r-distinguished by $\{ab, aab\}$. Thus $W = \{b, ab, aab\}$ is a characterization set of M.

3.4   Test Suite Generation

This section describes the generation of a test suite from a FSM using state counting. The method is from [5] and is essentially based on the results given in [12] for the case in which the IUT is known to be deterministic.

Suppose a state cover S and a characterization set W have been constructed. Q_S is used to denote the set of all d-reachable states of M. Let $Q^1, \ldots, Q^j$ denote the maximal sets of pairwise r-distinguishable states of M. Let also $Q_S^i = Q^i \cap Q_S$, $1 \le i \le j$.

Recall that the scope of testing is to check language inclusion between the (unknown) implementation and the specification. Thus, the task is to find a state q' in the implementation such that the input/output exhibited from q' is not allowed from the corresponding state q of the specification. A test suite will be then constructed using a breadth-first search through input/output sequences from each d-reachable state of M, in which the termination criterion is based on the observation that if a pair of states $(q, q') \in Q \times Q'$, from which a failure may be exhibited, is reachable then it is reachable by some minimal input/output sequence. Such a minimal sequence will not have visited the same pair of states twice and, furthermore cannot contain pairs of states that have already been reached by the sequences in S. More specifically, the following two ideas are used:

- If an input/output sequence (s, g) visits states of some Q^i, a tester can use W after each prefix of (s, g) to distinguish between the corresponding states visited along (s, g) in the implementation. If states from Q^i are visited n_i times along a minimal sequence (s, g) in the specification, then n_i distinct states will be visited in the implementation. Thus, n_i cannot exceed m', the upper bound on the number of states of the implementation, by more than 1.
- There could be some d-reachable states among those in Q^i and the corresponding states in the implementation will also be reached by sequences from S; this leaves $|Q_S^i|$ less pairs of states to explore.

By combining these two ideas, the breadth-first search can be ended once it has been established that states from some Q^i have been visited $m' - |Q_S^i| + 1$ times.

More formally, given a state $q \in Q_S$, the set $Tr(q)$, called a *traversal set* in [11], is constructed in the following way:

- A set $TrIO(q)$ is defined to consist of all input/output sequences (s, g) for which there exists i, $1 \le i \le j$, such that (s, g) visits states from Q^i exactly $m' - |Q_S^i| + 1$ times when followed from q (the initial state of the path is not included in the counting) and this condition does not hold for any proper prefix of (s, g).

– $Tr(q)$ is the set of input sequences such that there is some corresponding input/output sequence in $TrIO(q)$, i.e. $Tr(q) = \{s \in \Sigma^* \mid \exists g \in \Gamma^* \cdot (s, g) \in TrIO(q)\}$.

Then the test suite produced is [5]:

$$Y = \bigcup_{s \in S} \{s\} pref(Tr(q_s))W$$

where for $s \in S$, q_s denotes the state reached by s.

When all the states of M are d-reachable and pairwise r-distinguishable, the test suite reduces to the set $S\Sigma[m' - m + 1]W$. This is equivalent to the test suite produced by the W-method when testing from a deterministic FSM. Where the specification does not satisfy these conditions, a larger test suite is required. Clearly, every sequence in $Tr(q)$ has length at most $m' + 1$. Based on the above definition, an algorithm for constructing $Tr(q)$ is provided in [5].

Consider the specification M as represented in Fig. 1 and the upper bound on the number of states of the implementation $m' = 4$. There is a single maximal set of pairwise r-distinguishable states, $Q^1 = \{0, 1, 2, 3\}$. Since $Q_S = \{0, 1, 2\}$, $Q_S^1 = \{0, 1, 2\}$. Thus the termination criterion for $TrIO(q)$ gives $m' - |Q_S^1| + 1 = 4 - 3 + 1 = 2$. Hence $Y = S\Sigma[2]W$.

4   Bounded Sequence Testing from Non-deterministic FSMs

This section shows how the above test generation method can be extended to the case of bounded sequences. In this case, the test suite will contain only sequences of length less than or equal to an upper bound $l \geq 1$ and will have to establish if the IUT behaves as specified for all sequences in $\Sigma[l]$. More formally, $Y \subseteq \Sigma[l]$ is a test suite if and only if for every M' in the fault model, $M' \leq_{\Sigma[l]} M$ if and only if $M' \leq_Y M$.

The extension is not straightforward, as it is not sufficient to extract the prefixes of length at most l from the test suite produced in the unbounded case. Consider, for example, M_n, $n \geq 2$, as represented in Fig. 2 (a), $m' = n + 2$ and $l = n + 1$. All states of M_n are d-reachable and pairwise r-distinguishable, $S = \{\epsilon, a, \ldots, a^n, b\}$ is a state cover of M_n and $W = \{a^n b\}$ is a characterization set of M_n. Thus $Y = S\Sigma[1]W = \{\epsilon, a, \ldots, a^n, b\}\{\epsilon, a, b\}\{a^n b\}$ and $pref(Y) \cap \Sigma[n + 1] = pref(a^{n+1}) \cup pref(\{a^i b a^{n-i} \mid 0 \leq i \leq n\}) \cup pref(bba^{n-1})$. Consider M'_n as represented in Fig. 2 (b). Let $D = \{axbybz \mid x, y, z \in \Sigma^*, \|x\| + \|y\| + \|z\| \leq n - 2\} \subseteq \Sigma[n + 1]$. It can be observed that $M'_n \leq_{\Sigma[n+1]} D M_n$, but $M'_n \leq_s M_n$ does not hold for any sequence $s \in D$. Since $pref(Y) \cap D = \emptyset$, $M'_n \leq_{pref(Y)} \Sigma[n+1] M_n$.

In what follows, it will be shown that state counting can be used to generate tests for bounded sequences, provided that the sets S and W will contain sequences of *minimum length* that reach or distinguish states of M; these sets will be called a proper state cover and a strong l-characterization set, respectively.

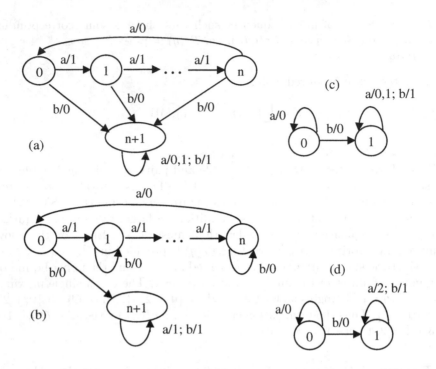

Fig. 2. The state-transition diagrams of M_n (a) and M'_n (b), M_0 (c) and M'_0 (d)

A few preliminary concepts are defined first. Without loss of generality, all FSMs considered are assumed to be initially connected, completely specified and observable and, furthermore, it will be assumed that every state can be reached by some sequence of length less than or equal to l.

For each state $q \in Q$, we define $level_M(q)$ as the length of the shortest path(s) from q_0 to q, i.e. $level_M(q) = min\{\|s\| \mid s \in \Sigma^*, q \in h_1(q_0, s)\}$. For M as represented in Fig. 1, $level_M(i) = i$, $0 \leq i \leq 3$.

States p and q of M are said to be l-dissimilar if p and q are r(k)-distinguishable for some $k \leq l - max\{level_M(p), level_M(q)\}$. The notion of l-dissimilar (l-similar) states is originally introduced in [1] and is used in [1] and [7] for constructing a minimal deterministic automaton and a minimal deterministic stream X-machine for a finite language. For M as represented in Fig. 1 and $l = 4$, states 2 and 3 are not l-dissimilar since they are not r(1)-distinguishable. On the other hand, every two other states of M are l-dissimilar.

Definition 1. *A set $S \subseteq \Sigma^*$ of input sequences is called a* proper state cover *of M if S is a minimal set such that every state q of M that is d-reachable by some sequence of length $level_M(q)$ is d-reached by some sequence s_q from S and $\|s_q\| = level_M(q)$.*

For M as represented in Fig. 1, $S = \{\epsilon, a, aa\}$ is a proper state cover of M.

The definition of a strong l-characterization set and the construction of the test suite are first given for a particular class of FSM specifications (quasi-deterministic FSMs) and then extended to the general type of FSM.

4.1   Quasi-deterministic FSMs

A *quasi-deterministic FSM* is a FSM in which for every $k > 0$, every pair of states that are not $\Sigma[k]$-equivalent are r(k)-distinguishable. In particular, this condition is satisfied by any deterministic FSM.

Definition 2. *Suppose M is a quasi-deterministic FSM. A set $W \subseteq \Sigma^*$ of input sequences is a called a strong l-characterization set of M, $l \geq 1$, if for every states p and q of M and every k, $0 < k \leq l - max\{level_M(p), level_M(q)\}$, for which p and q are r(k)-distinguishable, W r(k)-distinguishes p and q.*

Obviously, it is sufficient to check that W r(k)-distinguishes p and q for the minimum integer $k \leq l - max\{level_M(p), level_M(q)\}$ for which p and q are r(k)-distinguishable. That is, the shortest possible sequences are included in W. Naturally, W will r-distinguish any two l-dissimilar states of M.

Consider again M_n as represented in Fig. 2 (a), $n \geq 2$. For every pair (i, j), $0 \leq i < j \leq n$, i and j are $\Sigma[n - j]$-equivalent and r$(n - j + 1)$-distinguishable. Furthermore, $n + 1$ is r(1)-distinguishable from any other state. Thus M_n is quasi-deterministic, but not deterministic. $W = \{a, \ldots, a^n, b\}$ is a strong l-characterization set of M_n. On the other hand, M in Fig. 1 is not quasi-deterministic since states 2 and 3 are neither $\Sigma[2]$-equivalent nor r(2)-distinguishable.

4.2   Test Suite Generation

Suppose that the specification M is a quasi-deterministic FSM, S is a proper state cover of M and W is a strong l-characterization set of M. Q_S is used to denote the set of all states of M reached by sequences in S.

Let $Q^1, \ldots, Q^j$ denote the maximal sets of pairwise l-dissimilar states of M and let $Q_S^i = Q^i \cap Q_S$, $1 \leq i \leq j$. Under these conditions, the set $Tr(q_s)$ is defined analogously to section 3.4.

Then the test suite for bounded sequences is:

$$Z = (\bigcup_{s \in S} \{s\} pref(Tr(q_s)) W_\epsilon) \cap \Sigma[l] \setminus \{\epsilon\}$$

where $W_\epsilon = W \cup \{\epsilon\}$.

When $Q_S = Q$ and all states of M are pairwise l-dissimilar, the test suite reduces to the set $S\Sigma[m' - m + 1]W_\epsilon \cap \Sigma[l] \setminus \{\epsilon\}$. This is equivalent to the test suite produced in [8] for deterministic FSMs.

Consider again M_n, $n \geq 2$, as represented in Fig. 2 (a), $m' = n + 2$, $l = n + 1$ and the IUT M_n' as represented in Fig. 2 (b). $S = \{\epsilon, a, \ldots, a^n, b\}$ is a proper state cover of M_n and $W = \{a, \ldots, a^n, b\}$ is a strong l-characterization set of M_n.

There is a single maximal set of pairwise l-dissimilar states, $Q^1 = \{0, \ldots, n+1\}$. Since $Q_S = \{0, \ldots, n+1\}$, $Q_S^1 = \{0, \ldots, n+1\}$. Thus $Z = S\Sigma[1]W_\epsilon \cap \Sigma[n+1] \setminus \{\epsilon\}$ $= \{\epsilon, a, \ldots, a^n, b\}\{\epsilon, a, b\}\{\epsilon, a, \ldots, a^n, b\} \cap \Sigma[n+1] \setminus \{\epsilon\} = \{a^i \mid 1 \le i \le n+1\} \cup \{a^i b a^j \mid 0 \le i \le n, 0 \le j \le n-i\} \cup \{a^i bb \mid 0 \le i \le n-1\} \cup \{bba^i \mid 1 \le i \le n-1\} \cup \{bab, bbb\}$. As $abb \in Z$, $M_n' \le_Z M_n$ does not hold.

Note that W_ϵ, rather than only W, is needed in the definition of Z. Consider the specification M_0 as represented in Fig. 2 (c), $l = 2$, $m' = 2$ and the faulty implementation M_0' as represented in Fig. 2 (d). The only sequence that detects the fault in the IUT is ba. $S = \{\epsilon, b\}$ is a proper state cover of M_0 and $W = \{b\}$ is a strong l-characterization set of M_0. Thus $Z = S\Sigma[1]W_\epsilon \cap \Sigma[2] \setminus \{\epsilon\} = \{a, b, ab, ba, bb\}$. As $ba \in Z$, $M_0' \le_Z M_0$ does not hold. On the other hand, if W was used instead of W_ϵ in the definition of the test suite, then ba would not be contained in Z, so no fault would be detected.

4.3   General Type of FSMs

We now consider the general type of FSM specifications. First, note that the test suite given above may not be valid when the specification is not quasi-deterministic. Consider, for example, M_1 as represented in Fig. 3 (a), $m' = 3$ and $l = 4$. M_1 is not quasi-deterministic since states 0 and 1 are neither r(1)-distinguishable nor Σ-equivalent. All states of M_1 are d-reachable and pairwise l-dissimilar. Then, according to the above definitions, $S = \{\epsilon, a, aa\}$, $W = \{a, aa\}$ and $Z = S\Sigma[1]W_\epsilon \cap \Sigma[4] \setminus \{\epsilon\}$. Consider M_1' as defined in Fig. 3 (b). It can be observed that $M_1' \le_{\Sigma[4]} {}_{\Sigma^2 bb} M_1$, but $M_1' \le_s M_1$ does not hold for any sequence $s \in \Sigma^2\{bb\}$. Since $Z \cap \Sigma^2\{bb\} = \emptyset$, Z will detect no fault in M_1'.

Intuitively, this happens because, when M is not quasi-deterministic, there may be states p' and q' in the implementation M' that are r-distinguished by shorter sequences than those that r-distinguish the corresponding states p and q of the specification M. In our example, states 0 and 1 of M_1' are r-distinguished by $\{b\}$, whereas states 0 and 1 of M_1 are not r(1)-distinguishable and a longer sequence, aa, is used to r-distinguish between them. Consequently, the incorrect transition $h'(2, b) = (0, 1)$ cannot be detected by the above Z, since b was not included in W. On the other hand, the sequence $aabaa \in S\Sigma[1]W$, which results from the inclusion in W of the distinguishing sequence aa, has length 5 and, consequently, will not be contained in the test suite.

Now, observe that a sequence s can r-distinguish states p' and q' of the implementation only if the corresponding states p and q of the specification are not $\{s\}$-equivalent, i.e. there exists $g \in \Gamma^*$ such that $(s, g) \in (L_M(p) \setminus L_M(q)) \cup (L_M(q) \setminus L_M(p))$. Thus, the problem can be addressed by extending W to include any sequence s for which there exist states of M that are neither $\{s\}$-equivalent, nor r($\|s\|$)-distinguishable. Then the definition of a strong l-characterization set can be extended to the general type of FSM as follows:

Definition 3. *A set $W \subseteq \Sigma^*$ of input sequences is a called a strong l-characterization set of M, $l \ge 1$, if the following two conditions hold:*

- *For every states p and q of M and every k, $0 < k \leq l - max\{level_M(p), level_M(q)\}$, for which p and q are $r(k)$-distinguishable, W $r(k)$-distinguishes p and q.*
- *$s \in W$ for every $s \in \Sigma^*$ for which there exist states p and q of M with $\|s\| \leq l - max\{level_M(p), level_M(q)\}$ such that p and q are neither $\{s\}$-equivalent nor $r(\|s\|)$-distinguishable.*

With this revised definition of W, the construction of the suite remains the same as for quasi-deterministic FSM specifications. The following two sections of the paper provide the formal proofs to validate this construction.

For M_1 in the above example, states 0 and 1 are neither $\{b\}$-equivalent, nor r(1)-distinguishable. Thus $W = \{a, b, aa\}$. Then $aabb \in Z = S\Sigma[1]W_\epsilon \cap \Sigma[4] \setminus \{\epsilon\}$, so $M' \leq_Z M$ does not hold.

Consider again M as represented in Fig. 1, $m' = 4$ and $l = 4$. $S = \{\epsilon, a, aa\}$ is a proper state cover of M and $Q_S = \{0, 1, 2\}$. The pairs of states $(0, 2)$, $(0, 3)$, $(1, 2)$ and $(1, 3)$ are r-distinguished by $\{b\}$; 0 and 1 are r-distinguished by $\{ab\}$ and are Σ-equivalent. Since states 2 and 3 are Σ-equivalent, no other sequence needs to be included in W. Thus $W = \{b, ab\}$ is a strong l-characterization set of M. The maximal sets of pairwise l-dissimilar states of M are $Q^1 = \{0, 1, 2\}$ and $Q^2 = \{0, 1, 3\}$. Thus $Q_S^1 = \{0, 1, 2\}$ and $Q_S^2 = \{0, 1\}$ and the two termination criteria for $TrIO(q)$ give $m' - |Q_S^1| + 1 = 4 - 3 + 1 = 2$ and $m' - |Q_S^2| + 1 = 4 - 2 + 1 = 3$, respectively. The tree generated in the construction of $TrIO(1)$ is represented in Fig. 4. A node is a leaf if the path from the root to it has visited (after the root) $n_1 = 2$ states from Q^1 or $n_2 = 3$ states from Q^2. On the other hand, only paths of length at most $l - level_M(1) = 4 - 1 = 3$ need to be constructed; in Fig. 4, the remaining branches are drawn with dashed line.

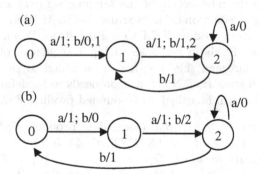

Fig. 3. The state-transition diagrams of M_1 (a) and M_1' (b)

5   The l-Bounded Product FSM

In order to compare the languages associated with two observable FSMs M and M', one can build a cross-product of their states, such that states (q, q') of the cross-product FSM correspond to pairs of states q, q' in the two FSMs.

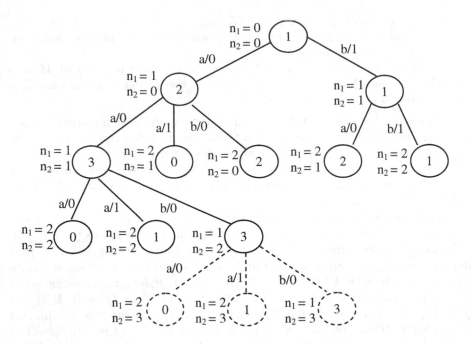

Fig. 4. The tree associated with $TrIO(2)$

A transition on input σ and output γ between states (q, q') and (p, p') exists in the cross-product FSM if and only if the transitions $(p, \gamma) \in h(q, \sigma)$ and $(p', \gamma) \in h'(q', \sigma)$ exist in M and M', respectively. The result of such a construction corresponds to the intersection of the languages $L(M)$ and $L(M')$. When checking that M' is a reduction of M, a transition in M' that is not allowed by M will lead in the cross-product FSM to a *Fail* state. When only the results produced by the two FSMs in response to input sequences of length at most l are compared, an integer i, $1 \leq i \leq l$, can be added to the state space and incremented by each transition. No transition needs to be defined for $i = l$. The resulting construction will be called an l-bounded product FSM of M and M'.

Definition 4. *Given* $l \geq 1$, *the* l-bounded product FSM *formed from* $M = (\Sigma, \Gamma, Q, h, q_0)$ *and* $M' = (\Sigma, \Gamma, Q', h', q'_0)$ *is the FSM* $P_l(M, M') = (\Sigma, \Gamma, Q_P, H, (q_0, q'_0, 0))$ *in which* $Q_P = Q \times Q' \times \{0, \ldots, l\} \cup \{Fail\}$ *with* $Fail \notin Q \times Q' \times \{0, \ldots, l\}$ *and* H *is defined by the following rules for all* $(q, q'), (p, p') \in Q \times Q', i \in \{0, \ldots, l-1\}, \sigma \in \Sigma$ *and* $\gamma \in \Gamma$:

 - *if* $(p, \gamma) \in h(q, \sigma)$ *and* $(p', \gamma) \in h'(q', \sigma)$ *then* $((p, p', i+1), \gamma) \in H((q, q', i), \sigma)$.
 - *if* $(p', \gamma) \in h'(q', \sigma)$ *and* $\gamma \notin h_2(q, \sigma)$ *then* $(Fail, \gamma) \in H((q, q', i), \sigma)$

and is undefined elsewhere.

As M and M' are observable, $P_l(M, M')$ is also observable (when M and M' are both deterministic, $P_l(M, M')$ is also deterministic [8]). On the other hand,

$P_l(M, M')$ is not completely specified even though M and M' are completely specified. More importantly, checking $M' \leq_{\Sigma[l]} M$ corresponds to establishing if the $Fail$ state of $P_l(M, M')$ is reachable.

Lemma 1. *The Fail state of $P_l(M, M')$ is not reachable if and only if $M' \leq_{\Sigma[l]} M$.*

Proof: From Definition 4, it follows that, for every $s \in \Sigma^*$ and $g \in \Gamma^*$, $H_g((q_0, q'_0, 0), s) = Fail$ if and only if $s = s_1\sigma$ with $s_1 \in \Sigma[l-1]$ and $\sigma \in \Sigma$ and $g = g_1\gamma$ with $g_1 \in \Gamma[l-1]$ and $\gamma \in \Gamma$ such that $g_1 \in h_2(q_0, s_1) \cap h'_2(q'_0, s_1)$ and $g_1\gamma \in h'_2(q'_0, s_1\sigma) \setminus h_2(q_0, s_1\sigma)$.

6 State Counting for Bounded Sequences

State counting can now be used to prove that, whenever the $Fail$ state is reachable, it will be reached by some sequence in the test suite. As in the unbounded case, it will be shown that the test suite contains all "minimal" input sequences that could reach $Fail$. Among the shortest sequences, the minimal sequences are those for which also the "distance" (defined in what follows) to the set S is the shortest. The basic idea is similar to that used in bounded sequence testing from deterministic FSM specifications [8]; however, when considering non-deterministic FSMs, we have to take into account that non-equivalent states may not necessarily be r-distinguishable (see Lemma 3).

Given $x \in \Sigma^*$ and $A \subseteq \Sigma^*$ with $\epsilon \in A$, the length of the shortest sequences(s) $t \in \Sigma^*$ for which there exists a sequence $s \in A$ such that $st = x$ is denoted by $d(x, A)$, i.e. $d(x, A) = min(\{\|t\| \mid t \in \Sigma^*, \exists s \in A \cdot st = x\}$. Since $\epsilon \in A$, the set $\{t \in \Sigma^* \mid \exists s \in A \cdot st = x\}$ is not empty, so $d(x, A)$ is well defined.

Lemma 2. *Let $p, q \in Q$, $p', q' \in Q'$, $U \subseteq \Sigma^*$ and $k > 0$. If $p' \leq_{U \ \Sigma[k]} p$, $q' \leq_{U \ \Sigma[k]} q$ and U $r(k)$-distinguishes p and q then U $r(k)$-distinguishes p' and q'.*

Proof: Follows by induction on k.

Lemma 3. *Let $s \in S$ and $t \in Tr(q_s)$ such that $\|st\| \leq l$ and s is the longest prefix of st that is in S. If $M' \leq_{(S \ \ s \ pref(t)W_\epsilon) \ \Sigma[l]} M$ then there exist $y_1 \in \{s\}pref(t) \setminus \{s\}$, $y_2 \in S \cup pref(y_1) \setminus \{y_1\}$ and $w_1, w_2 \in \Gamma^*$ such that the following two conditions hold:*

- $\|y_2\| < \|y_1\|$ *or* $\|y_2\| \leq \|y_1\|$ *and* $d(y_2, S) < d(y_1, S)$
- $H_{w_1}((q_0, q'_0, 0), y_1) = (q_1, q', \|y_1\|)$ *and* $H_{w_2}((q_0, q'_0, 0), y_2) = (q_2, q', \|y_2\|)$ *for some states $q_1, q_2 \in Q$ and $q' \in Q'$ such that $L_M(q_1) \cap L_{M'}(q') \cap (\Sigma \times \Gamma)[l - \|y_1\|] = L_M(q_2) \cap L_{M'}(q') \cap (\Sigma \times \Gamma)[l - \|y_1\|]$.*

Proof: Let i, $1 \leq i \leq j$. Suppose s_1 and s_2 are two distinct elements of S such that $q_{s_1}, q_{s_2} \in Q^i$ and let $q'_{s_1} \in h_1(q_0, s_1)$ and $q'_{s_2} \in h_1(q_0, s_2)$. Since q_{s_1} and q_{s_2} are l-dissimilar there exists k, $0 < k \leq l - max\{level_M(q_{s_1}), level_M(q_{s_2})\}$, such that q_{s_1} and q_{s_2} are $r(k)$-distinguishable. As W is a strong l-characterization set of M, W r(k)-distinguishes q_{s_1} and q_{s_2}. Since $M' \leq_{SW} {}_{\Sigma[l]} M$, $q'_{s_1} \leq_W {}_{\Sigma[k]} q_{s_1}$ and $q'_{s_2} \leq_W {}_{\Sigma[k]} q_{s_2}$. Thus, by Lemma 2, q'_{s_1} and q'_{s_2} are r(k)-distinguishable. Consequently, the sequences in S will reach at least $|Q^i_S|$ distinct states of M'.

On the other hand, since $t \in Tr(q_s)$, there is some $g \in \Gamma^*$ and i, $1 \leq i \leq j$, such that (t, g) visits states from Q^i exactly $m' - card(Q^i_S) + 1$ times when followed from q_s. Since S has already reached at least $|Q^i_S|$ states, there will be a state q' of M' that either has been visited twice by (t, g) or has been reached by some sequence in S. Thus, there exist $y_1 \in \{s\}pref(t) \setminus \{s\}$, $y_2 \in S \cup \{s\}pref(y_1) \setminus \{y_1\}$ and $w_1, w_2 \in \Gamma^*$ such that $h_{w_1}(q_0, y_1) = q_1$, $h_{w_2}(q_0, y_2) = q_2$, $h'_{w_1}(q'_0, y_1) = q'$ and $h'_{w_2}(q'_0, y_2) = q'$. for some states $q_1, q_2 \in Q^i$ and $q' \in Q'$. Then $H_{w_1}((q_0, q'_0, 0), y_1) = (q_1, q', \|y_1\|)$ and $H_{w_2}((q_0, q'_0, 0), y_2) = (q_2, q', \|y_2\|)$.

Let $\mu = max\{\|y_1\|, \|y_2\|\}$. We prove by contradiction that q_1 and q_2 are not r($l - \mu$)-distinguishable. Assume q_1 and q_2 are r($l - \mu$)-distinguishable, $\mu < l$. Since W is a strong l-characterization set of M, W r($l - \mu$)-distinguishes q_1 and q_2. On the other hand, since $M' \leq_{y_1, y_2} W {}_{\Sigma[l]} M$, $q' \leq_W {}_{\Sigma[l-\mu]} q_1$ and $q' \leq_W {}_{\Sigma[l-\mu]} q_2$. Thus, by Lemma 2, W would r($l-\mu$)-distinguish q' from itself. This is obviously a contradiction.

We now show that $\|y_2\| < \|y_1\|$ or $\|y_2\| \leq \|y_1\|$ and $d(y_2, S) < d(y_1, S)$. If $y_2 \in pref(y_1) \setminus \{y_1\}$ then $\|y_2\| < \|y_1\|$. Otherwise $y_2 \in S \setminus \{s\}$, so $\|y_2\| = level_M(q_2)$. Then there are two cases:

- $q_1 = q_2$. Then $level_M(q_2) \leq \|y_1\|$ so $\|y_2\| \leq \|y_1\|$. Since $y_1 \notin S$ and $y_2 \in S$, $d(y_2, S) < d(y_1, S)$.
- $q_1 \neq q_2$. We prove by contradiction that $\|y_2\| < \|y_1\|$. Assume $\|y_1\| \leq \|y_2\|$. Then $level_M(q_1) \leq \|y_1\| \leq \|y_2\| = level_M(q_2)$. Hence $level_M(q_1) \leq level_M(q_2) = \|y_2\|$. As $q_1, q_2 \in Q^i$, q_1 and q_2 are r($l - \|y_2\|$)-distinguishable. On the other hand, we have shown that q_1 and q_2 are not r($l - \mu$)-distinguishable. Since $\mu = \|y_2\|$, this is a contradiction.

Thus $\|y_2\| < \|y_1\|$ or $\|y_2\| \leq \|y_1\|$ and $d(y_2, S) < d(y_1, S)$. Since $\|y_2\| \leq \|y_1\|$, $\mu = \|y_1\|$, so q_1 and q_2 are not r($l - \|y_1\|$)-distinguishable.

Finally, we prove by contradiction that $L_M(q_1) \cap L_{M'}(q') \cap (\Sigma \times \Gamma)[l - \|y_1\|] = L_M(q_2) \cap L_{M'}(q') \cap (\Sigma \times \Gamma)[l - \|y_1\|]$. Assume otherwise and let $(s_0, g_0) \in L_{M'}(q') \cap (\Sigma \times \Gamma)[l - \|y_1\|] \cap ((L_M(q_2) \setminus L_M(q_1)) \cup (L_M(q_1) \setminus L_M(q_2)))$. Since W is a strong l-characterization set of M and q_1 and q_2 are not r($l - \|y_1\|$)-distinguishable, $s_0 \in W$. As $M' \leq_{y_1, y_2} W {}_{\Sigma[l]} M$ and $\|y_2 s_0\| \leq \|y_1 s_0\| \leq l$, it follows that $M' \leq_{y_1 s_0, y_2 s_0} M$. Thus, since $(s_0, g_0) \in L_{M'}(q')$, $(s_0, g_0) \in L_M(q_1)$ and $(s_0, g_0) \in L_M(q_2)$. This provides a contradiction, as required.

Lemma 4. Let $(q_1, q', j_1), (q_2, q', j_2) \in Q \times Q' \times \{0, \ldots, l\}$, $0 \leq j_2 \leq j_1 \leq l - 1$, and $(x, w) \in (\Sigma \times \Gamma)[l - j_1]$. Suppose $L_M(q_1) \cap L_{M'}(q') \cap (\Sigma \times \Gamma)[l - j_1] = L_M(q_2) \cap L_{M'}(q') \cap (\Sigma \times \Gamma)[l - j_1]$. If $H_w((q_1, q', j_1), x) = Fail$ then $H_w((q_2, q', j_2), x) = Fail$.

Proof: If $H_w((q_1, q', j_1), x) = Fail$ then $x = s\sigma$ with $s \in \Sigma[l - j_1 - 1]$, $\sigma \in \Sigma$ and $w = g\gamma$ with $g \in \Gamma[l - j_1 - 1]$, $\gamma \in \Gamma$ such that $g \in h_2(q_1, s) \cap h'_2(q', s)$ and $g\gamma \in h'_2(q', s\sigma) \setminus h_2(q_1, s\sigma)$. Since $L_M(q_1) \cap L_{M'}(q') \cap (\Sigma \times \Gamma)[l - j_1] = L_M(q_2) \cap$

$L_{M'}(q') \cap (\Sigma \times \Gamma)[l - j_1]$, $g \in h_2(q_2, s) \cap h'_2(q', s)$ and $g\gamma \in h'_2(q', s\sigma) \setminus h_2(q_2, s\sigma)$. As $j_2 \leq j_1$, it follows that $H_w((q_2, q', j_2), x) = Fail$.

Lemma 5. *If $M' \leq_Z M$ then the Fail state of $P_l(M, M')$ is not reachable.*

Proof: We provide a proof by contradiction. Assume $Fail$ is reachable and let X be the set of all sequences of minimum length that reach $Fail$. Let $\mu = min\{d(x, S) \mid x \in X\}$ and $X_\mu = \{x \in X \mid d(x, S) = \mu\}$.

We prove by contradiction that $X_\mu \cap (\bigcup_{s \in S}\{s\}pref(Tr(q_s))) \neq \emptyset$. Assume $X_\mu \cap (\bigcup_{s \in S}\{s\}pref(Tr(q_s))) = \emptyset$ and let $x \in X_\mu$. Then $x \notin \bigcup_{s \in S}\{s\}pref(Tr(q_s))$. Since $\epsilon \in S$, $x \in S\Sigma^*$. Let $s \in S$ be the longest prefix of x that is in S. Then $x = stu$, for some $t \in Tr(q_s)$ and $u \in \Sigma^* \setminus \{\epsilon\}$ with $\|stu\| \leq l$ and there exist $g, v \in \Gamma^*$ such that $g \in h_2(q_0, st) \cap h'_2(q'_0, st)$ and $gv \in h'_2(q'_0, stu) \setminus h_2(q_0, stu)$. Since $M' \leq_Z M$ and $(S \cup \{s\}pref(t)W_\epsilon) \cap \Sigma[l] \subseteq Z$, by Lemma 3, there exist $y_1 \in \{s\}pref(t) \setminus \{s\}$, $y_2 \in S \cup pref(y_1) \setminus \{y_1\}$ and $w_1, w_2 \in \Gamma^*$ such that the following two conditions hold:

- $\|y_2\| < \|y_1\|$ or $\|y_2\| \leq \|y_1\|$ and $d(y_2, S) < d(y_1, S)$
- $H_{w_1}((q_0, q'_0, 0), y_1) = (q_1, q', \|y_1\|)$ and $H_{w_2}((q_0, q'_0, 0), y_2) = (q_2, q', \|y_2\|)$ for some states $q_1, q_2 \in Q$ and $q' \in Q'$ such that $L_M(q_1) \cap L_{M'}(q') \cap (\Sigma \times \Gamma)[l - \|y_1\|] = L_M(q_2) \cap L_{M'}(q') \cap (\Sigma \times \Gamma)[l - \|y_1\|]$.

Let $z \in \Sigma^*$ such that $st = y_1 z$ and $w_z \in \Gamma^*$ such that $g = w_1 w_z$. As $H_{gv}((q_0, q'_0, 0), x) = Fail$, $H_{w_z v}((q_1, q', \|y_1\|), zu) = Fail$. Then, by Lemma 4, $H_{w_z v}((q_2, q', \|y_2\|), zu) = Fail$. Thus $H_{w_2 w_z v}((q_0, q'_0, 0), y_2 zu) = Fail$. If $\|y_2\| < \|y_1\|$ then $y_2 zu$ is a sequence shorter than x that reaches $Fail$. Thus $x \notin X$, which is a contradiction. Otherwise, $\|y_2\| = \|y_1\|$ and $d(y_2, S) < d(y_1, S)$. Since no sequence in $\{y_1\}pref(zu)$ is contained in S, $d(y_2 zu, S) < d(y_1 zu, S)$. Consequently $\|y_2 zu\| = \|x\|$ and $d(y_2 zu, S) < d(x, S)$. Thus $x \notin X_\mu$, which provides a contradiction, as required. Hence $X_\mu \cap (\bigcup_{s \in S}\{s\}pref(Tr(q_s))) \neq \emptyset$.

On the other hand, since $M' \leq_Z M$, no sequence in Z will reach $Fail$. Thus $X_\mu \cap Z = \emptyset$. As $\bigcup_{s \in S}\{s\}pref(Tr(q_s)) \subseteq Z$, this provides a contradiction, as required. Hence $Fail$ is not reachable.

Theorem 1. *$M' \leq_{\Sigma[l]} M$ if and only if $M' \leq_Z M$.*

Proof: "$\Rightarrow$": Obvious, since $Z \subseteq \Sigma[l]$. "$\Leftarrow$": Follows from Lemmas 5 and 1.

7   Conclusions

This paper extends the state counting based method of deriving tests from a non-deterministic FSM to the case of bounded sequences. The method for bounded sequences has practical value, as many applications of finite state machines actually use only input sequences of limited length. In such applications, the test suite produced may contain only a small fraction of all sequences of length less than or equal to the upper bound. The test suite for M_n in our example (Fig. 2 (a)), $m' = n + 2$ and $l = n + 1$ will contain only $(n^2 + 9n + 6)/2$ sequences out of a total of $2^{n+2} - 2$ sequences.

Improvements in the size of the test suite may be obtained by using only subsets of W to identify the states reached by the sequences in $Tr(q_s)$, in a way similar to the Wp method for unbounded [3] and bounded [8] sequences. This will be the subject of a future paper. Possible future work also involves the generalization of these bounded sequence testing methods to classes of extended finite state machines, such as stream X-machines [6].

References

1. Campeanu, C., Santean, N., Yu, S. Minimal cover automata for finite languages. *Theoretical Computer Science*, 267, 3-16 (1999)
2. Chow, T. S. Testing software design modeled by finite state machines, *IEEE Transactions on Software Engineering*, 4(3), 178-187 (1978)
3. Fujiwara, S., von Bochmann, G., Khendek, F., Amalou, M. and Ghedamsi A. Test Selection Based on Finite State Models. *IEEE Transactions on Software Engineering*, 17(6), 591-603 (1991)
4. Hierons, R. M. Adaptive testing of a deterministic implementation against a nondeterministic finite state machine. *The Computer Journal*, 41(5), 349-355 (1998)
5. Hierons, R. M. Testing from a Non-Deterministic Finite State Machine Using Adaptive State Counting. *IEEE Transactions on Computers* 53(10), 1330-1342 (2004)
6. Holcombe, M., Ipate, F. *Correct Systems: Building a Business Process Solution. Springer Verlag*, Berlin (1998)
7. Ipate, F. On the Minimality of Finite Automata and Stream X-machines for Finite Languages, *The Computer Journal*, 48(2), 157-167 (2005)
8. Ipate, F. Bounded Sequence Test Selection from Finite State Machines, submitted.
9. Lee, D. and Yannakakis, M. Principles and Methods of Testing Finite State Machines - A Survey. *Proceedings of the IEEE*, 84(8), 1090-1123 (1996)
10. Luo, G. L., Bochmann, G. v. and Petrenko, A. Test selection based on communicating nondeterministic finite-state machines using a generalized Wp-method. *IEEE Transactions on Software Engineering*, 20(2), 149-161 (1994)
11. Petrenko, A., Yevtushenko, N., Lebedev, A. and Das, A. Nondeterministic state machines in protocol conformance testing. In *Proc. of Protocol Test Systems, VI (C-19)*, Pau, France, 28-30 September, Elsevier Science, 363-378 (1994)
12. Petrenko, A., Yevtushenko, N., Bochmann G.v. Testing deterministic implementations from nondeterministic FSM specifications. In *Proc. of 9th International Workshop on Testing of Communicating Systems (IWTCS'96)*, Darmstadt, Germany, 9-11 September 1996, Chapman and Hall, 125-140 (1996)
13. Sidhu, D. and Leung, T. Formal methods for protocol testing: A detailed study. *IEEE Transactions on Software Engineering*, 15(4), 413-426, 1989.
14. Yevtushenko, N. V., Lebedev, A. V. and Petrenko, A. F. On checking experiments with nondeterministic automata. *Automatic Control and Computer Sciences*, 6, 81-85 (1991)

LaTe, a Non-fully Deterministic Testing Language

Emmanuel Donin de Rosière[1], Claude Jard[2], and Benoît Parreaux[1]

[1] France Télécom R&D,
2 Avenue Pierre Marzin 22307 Lannion, France
emmanuel.doninderosiere@francetelecom.com,
benoit.parreaux@francetelecom.com
[2] ENS Cachan,
Campus de Kerlann, 35 170 Bruz, France
claude.jard@bretagne.ens-cachan.fr

Abstract. This paper presents a case study which is the test of a voice-based service. To develop this application, we propose new functionalities for testing languages and a new language called *LaTe* that implements them.

With LaTe, one testing scenario can describe several different executions and the interpreter tries to find the execution that best fits with the real behavior of the System Under Testing (SUT).

We propose an operational semantics of these non-deterministic operators. Experimental results of the test of the voice-based service are also included.

1 Introduction

The world of testing languages remains complex and dense: there are often more than one language by application domain, e.g. hardware testing [1, 2], protocol testing [3], component testing [4, 5]... Several systems take programming languages in order to use them for testing purpose [6, 7]. One objective of this paper is to experiment with a paradigm that is non-deterministic testing. Despite a more complex interpretation, we will prove that this can increase the quality of black box testing languages (also called functional testing languages) on complex SUT.

This paper focuses on black box testing, i.e. the fact of testing a system where inputs and outputs of functions are known, but internal code structure is irrelevant. It is a form of testing which cannot target specific software components or portions of the code. So, in the rest of this article, we will use the same definitions than TTCN [8] which is the reference in this domain. This language, in its version 3, tries to be as generalist as possible and the most independent of the SUT.

SUT are more and more complex (and sometimes non-deterministic), so we need a testing language that has to be as powerful and expressive as a programming language. This is exemplified by the evolution of TTCN. Another instance

M.Ü. Uyar, A.Y. Duale, and M.A. Fecko (Eds.): TestCom 2006, LNCS 3964, pp. 71–86, 2006.
© IFIP International Federation for Information Processing 2006

is Tela (an UML 1.4 based testing language) [9], which gives lots of control structures like loops, branches, guards, interleaving and so on. However, Tela is more
a test description language than a test execution language.

New operators which are resource-greedy can be proposed if they have good
qualities because the time of test execution is not something important in our
context. For decreasing the cost of learning a new testing language, they must
be as easy and generalist as possible in order to test several different SUT with
the same testing language (e.g. web services, java SUT etc).

This paper is organized as follows. In the next section, we present *LaTe*, a
testing language which implements non deterministic operators. We also give
the semantics of these operators. Then, we give some information about the
case study: *testing a voice-based service* and the actual difficulties. Section 4 is
dedicated to the test architecture of the experiments and to the examples of test
cases in order to show the pros and cons of the constructions proposed here.
Finally, we discuss the results and conclude.

2   Presentation of *LaTe*

One of the aims of this study is to show that non-deterministic operators can
be very useful in testing languages. In order to evaluate these new operators,
we have created a new language, but they could be added in a more complete
language like TTCN. In this section, the main characteristics of LaTe will be
presented. Then a description of the non-deterministic operators will be given.
Finally, we will see the power of these operators and LaTe through a small
example of unanimity vote.

2.1   Main Requirements

In Late, we try to select some important features:

Genericity: The language should be able to test different SUT. It is unfortunate
to have to learn a different testing language for each system you want to test.
So, it is better if an unique language can test every SUT (may be through
an adaptation layer). As in TTCN, the use of *SUT adapters* can be useful.
Nevertheless, unlike TTCN, it is interesting to write only one *SUT adapter*
to test different SUT of the same type, e.g. one SUT adapter for testing
all the SUT written in JAVA, one SUT adapter for Web services. Actually,
three *SUT adaptor factories* have been written: the Java one, the C one and
the socket one and. They contain respectively 250, 400 and 100 lines of code.
It was very easy to write them because of the use of reflection in Java and
the Java Native Interface that allows use C functions in Java code.

Using stubs: It is sometimes useful to develop stubs in order to test some SUT.
But, in most languages, the user needs to write a specific stub for each
test case. In addition, the interactions between the SUT and the tester and
those between the stub and the SUT are often described separately. Thus,
it is necessary to add synchronization points in the scenario in order to

express precedence between an action of the tester and one of the stub. With LaTe, like for SUT adapters, you need to write only one *stub constructor* to create stubs for a specific type of SUT and all the interactions between stubs and the SUT are directly written in the global scenario. For example, the following code specifies the creation of a Java stub that implements the interface `MyInterface`.

```
mystub:=createStub("Java","MyInterface");
//You create a stub
callSut(mysut,"register",mystub);
//You send it to the previously created sut
stubcall:=getCalled(mystub);
//You verify that the stub is called by the sut
```

This example shows that you can easily describe the general behavior of the system. In a system where you must separate the behavior of the stub and the tester component, the tester has to send a message to the stub in order to inform it that the call to the SUT have been made and it will receive a call from it. With our scenario, it is more compact and we can easily show the event succession.

Powerful API: SUT become more and more complex and non deterministic. A large collection of APIs is then needed to check that the value returned by the SUT is in accordance with the specification. The easiest way to have a large collection of API is to allow testers to use API from a programming language. In our language, we can easily use the Java APIs.

Dynamic language: Just like above, some SUT can create dynamically PCO (Point of Control and Observation) so we need to discover them during the execution of the scenario and dynamically connect to these PCO.

Something like a scripting language: Sometimes, the tester needs to test in real-time the SUT. For example, when the tester debugs a test scenario, he may need to have a console to execute his own commands. So an interpreted language will therefore be more useful than a compiled one. Moreover, it will be easier to have a dynamically typed language in order to write quickly the scenario and avoid the check and cast of values each line.

2.2 Non-deterministic Operators

As mentioned above, SUT become more complex, and this complexity leads to more non determinism. However the non determinism in a SUT is something quite difficult for a test writer because he has to infer the possible state of the SUT. So he has to add lots of if ... then ... elsif ... and if it cannot distinguish two cases, he may have to indicate an inconclusive verdict. Unfortunately, some SUT are intrinsically non deterministic: there are not fully observable, so we cannot know their internal state just from their outputs. To succeed in testing non deterministic SUT, we have hadded two non deterministic operators in the language. These operators are the non deterministic interleaving and the non deterministic choice. They are presented in details in this section. Note that a similar paradigm has already been used in the procol testing domain in [10].

In a nutshell, the solution of the non deterministic SUT problem is to have different executions at the same time for the same scenario. When a divergence point is found in an execution, several executions are created in order to cover all the possibilities. When a contradiction is detected in an execution, then this execution is stopped and destroyed. The test passes when there is at least one execution that arrives at the end of the scenario, otherwise, it fails.

The first non-deterministic operator we define is noted ||. It represents a choice in a scenario. For instance, a:=3||a:=5 means that we have two executions: one where a equals 3 and another where a equals 5. All statements afterwards will be executed twice, once for each execution.

It can be done because the statements in one execution are independent of the ones of another execution. However, communications between the tester and the SUT do modify the state of the SUT, so precautions must be taken before executing these instructions.

In *LaTe*, when an execution wants to send a message, it always requests it to a component that can view all the current executions.

The other non-deterministic operator is noted &&. It represents a non deterministic interleaving. It executes all of the possible interleavings of two given branches. For example, {A;B}&&{C;D} is equivalent to:

$$\{A;B;C;D\}||\{A;C;B;D\}||\{A;C;D;B\}||\{C;D;A;B\}||\{C;A;B;D\}||\{C;A;D;B\}.$$

This operator helps to easily describe two independent behaviors. For example, when stubs are used in a scenario, it allows to specify that the behavior of one is independent of the other. Nevertheless, the longer the branches are, the more different executions are evaluated. Thus, in order to decrease this number, we decide that only communication with the SUT statements will cause the evaluation of new execution. For instance, with the following scenario: { ?a; assert(a==5) } && { ?b;assert(b==6) }, only two executions will be evaluated: ?a;assert(a==5);?b;assert(b==6) and ?b;assert(b==6); ?a;assert(a==5) instead of the 6 possible interleavings. We can do this because we suppose the instructions that do not communicate with the SUT can be executed in any order without changing anything. This can be true only if a statement of a particular branch cannot influence other branches. In order to prevent this, we implement a locking variable system: if a variable is modified in a particular branch, it cannot be read or written in other branches. With all these restrictions, we can verify the initial hypothesis.

As we said before, we need a component that can view all the executions in order to decide which messages can be sent and when. This component will be called *top level* in this article. Each time an execution wants to send a message, it sends a request to the top level. If all the executions want to send the same message, the top level emits it and wakes up all the executions. If all the executions want to emit different messages then the top level may choose randomly an execution, sends the corresponding message and destroys the other messages or may stop the evaluation of the scenario and returns an error depending of the configuration of the *LaTe* interpreter.

2.3   Operational Semantics

In the following equations, $\mathfrak{P}(S)$ denotes the set of all subsets of S, $\bullet$ the concatenation operator and $\sqcup$ the shuffle product.

Let $[\![]\!]$ be the following semantic operator:

$$[\![]\!] : \mathbb{P} \times (\Sigma_r \times \mathbb{R}^+) \quad \times \mathbb{R}^+ \times \mathfrak{P}(\Sigma_e \times \mathbb{R}^+) \rightarrow \tag{1}$$
$$\mathfrak{P}(\mathbb{P} \times (\Sigma_r \times \mathbb{R}^+) \quad \times \mathbb{R}^+ \times \mathfrak{P}(\Sigma_e \times \mathbb{R}^+) \times \mathfrak{P}(\Sigma_e \times \mathbb{R}^+))$$

$\mathbb{P}$ corresponds to the set of all programs formed by waiting, sending and receiving statements and interleaving, alternative and sequence operators. It also contains the null program (P_0).

Σ_r is the set of messages that can be sent by the SUT and Σ_e is the set of all messages sent by the tester.

This operator expresses the set of all possible futures from a program P, a trace σ^i containing the received messages and their arriving times, the current time, and a list of sendable messages. Each future is made up by the remaining program to execute (which can be null), the current time, the list of sendable messages and the set of messages to send. If the remaining program is null, then all the statements were executed else the execution is waiting for the authorization to send a message to the SUT.

The null program can be executed but does not modify the context:

$$[\![P_0, \sigma^i, t, a]\!] = \{(P_0, \sigma^i, t, a, \phi)\} \tag{2}$$

Let $?e^{<w}$ be the statement meaning that the next message received must be e and that it must arrive before w seconds. If it is true ($\sigma^i(0) = e^\tau$ and $t + w > \tau$), then e is deleted from the trace and the clock is moved forward by τ.

$$[\![?e^{<w}, \sigma^i, t, a]\!] = \begin{cases} \{(P_0, \sigma^{i+1}, t+\tau, a, \phi)\} & \text{if } \sigma^i(0) = e^\tau \wedge t + w > \tau \\ \phi & \text{otherwise} \end{cases} \tag{3}$$

Let $!E$ be the statement that means **E** must be sent to the SUT. First, if E belongs to the set of sendable messages, it is sent. If not, the execution is stopped, and **E** (associated with the current time) is added to the list of messages we want to send. We also verify that any more incoming message does not arrive.

$$[\![!e, \sigma^i, t, a]\!] = \begin{cases} \{(P_0, \sigma^i, t', a \setminus \{e\}, \phi)\} & \text{if } \exists t', t'' \in \mathbb{R}^+, \exists f \in \Sigma | \\ & e^{t'} \in a \wedge \sigma^i(0) = f^{t''} \wedge t'' \geq t' \\ \phi & \text{if } \forall t' \in \mathbb{R}^+, \exists t'' \in \mathbb{R}^+, \exists f \in \Sigma | \\ & e^{t'} \in a \wedge \sigma^i(0) = f^{t''} \wedge t'' < t' \\ \{(!e, \sigma^i, t, a, \{e^t\})\} & \text{otherwise} \end{cases} \tag{4}$$

In order to execute $P; Q$, P is executed first with the current environment. Then, for each possible future where all the statements were executed, Q is executed with the new environment. For each future where a sending authorization is awaiting, the remaining program is rebuilt.

$$\llbracket P; Q, \sigma^i, t, a \rrbracket = \bigcup \{ \llbracket Q, \sigma^j, t', a' \rrbracket | (P_0, \sigma^j, t', a', \phi) \in \llbracket P, \sigma^i, t, a \rrbracket \}$$
$$\cup \bigcup \{ (P'; Q, \sigma^j, t', a', d) | (P', \sigma^j, t', a', d) \in \llbracket P, \sigma^i, t, a \rrbracket \wedge P' \neq P_0 \}$$
$$(5)$$

The possible futures of $P||Q$ are the union of possible futures of P with those of Q:

$$\llbracket P||Q, \sigma^i, t, a \rrbracket = \llbracket P, \sigma^i, t, a \rrbracket \cup \llbracket Q, \sigma^i, t, a \rrbracket \qquad (6)$$

The interleaving operator (&&) is the most complex operator in these semantics.

In order to execute $P\&\&Q$, P and Q are executed separately. To do this, two disjoint sets a_p and a_q are extracted from a. Each set corresponds to the part of a used by each process. $\sigma^i \setminus \sigma^j$ is also divided into two parts thanks to two projections h_p and h_q. If the two processes are fully executed, the execution is well finished and the new current time is the maximum of the two final times. If at least one of them is waiting for a sending authorization, the state of the program is rebuilt at the moment of the send of the message. Therefore, waiting times ($S(t' - t)$ and $S(t'' - t)$) have to be added in order to take the passing time into account.

$$\llbracket P\&\&Q, \sigma^i, t, a \rrbracket = \{ (P_0, \sigma^j, max(t', t''), a \setminus (a_p \cup a_q), \phi), \exists j \geq i,$$
$$\exists h_p, h_q \in \text{Projection}, \exists t', t'' \in \mathbb{R}^+, \exists a_p, a_q \in \mathfrak{P}(\Sigma_e \times \mathbb{R}^+)$$
$$|\sigma^i \setminus \sigma^j \in h_p(\sigma^i \setminus \sigma^j) \sqcup h_q(\sigma^i \setminus \sigma^j)$$
$$\wedge (P_0, \sigma^j, t, \phi, \phi) \in \llbracket P, h_p(\sigma^i \setminus \sigma^j) \bullet \sigma^j, t', a_p \rrbracket$$
$$\wedge (P_0, \sigma^j, t, \phi, \phi) \in \llbracket Q, h_q(\sigma^i \setminus \sigma^j) \bullet \sigma^j, t'', a_p \rrbracket$$
$$\wedge a_p \cap a_q = \phi \wedge a_p \subset a \wedge a_q \subset a \}$$
$$\cup \{ ((S(t' - t); P')\&\&(S(t'' - t); Q'), \sigma^j, t, \phi, d_p \cup d_q) \setminus \exists j \geq i,$$
$$\exists h_p, h_q \in \text{Projection}, \exists t', t'' \in \mathbb{R}^+, \exists a_p, a_q \in \mathfrak{P}(\Sigma_e \times \mathbb{R}^+)$$
$$|\sigma^i \setminus \sigma^j \in h_p(\sigma^i \setminus \sigma^j) \sqcup h_q(\sigma^i \setminus \sigma^j)$$
$$\wedge (P', \sigma^j, t', \phi, d_p) \in \llbracket P, h_p(\sigma^i \setminus \sigma^j) \bullet \sigma^j, t, a_p \rrbracket$$
$$\wedge (Q', \sigma^j, t'', \phi, d_q) \in \llbracket Q, h_q(\sigma^i \setminus \sigma^j) \bullet \sigma^j, t, a_q \rrbracket$$
$$\wedge a_p \cap a_q = \phi \wedge a_p \cup a_q = a \wedge d_p \cup d_q \neq \phi \}$$
$$(7)$$

We have just seen that some executions may enter in a waiting state. In fact, an emission is something that cannot be cancelled. As a result, we have to check that all executions want to send the same message. Indeed, in order to choose what and when an emission can be made, we must have a total view of all the executions. It is why the use of a *top level* is unavoidable. It is the only element that can see the global state of the tester. This component gets back the emission requests, computes them in order to find if the executions are determinable. After that, it gives the authorization to send messages.

Let TL be semantics of the *top level*:

$$TL : \mathfrak{P}(\mathbb{P} \times (\Sigma_r \times \mathbb{R}^+) \quad \times \mathbb{R}^+ \times \mathfrak{P}(\Sigma_e \times \mathbb{R}^+) \times \mathfrak{P}(\Sigma_e \times \mathbb{R}^+)) \to \text{Bool}$$

TL is defined as follow:

$$TL(P_g) = \begin{cases} true \text{ if } \exists \sigma^i \in (\Sigma_r \times \mathbb{R}^+) \quad , \exists t \in \mathbb{R}^+ | (P_0, \sigma^i, t, \phi, \phi) \in P_g \\ TL(\bigcup \{[\![P, \sigma, t, d]\!] | \exists a \in \mathfrak{P}(\Sigma_e \times \mathbb{R}^+), (P, \sigma, t, \phi, a) \in P_g\}) \text{ if } \exists d \\ \in \mathfrak{P}(\Sigma_e \times \mathbb{R}^+) \, , \, d = \{e^{max(t_i)} | \forall d_i, \exists t_i \ e^{t_i} \in d_i\} \wedge d \neq \phi \wedge P_g \neq \phi \\ false \text{ otherwise} \end{cases}$$

$$(8)$$

If one possible future finishes its execution, the trace conforms to the testing scenario. Otherwise, all the executions are waiting at least one sending authorization. If the intersection of all of these requests is not empty, we authorize their emission. In any other cases, we cannot find a consensus. Thus, the test is declared false. Denote that the same emission for different executions must be done in the same time (because there can be only one real emission). We emit therefore this message at the maximum time given by all executions.

With these semantics, we do not assure that the emission will immediately take place when the request is made. We use a *best effort* policy. Nevertheless, these semantics can be easily modified in order to emit the message immediately, but it risks compelling too much the executions of the scenario.

Finally, we just have to compute $TL([\![P, \sigma^0, 0, \phi]\!])$ to find if a trace σ satisfies a program P.

2.4 An Example: Unanimity Vote

We point out some of the advantages of these operators through a small example: a system of unanimity vote. Here, the SUT is a java class that can be called to register to the electoral list an object implementing a particular interface. When a question is asked to the SUT, it transmits the question to some electors. These calls can be executed concurrently and if one of the electors replies `false`, the returned value of the SUT will be `false` and all the electors may not be asked to vote (it is a unanimity vote).

This example is quite difficult to test using classical systems because:

- it has to use stubs for electors and has to control each stub;
- some of the interactions are concurrent. We have to verify if the test scenario agrees with all the possibilities of interleaving the SUT interactions;
- it contains lots of non-determinism: if one of the electors replies `false`, the SUT may continue to call all of the electors but it also can stop and directly replies to the tester. The test scenario must describe all of these cases.

With LaTe, it is quite simple to write a testing case for this SUT. For example, for 2 electors, the scenario may be the following:

```
1   mysut:=createStub("Java","SUT");
2   elector1:=createStub("Java","ElectorInterface");
3   callSut(mysut,"register",elector1);
4   //We create a stub and register it to the SUT
5   elector2:=createStub("Java","ElectorInterface");
6   callSut(mysut,"register",elector2);
7   //We create a stub and register it to the SUT
8
9   sutcall:=callSUT(mysut,"ask","Is 42 prime ?");
10  {
11    @<10*sec{stubcall1:=getCalled(elector1)};
12    replySUT(stubcall1,false);
13    //The SUT asks the question to the elector 1 and we reply false
14  }
15  &&
16  {
17    {
18      @<10*sec{stubcall2:=getCalled(elector2)};
19      replySUT(stubcall2,true);
20      //The SUT may ask the question to the elector 2 and we reply true
21    }
22    ||
23    {
24    }
25    //But it is optional
26  };
27  assert(getReply(callSut)==false);
28  //The answer is "false"
```

Listing 1.1. LaTe scenario for the unanimity vote

As mentioned above, LaTe evaluates all the possible interleavings of the scenario. In other words, whatever the order of stub calls by the SUT and whatever the number of called stubs, LaTe is capable of verifying that the execution fits the specification described in the scenario. Time specification can also be easily added by using the @ operator.

Nevertheless, LaTe may have difficulties to evaluate particular scenarios. For instance, the pseudo code { ?a;!A } && { ?b;!B } is problematic if the SUT sends the two messages a and b, without waiting for the replies A or B, because this behavior is extremely non deterministic. All the executions are presented in Figure 1.

In this example, if the SUT sends a and waits for A before sending b, there is also a problem. All the branches beginning by ?b are not executable because

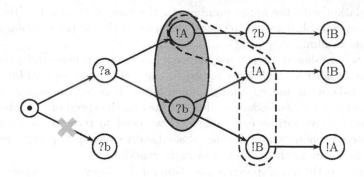

Fig. 1. All the possible interleavings

the first message received by the tester is **a** and the assertion **?b**, which signifies that the next message receive by the tester is **b**, will be broken. Thus, in the next step, two executions are possible: **?b** and **!A**. The second execution (with **?b**) is in a waiting state, because the SUT does not send an additional message and the first execution cannot emit **A**, because there is an execution that does not want to send something. We are in a livelock. In order to resolve this conflict, the scenario must contain information about timeouts in all the *receive* instructions, e.g. if the **?b** lasts more than 5 seconds, this execution will be destroyed, and there will be only one possible execution (the first one). The evaluation of the scenario will continue.

In other case, if the SUT sends the two messages **a** and then **b** without waiting for the replies **A** or **B**, there will be another problem. Three executions will be then possible. They are displayed with dash in Figure 1. Two of these executions want to send **A** and the last intends to emit **B**. This is because of the exact symmetry of the test scenario (there is no difference between the two stubs). In this case, we decide that LaTe may choose randomly an execution and continues it. Then, LaTe prints a warning message in order to warn the tester that the scenario contains conflicts. LaTe may also stop the execution according to its configuration.

Through this example, we can see some of the advantages of the non-determinism operators of LaTe: they permit to easily describe the parallelism and the non-determinism of the SUT. Although this example is an extreme case of non-determinism associated with concurrency, we can test quite easily this behavior. Nevertheless, the tester has to think of all possible executions. In the next sections, a more complex example will be studied in details, the voice-based service that we will just study: the vocal-based telephone directory.

3   Description of the Case Study: Testing a Voice-Based Service

We decided to validate our approach on the test execution of voice-based services and particularly a vocal-activated telephone directory. In this service, the user gives a name and the service seeks this name and then proposes to put the user in relation with the found number. In the case of homonyms, the service proposes several solutions and requests the user to choose the solution which is appropriate to him.

The new voice-based services use intensively speech synthesis and recognition. These functionalities simplify the access to the voice-based services but complicate their validation. Indeed, they produce lots of non-determinism if we try to automatically test it. For example, the volume and the speed of the voice during two different conversations can change. If we need to recognize automatically the sentences pronounced by the voice-based services using a speech recognition tool, we should make the verdict even more random.

Furthermore, the use of speech recognition for the tester also increases the non-determinism of all the system. Many factors can affect the result of the speech recognition such as the quality of the transmitted messages and the quality of the line and so on. It is possible for a same message to be correctly recognized the first time but not the following times. Thus, the presence of speech recognition and synthesis causes a significant number of *inconclusive* verdicts during automatic tests.

One current solution for this problem is to replace all speech signals emitted and expedted by the platform by DTMF (Dual Tone Multi Frequency) signals. These signals are the sounds produced when you dial a number with your phone. This solution is not completely satisfactory because it requires the modification of the voice-based service. The verdict of the tests can also be deteriorated by these modifications. The other solution is to carry out the tests, taking the risk of obtaining a significant number of *inconclusives* verdicts. No other solution are possible with common testing languages. The language that we propose in this article enables us to bring a new solution to this problem by using non-deterministic operators in the test scenario. So this vocal-based phone directory, although very simple, is enough to clarify the interests of our language.

4   Methodology of the Experiment

In this section, we will see in details how the vocal-based telephone directory was tested using LaTe. First, we will see the test architecture of this experiment, then we study the communication protocol between the tester and the calling platform. Finally, some test scenarios will be presented.

4.1   Test Architecture

In section 3, we discussed in details how works the vocal-based telephone directory testing here. Thus, we develop a special test architecture in order to allow the tester to connect to the voiceXML service. This architecture is represented in Figure 2.

The tester represents the machine where LaTe and the scenarios are executed. It communicates through a socket with the call API. This computer is linked to a call card which can makes calls and conversation on an analogic line. Therefore, this computer can call the voice-based service. As we just said, there is a dialog

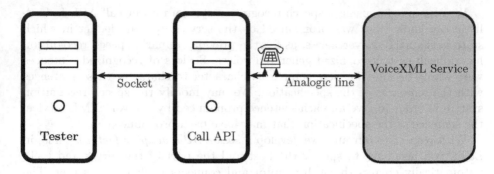

Fig. 2. Test Architecture

between the tester and the call API through a socket. A special protocol has been defined to allow the tester to test the service and obtain information about the conversation. This protocol uses all of these messages:

BECMDCall num ENDCMD: this message asks the call API to call the number *num*.

BECMDDTMF num ENDCMD: it asks the call API to simulate the pushing of a particular touch (defined by num) in the telephone keyboard. This is done by emiting a special sound called *DTMF* during the conversation.

BECMDWaitDTMF num ENDCMD: this message asks the API to wait a particular DTMF sent by the directory.

BECMDTalk file ENDCMD: it asks to play a particular file during the conversation. This file must be in a PCM format at a good frequency.

BECMDRecordAll file ENDCMD: it asks the API to record all of the conversation in the file file. This conversation will be saved in a PCM format.

BECMDHangUp ENDCMD: sent when the tester wants to hang up.

BECMDEnd ENDCMD: sent when the tester wants to stop the communication between it and the calling computer.

For each of this message, the call API sends a corresponding **done** message when the operation executed without any problem or a **not started** when a problem occurs. Moreover, the call API can emit particular messages:

BEGINFDetect Speech ENDINF: it is sent when the call API discovers that there is someone who speak during the conversation.

BEGINFEnd Speech ENDINF: it is sent when the call API discovers that the speech finishes.

BEGINFHangUp ENDINF: sent when the call API discovers that the line has been hanged up.

At first, we tried to allow the API to use vocal synthesis in order to send every possible message. However, we discovered that the speech recognition system of the voiceXML service has lots of difficulties to recognize generated voice. These problems increase the number of cases we have to manage in our scenarios, so we finally preferred to record the sentences we will play during the tests for this reason.

We also imagine using a speech recognition system for our call API. Indeed, if we can know what was pronounced by the service, we can deduct in which state is the SUT. Nevertheless, as for the previous remark, speech recognition has difficulties to recognized generated voices. So lots of recognized sentences were wrong because we were comparing character to character these sentences with the sentences of the specification. We may modify the speech recognition system in order to give for each sentence pronounced by the voiceXML service, the sentence of the specification that may have been pronounced.

With respect to our aim, we develop a *socket SUT adaptor factory*. Thus, in LaTe, you just have to specify the host and the port of the server and LaTe automatically creates the stub adaptor and connects itself to the server. The code of this factory is quite simple but we will not show it here because we are not specially interested by this information in this paper. Just with these information, we allow LaTe to easily test the vocal-based directory. So in the next section, we will see in details some scenarios and further the advantages of the non-deterministic operators on this particular test case.

4.2   Some Test Cases

For all of the following test cases, we defined several LaTe functions to simplify the writing of scenarios. These functions are:

connection(host,port): this function initializes the communication between the tester and the call API.

call(sut,num): for a giving SUT, it send a message for calling the phone number num.

sendCommand(sut,command): it sends the corresponding command to the SUT.

listenSpeech(sut,t1,t2): it verifies that a sentence is pronanced before the time t1 and during at most the time t2. Otherwise, the execution is destroyed.

getMessageIn(sut,message,t): it waits at most the time t for a particular message. If any message arrives or the first message was not message, the execution is destroyed.

getMessage(sut1,message): it makes the same thing than the previous function, but without any timeout.

It is very easy to write thess fonctions. For example, the code of the function listenSpeech is the following:

```
1   function listenSpeech(sut1,t1,t2)
2   {
3     @<t1{ assert(getMessage(sut)=="BEGINFDetect SpeechENDINF") };
4     @<t2{ assert(getMessage(sut)=="BEGINFEnd SpeechENDINF") };
5   };
```

Listing 1.2. User-defined function

For the first test case, our aim is to connect to the call API, call the directory, pronounce a name and verify that people picks up.

```
1    sut:=connection("1-at7290",4442);
2    call(sut,"123");
3    sendCommand(sut,"RecordAll communication.pcm");
4    listenSpeech(sut,30*sec,30*sec);
5    sendCommand(sut,"Talk testername");
6      {   listenSpeech(sut,50*sec,30*sec); }
7    &&
8      {   getCommandeTimer(sut,"Send talk done",20*sec); };
9    wait(5000);
10   sendCommand(sut,"Talk yes");
11     {   listenSpeech(sut,50*sec,30*sec); }
12   &&
13     {   getMessageIn(sut,"Send talk done",15*sec); };
14   sendCommand(sut,"HangUp");
15     {
16         { getMessage(sut,"HangUp detected"); }
17       &&
18         { listenSpeech(sut,50*sec,30*sec); }
19     }
20   ||
21     {   getMessage(sut,"HangUp detected"); };
22     {   getMessageIn(sut,"RecordAll done",15*sec); }
23   &&
24     {   getMessageIn(sut,"HangUp done",15*sec); };
```

Listing 1.3. LaTe scenario for a simple test

The fact that we use an analogic line to connect to the service adds randomness in the receipt order of messages. For example, we do not know in advance if the message **Send talk done** will be received before the beginning of the speech. So we have to set that all of **Send talk done** messages can be interleaved with the listenSpeech function. A lot of executions will be evaluated, but only one will fit the real events.

Moreover, when the tester executes manually the test, he can observe that sometimes the call API detects two different speeches and other times, only one. Thus, in the scenario (lines 15 to 21) we specify that these two cases can occur by using the || operator. We have either an **HangUp detected** message, or a **HangUp detected** message interleaved with a speech detection. At the end of the scenario, we also specify that a **HangUp done** is interleaved with a **RecordAll done** message.

5   Experimental Results and Discussion

5.1   Results and Pros

The use of LaTe in this case allowed the tester to semi automatically test the voiceXML service. It was not fully automatic because the test needs an operator which verifies that his phone rings and picks up for some test cases. But compared to the manual testing, this solution reduces the interactions between the tester and the SUT.

If we compare this solution to one based on TTCN, we can observe that TTCN contains an `interleave` statement that specifies that different branches must be executed concurrently. TTCN allows user to write its own functions (internally or externally), nevertheless, they can be used in an `interleave` branch. In this particular case, LaTe is more powerfull than TTCN, because we don't have to inline the `getMessage` and `listenSpeech` functions. Moreover, TTCN does not allow to specify in a test scenario that something is optional. The only statement that can be used for that is the `alternative` one, but the user have to give a guard for each branches of the `alternative` statement which is very difficult in our example.

Another advantage of this testing architecture is that we have to our disposal all the traces of the conversations between the tester and the voiceXML service. So when we discover an error, we can easily locate it thanks to these traces.

5.2   Discussion

As we said previously, this method for testing the vocal-based phone directory is not perfect: the tester does not know exactly what is pronounced during the conversation and he has to pre-record all of the sentences before using it. Thus, one possible evolution of this technique is to use speech recognition. We saw in section 4.1 that a normal system may not work for our example. The recognition is not something perfect and may make mistakes. So some verdicts may be `fail` with any difference between the specification and the SUT. One possible solution is to modify the speech recognition system so that it gives a set of possible sentences that have been pronounced. With this modification, the speech recognition system will someway be non deterministic: several possible verdicts will be returned. Associated with our non-deterministic operators, it can easily find after few steps which sentence was pronounced thanks to the following sentences. Thus, with only few modifications of the scenarios, this system will explore more deeply the real behavior of the SUT and will be more capable of detecting mistakes.

Another possible improvement of this system is to generate directly the *LaTe* scenarios from the specification. One of our perspective is to modify *TGV* [11] for this aim. TGV allows the generation of an abstract test case from a specification and a test purpose. The generation is done "on-the-fly" on the synchronous product of the specification with the test purpose. It is based on Tarjan's algorithm. During the depth-first search (DFS), TGV performs abstraction and determinization of this product. The DFS stops when an accepting state of test purpose is reached. During the backtracking, TGV synthesizes the transitions of the test case.

Currently, TGV generates test cases in both BCG [12] and Aut [13] formats, so if it can be modified to directly generate test scenarios in *LaTe* format, we will be able to reduce the work of writing these scenarios.

6 Conclusion

Some SUT are so complex and non-deterministic that usual testing systems and languages have difficulties to evaluate the state of the SUT. Thus, we have defined two particular operators for testing languages, the non-deterministic choice and the non-deterministic interleaving. These operators allow the tester to maintain several different executions at the same time. Each execution is independent of the others and is destroyed when a contradiction is found. Nevertheless, communications between these executions and the SUT must be managed because of the communication cannot be undone and modify the environment of all executions.

In order to show that these non-deterministic constructions may be useful, we have implemented them in a new language, LaTe and we have applied them on a particular case study: testing a vocal-based telephone directory. They were particularly useful on this case because the SUT contains lots of non-deterministic behavior like interleaved events, optional messages etc. Thus, these constructions allowed to increase the automation of this task and also allowed the scenarios to test deeper behavior than usual test scenarios. Finally, we have proposed several enhancements for this particular case study, like adding a speech recognition system in order to increase the power of the system and test deeper voice-based services.

References

1. Thomas, D.E., Moorby, P.R.: The Verilog Hardware Description Language. 3rd edn. Kluwer Academic Publishers (1996)
2. Offerman, A., Goor, A.: An experimental user level implementation of tcp. Technical Report 1-68340-44(1997)07, Delft University of Technology (1997)
3. Aho, A.V., Dahbura, A.T., Lee, D., Uyar, M.U.: An optimization technique for protocol conformance test generation based on UIO sequences and rural chinese postman tours. IEEE Transactions on Communications **39** (1991) 1604–1615
4. Doong, R.K., Frankl, P.G.: The ASTOOT approach to testing object-oriented programs. ACM Transactions on Software Engineering and Methodology **3** (1994) 101–130
5. Cheon, Y., Leavens, G.T.: A Simple and Practical Approach to Unit Testing: The JML and JUnit Way. In: ecoop. (2002)
6. Beck, K., Gamma, E.: Junit test infected: Programmers love writing tests. Technical report, Java Report (1998)
7. Massol, V., Husted, T.: JUnit In Action. Manning (2003)

8. ITU-T Z.140: The Tree and Tabular Combined Notation Version 3 (TTCN-3): Core Language. (2001)
9. Pickin, S., Jard, C., Le Traon, Y., Jézéquel, J., Le Guennec, A.: System test synthesis from uml models of distributed software. In: FORTE'2002, IFIP Int. Conf. on Formal description techniques, Houston, Texas (2002)
10. Ghriga, M., Frankl, P.G.: Adaptive testing of non-deterministic communication protocols. In: Protocol Test Systems. (1993) 347–362
11. J. C. Fernandez, C. Jard, T. Jéron, G. Viho: Using on-the-fly verification techniques for the generation of test suites. In Rajeev Alur, Thomas A. Henzinger, eds.: Proceedings of the Eighth International Conference on Computer Aided Verification CAV. Volume 1102., New Brunswick, NJ, USA, Springer–Verlag (1996) 348–359
12. Tock, L.P.: The bcg postscript format. Technical report, INRIA Rhône-Alpes (1995)
13. Fernandez, J.C.: Aldebaran user's manual. Technical report, Laboratoire de Génie Informatique - Institut IMAG (1989)

Customized Testing for Probabilistic Systems[*]

Luis F. Llana-Díaz, Manuel Núñez, and Ismael Rodríguez

Dept. Sistemas Informáticos y Programación,
Universidad Complutense de Madrid, 28040 Madrid, Spain
{llana, mn, isrodrig}@sip.ucm.es

Abstract. In order to test the correctness of an IUT (*implementation under test*) with respect to a specification, testing its whole behavior is desirable but unfeasible. In some situations, testing the behavior of the IUT assuming that it is stimulated by a given *usage model* is more appropriate. Though considering this approach to test functional behaviors consists simply in testing a subset of the IUT, to study the *probabilistic* behavior of systems by using this *customized testing* approach leads to some new possibilities. If usage models specify the probabilistic behavior of stimuli and specifications define the probabilistic behavior of reactions to these stimuli, then, by composing them, the probabilistic behavior of *any* behavior is completely specified. So, after a finite set of behaviors of the IUT is checked, we can compute an *upper bound* of the probability that a user following the usage model finds an error in the IUT. This can be done by considering the *worst case* scenario, that is, that any unchecked behavior is *wrong*.

1 Introduction

Even though testing the whole behavior of a system is desirable, this implies, in general, applying an infinite number of tests. So, formal testing methodologies usually focus on critical parts or aspects of the system. In particular, it is specially important to check that systems provide some minimal functionalities, even if other less relevant functionalities fail. That is, we check that some critical *usage modes* are correct and remain available as expected. More generally, we can group and abstract a set of usage modes in terms of a (probably abstract) user that produces them, that is, in terms of a user model that represents a subset of manners to use the system. Once we are provided with a suitable user model, this model can be used to particularize the goals of the testing procedure. In other words, we can test the correctness of an implementation under test (IUT) with respect to a specification *under the assumption* that the system is stimulated according to the user model. Let us note that if our testing methodology focuses on checking the *functional* behavior of the IUT (i.e., what it does and what it does not), then testing the IUT with respect to a user model may be

---

[*] Research partially supported by the Spanish MCYT project TIC2003-07848-C02-01, the Junta de Castilla-La Mancha project PAC-03-001, and the Marie Curie project MRTN-CT-2003-505121/TAROT.

M.Ü. Uyar, A.Y. Duale, and M.A. Fecko (Eds.): TestCom 2006, LNCS 3964, pp. 87–102, 2006.

easy. In particular, it is enough that, among all the stimulation sequences that can be proposed to the system (i.e., among all stimulation sequences considered in the specification), we consider only those that can be produced by the user model as well. That is, testing the IUT with respect to a specification and a user model actually consists in testing the IUT with respect to a *subset* of the specification.

However, the application of this *user customized* approach to test other types of properties is less trivial and more interesting. This is the case if we consider the *probabilistic* behavior of entities. Let us suppose that the specification provides the desirable probabilistic behavior of the IUT and that, in addition, the user model explicitly defines the probabilistic propensity of each action stimulating the IUT. In this case, we have an environment where the ideal probabilistic behavior of the system, consisting of the IUT and the user model, is *completely specified*. That is, in any interaction between a user and a correct IUT, the probabilistic weight of each choice can be quantified. This extra information allows the testing methodology to go further than other methodologies by providing a relevant diagnosis result: After a finite set of tests is applied, we can compute, for a given *feasibility* degree, an *upper bound* of the probability that a user behaving according to the considered user model finds a wrong behavior in the IUT. This measure will be computed by considering that all the IUT traces that have not been produced yet behave *incorrectly*. Let us note that, after applying a finite set of tests, the number of traces that have not been studied yet is, in general, infinite. However, the cumulated probability of these traces (that is, the probability that any of them is produced) is, like any probability, finite, and we can compute it. Actually, this probability is the complementary of the probability that any *already* analyzed trace is performed. Let us remark that the (ideal) probability of each of these traces is given by the probability defined in the specification for that trace. Unfortunately, we cannot know with certainty whether the traces of the IUT that have been already analyzed actually follow the probabilities defined in the specification. However, by testing each nondeterministic choice of the IUT a high number of times and by applying a suitable contrast hypothesis, we can determine, for a given *feasibility* or *credibility* degree, whether the probabilistic behavior of the IUT corresponds to the one of the specification. For instance, if the specification indicates that, at a given point, the probabilities of performing a and b are the same, then, if the implementation has performed a 507 times and b 493 times then that credibility will be high. However, the credibility will be lower if they were recorded 614 and 386 times, respectively. Since our knowledge of the probabilistic behavior of the IUT will depend on a feasibility degree, the upper bound of the probability that a user finds an error in an IUT will be defined in probabilistic terms as well, that is, for a given feasibility degree.

Let us note that to provide an upper bound of the probability of error is not only useful for a (probabilistic) diagnostic of the IUT correctness. In fact, it may guide the testing process itself. The task of selecting, among an infinite set of tests, a finite set of tests to be applied during the (finite) time assigned to testing is not simple. Thus, we will be interested in tests with high discrimination power.

In other words, we should choose tests such that, when successfully passed, induce a high certainty about the correctness of the IUT. Actually, the upper bound of error probability implicitly provides a guide for selecting tests. We will prefer those sets of tests such that, when correctly passed by the IUT, provide a *lower* upper bound of error, that is, provide a higher certainty of the IUT correctness. That is, we will prefer those sets providing upper bounds with higher feasibility degrees.

In this paper we develop these ideas and we construct a testing methodology for testing probabilistic systems that interact with user models. We define two implementation relations. The first one directly compares the probabilities of the traces in the IUT and in the specification. The second one indirectly compares these probabilities by applying a contrast hypothesis to a sample collected by interacting with the IUT. Besides, we show how the measure that we commented before is computed from an IUT sample. In terms of related work, there is significant work on testing preorders and equivalences for probabilistic processes [2, 10, 12, 3, 1, 13, 7, 5]. Most of these proposals follow the *de Nicola and Hennessy's style* [6, 4], that is, two processes are equivalent if the application of any test belonging to a given set returns the same result. Instead, we are interested in checking whether an implementation conforms to a specification. In particular, our relations are more similar to the ones introduced in [14, 8]. Regarding probabilistic user models, it is worth to point out that these previous works do not explicitly consider this notion. User models have been used in specific software testing scenarios (e.g., to test C++ templates [11]). Other work deals with user models in the testing context [16, 15], but they do not consider formal conformance testing techniques.

The rest of the paper is structured as follows. In the next section we present some basic notions to denote specifications, IUTs, and user models. In Section 3 we show how these notions are related and we define tests. In Section 4 we present our (probabilistic) conformance relations. Then, in Section 5 we give the upper bound of the probability that a user finds an error in an IUT. Finally, in Section 6 we present our conclusions.

2   Basic Notions

In this section we present some basic notions used in the paper. First, we introduce some statistics notions. An *event* is any reaction we can detect from a system or environment; a random variable is a function associating each event with its probability.

Definition 1. Let $\mathcal{A}$ be a set of events and $\xi : \mathcal{A} \to [0, 1]$ be a function such that $\sum_{\alpha \in} \xi(\alpha) = 1$. We say that ξ is a *random variable* for the set of events $\mathcal{A}$.

If we observe that the event $\alpha \in \mathcal{A}$ is produced by a random source whose probabilistic behavior is given by ξ then we say that α *has been generated by* ξ. We extend this notion to sequences of events as expected: If we observe that the sequence of events $H = \langle \alpha_1, \ldots, \alpha_n \rangle$ is consecutively produced by a random source whose probabilistic behavior is given by ξ then we say that H *has been generated by* ξ or that H is a *sample* of ξ.

Given the random variable ξ and a sequence of events H, we denote the *confidence* that H is *generated by* ξ by $\gamma(\xi, H)$.   □

This definition introduces a simple version of discrete random variable where all the events are independent. The actual definition of a *random variable* is more complex but it is pointless to use its generality in our setting. In the previous definition, the application of a suitable *hypothesis contrast* is abstracted by the function γ. We have that $\gamma(\xi, H)$ takes a value in $[0, 1]$. Intuitively, a sample will be *rejected* if the probability of observing that sample from a given random variable is low. At the end of this section we present a working definition of the function γ. It is worth to point out that the results of this paper do not depend on the formulation of γ, being possible to *abstract* the actual definition.

Next we present the formalism we will use to define *specifications* and *implementations*. A *probabilistic finite state machine* is a finite state machine where each transition is equipped with a probability denoting its probabilistic propensity. Thus, a transition $s \xrightarrow{i/o}_p s'$ denotes that, when the machine is in state s and the input i is received, then, with probability p, it moves to the state s' and produces the output o. We will assume that the environment stimulates the machine with a single input at any time. Given an input, the machine probabilistically chooses the transition it takes from its current state. Hence, the probability of a transition allows to compare its propensity with the one of any other transition departing from the same state and receiving the *same* input. That is, given s and i, the addition of all values p such that there exist o, s' with $s \xrightarrow{i/o}_p s'$ must be equal to 1. In contrast, there is no requirement binding the probabilities departing from the same state and receiving different inputs because each one describes (part of) a different probabilistic choice of the machine.

Definition 2. A *Probabilistic Finite State Machine*, in short PFSM, is a tuple $M = (S, I, O, \delta, s_0)$ where

- S is the *set of states* and $s_0 \in S$ is the *initial state*.
- I and O, with $I \cap O = \varnothing$, denote the sets of *input* and *output* actions, respectively.
- $\delta \subseteq S \times I \times O \times (0, 1] \times S$ is the *set of transitions*. We will write $s \xrightarrow{i/o}_p s'$ to denote $(s, i, o, p, s') \in \delta$.

Transitions and states fulfill the following additional conditions:

- For all $s \in S$ and $i \in I$, the probabilities associated with outgoing transitions add up to 1, that is, $\sum \{p \mid \exists o \in O, s' \in S : s \xrightarrow{i/o}_p s'\} = 1$.
- PFSMs are *free of non-observable non-determinism*, that is, if whenever we have the transitions $s \xrightarrow{i/o}_{p_1} s_1$ and $s \xrightarrow{i/o}_{p_2} s_2$ then $p_1 = p_2$ and $s_1 = s_2$.
- In addition, we will assume that implementations are *input-enabled*, that is, for all state s and input i there exist o, p, s' such that $s \xrightarrow{i/o}_p s'$.   □

Although PFSMs will be used to define specifications, a different formalism will be used to define *user models*. Specifically, we will use *probabilistic labeled transition systems*. A user model represents the external environment of a system. User models actively produce inputs that stimulate the system, while passively receive outputs produced by the system as a response. The states of a user model are split into two categories: *Input states* and *output states*. In input states, all outgoing transitions denote a different input action. Since inputs are probabilistically chosen by user models, any input transition is endowed with a probability. In particular, $s \xrightarrow{i}_p s'$ denotes that, with probability p, in the input state s, the input i is produced and the state is moved to s'. Given an input state s, the addition of all probabilities p such that there exists i, s' with $s \xrightarrow{i}_p s'$ must be *lower* than or equal to 1. If it is lower then we consider that the remainder up to 1 implicitly denotes the probability that the interaction with the system *finishes* at the current state. Regarding output states, all transitions departing from an output state are labeled by a different output action. However, output transitions do not have any probability value (let us remind that outputs are chosen by the system). Input and output states will strictly alternate, that is, for any input state s, with $s \xrightarrow{i}_p s'$, s' is an output state, and for any output state s, with $s \xrightarrow{o} s'$, s' is an input state.

Definition 3. A *probabilistic labeled transition system*, in short PLTS, is a tuple $U = (S_I, S_O, I, O, \delta, s_0)$ where

- S_I and S_O, with $S_I \cap S_O = \varnothing$, are the sets of *input* and *output* states, respectively. $s_0 \in S_I$ is the *initial state*.
- I and O, with $I \cap O = \varnothing$, are the sets of *input* and *output* actions, respectively.
- $\delta \subseteq (S_I \times I \times (0,1] \times S_O) \cup (S_O \times O \times S_I)$ is the *transition relation*. We will write $s \xrightarrow{i}_p s'$ to denote $(s, i, p, s') \in S_I \times I \times (0,1] \times S_O$ and $s \xrightarrow{o} s'$ to denote $(s, o, s') \in S_O \times O \times S_I$.

Transitions and states fulfill the following additional conditions:

- For all input states $s \in S_I$ and input actions $i \in I$ there exists at most one outgoing transition from s: $|\{s \xrightarrow{i}_p s' \mid \exists\, p \in (0,1],\ s' \in S_O\}| \leq 1$.
- For all output states $s \in S_O$ and output actions $o \in O$ there exists exactly one outgoing transition labeled with o: $|\{s \xrightarrow{o} s' \mid \exists\, s' \in S_I\}| = 1$.
- For all input state $s \in S_I$ the addition of the probabilities associated with the outgoing transitions is lower than or equal to 1, that is, $\text{cont}(s) = \sum\{p \mid \exists\, s' \in S_O :\ s \xrightarrow{i}_p s'\} \leq 1$. So, the probability of stopping at that state s is $\text{stop}(s) = 1 - \text{cont}(s)$. □

By iteratively executing transitions, both PFSMs and PLTSs can produce sequences of inputs and outputs. The probabilities of these sequences are given by the probabilities of the transitions. Next we introduce some *trace* notions. A *probability trace* is a sequence of probabilities, a *trace* is a sequence of inputs and outputs, and a *probabilistic trace* is a tuple containing both.

Definition 4. A *probability trace* π is a finite sequence of probabilities, that is, a possibly empty sequence $\langle p_1, p_2, \ldots, p_n \rangle \in (0,1]^*$. The symbol ϵ denotes the empty probability trace. Let $\pi = \langle p_1, p_2, \ldots, p_n \rangle$ be a probability trace. We define its *sef-product*, denoted by $\prod \pi$, as $\prod_{1 \leq i \leq n} p_i$. Since $\prod_{a \in \varnothing} = 1$, we have $\prod \epsilon = 1$. Let $\pi = \langle p_1, p_2, \ldots, p_n \rangle$ and $\pi' = \langle p'_1, p'_2, \ldots, p'_m \rangle$ be probability traces. Then, $\pi \cdot \pi'$ denotes their concatenation that is, $\langle p_1, p_2, \ldots, p_n, p'_1, p'_2, \ldots, p'_m \rangle$, while $\pi * \pi'$ denotes its pairwise product, that is, $\langle p_1 * p'_1, p_2 * p'_2, \ldots, p_r * p'_r \rangle$, where $r = \min(n, m)$.

A *trace* ρ is a finite sequence of input/output actions $(i_1/o_1, i_2/o_2, \ldots, i_n/o_n)$. The symbol ϵ denotes the empty trace. Let ρ and ρ' be traces. Then, $\rho \cdot \rho'$ denotes their concatenation. A *probabilistic trace* is a pair (ρ, π) where ρ is a trace $(i_1/o_1, i_2/o_2, \ldots, i_n/o_n)$ and $\pi = \langle p_1, p_2, \ldots, p_n \rangle$ is a probability trace. If ρ and π are both empty then we have the *empty probabilistic trace*, written as (ϵ, ϵ). Let (ρ, π) and (ρ', π') be probabilistic traces. Then, $(\rho, \pi) \cdot (\rho', \pi')$ denotes their concatenation, that is, $(\rho \cdot \rho', \pi \cdot \pi')$. □

Next we define how to extract traces from PFSMs and PLTSs. First, we consider the reflexive and transitive closure of the transition relation, and we call it *generalized* transition. Then, probabilistic traces are constructed from generalized transitions by considering their sequences of actions and probabilities.

Definition 5. Let M be a PFSM. We inductively define the *generalized* transitions of M as follows:

- If $s \in S$ then $s \overset{\epsilon}{\Longrightarrow}_\epsilon s$ is a generalized transition of M.
- If $s \in S$, $s \overset{\rho}{\Longrightarrow}_\pi s'$, and $s' \overset{i/o}{\longrightarrow}_p s_1$ then $s \overset{\rho\, i/o}{\Longrightarrow}_{\pi\ p} s_1$ is a generalized transition of M.

We say that (ρ, π) is a *probabilistic trace* of M if there exists $s \in S$ such that $s_0 \overset{\rho}{\Longrightarrow}_\pi s$. In addition, we say that ρ is a *trace* of M. The sets $\mathtt{pTr}(M)$ and $\mathtt{tr}(M)$ denote the sets of *probabilistic traces* and *traces* of M, respectively. □

The previous notions can be defined for PLTSs. In order to obtain sequences of paired inputs and outputs, traces begin and end at input states; generalized transitions are constructed by taking *pairs* of consecutive PLTS transitions.

Definition 6. Let M be a PLTS. We inductively define the *generalized* transitions of U as follows:

- If $s \in S_I$ then $s \overset{\epsilon}{\Longrightarrow}_\epsilon s$ is a generalized transition of U.
- If $s \in S_I$, $s \overset{\rho}{\Longrightarrow}_\pi s'$, and $s' \overset{i}{\longrightarrow}_p s'' \overset{o}{\longrightarrow} s_1$ then $s \overset{\rho\, i/o}{\Longrightarrow}_{\pi\ p} s_1$ is a generalized transition of U.

We say that (ρ, π) is a *probabilistic trace* of U if there exists $s \in S_I$ such that $s_0 \overset{\rho}{\Longrightarrow}_\pi s$. In addition, we say that ρ is a *trace* of U. We define the probability of U to stop after ρ, denoted by $\mathtt{stop}_U(\rho)$, as $\mathtt{stop}(s)$. The sets $\mathtt{pTr}(U)$ and $\mathtt{tr}(U)$ denote the set of *probabilistic traces* and *traces* of U, respectively. □

2.1   Definition of a Hypothesis Contrast: Pearson's χ^2

In this paper we consider *Pearson's χ^2 contrast* but other contrasts could be used. The mechanism is the following. Once we have collected a sample of size n we perform the following steps:

- We split the sample into k classes covering all the possible range of values. We denote by O_i the *observed frequency* in class i (i.e., the number of elements belonging to the class i).
- We calculate, according to the proposed random variable, the probability p_i of each class i. We denote by E_i the *expected frequency* of class i, that is, $E_i = np_i$.
- We calculate the *discrepancy* between observed and expected frequencies as $X^2 = \sum_{i=1}^{n} \frac{(O_i - E_i)^2}{E_i}$. When the model is correct, this discrepancy is approximately distributed as a χ^2 random variable.
- The number of freedom degrees of χ^2 is $k - 1$. In general, this number is equal to $k - r - 1$, where r is the number of parameters of the model which have been estimated by maximal likelihood over the sample to estimate the values of p_i. In our framework we have $r = 0$ because the model completely specifies the values of p_i before the samples are observed.
- We will *accept* that the sample follows the proposed random variable if the probability to obtain a discrepancy greater than or equal to the detected discrepancy is high enough, that is, if $X^2 < \chi^2_\alpha(k-1)$ for some α high enough. Actually, as such margin to accept the sample decreases as α increases, we can obtain a measure of the validity of the sample as $\max\{\alpha | X^2 \leq \chi^2_\alpha(k-1)\}$.

According to the previous steps, next we present an operative definition of the function γ that was introduced in Definition 1. We will consider two sets of events $\mathcal{A}$ and $\mathcal{A}'$, with $\mathcal{A} \subseteq \mathcal{A}'$. The set $\mathcal{A}$ gives the domain of the random variable ξ, while the events denoted by H belong to $\mathcal{A}'$. If the sample includes any event a that is not considered by the random variable (i.e., $a \notin \mathcal{A}$) then the sample cannot be generated by the random variable and the minimal *feasibility*, that is 0, is returned. Otherwise, we return the maximal feasibility α such that the hypothesis contrast is passed.

Definition 7. Let $\mathcal{A}$ and $\mathcal{A}'$ be sets of events, with $\mathcal{A} \subseteq \mathcal{A}'$, and H be a sample of elements belonging to $\mathcal{A}'$. Let $\xi : \mathcal{A} \to (0,1]$ be a random variable. We define the confidence of ξ on H, denoted by $\gamma(\xi, H)$, as follows:

$$\gamma(\xi, H) = \begin{cases} 0 & \text{if } H \cap \bar{\mathcal{A}} \neq \varnothing \\ \max\{\alpha \mid X^2_\xi \leq \chi^2_\alpha(k-1)\} & \text{otherwise} \end{cases}$$

where X^2_ξ denotes the discrepancy level of the sample H on ξ, calculated as explained above by considering that the sampling space is $\mathcal{A}$.   □

3   Tests and Composition of Machines

In this section we define our tests as well as the interaction between the notions introduced in the previous section (PFSMs and PLTSs). As we said before, we will use PLTSs to define the behavior of the external environment of a system, that is, a user model. Moreover, PLTSs are also appropriate to define the *tests* we will apply to an IUT. Tests are PLTSs fulfilling some additional conditions. Basically, a test defines a finite sequence of inputs that can be interrupted depending on the outputs produced by the IUT as response: If one of the expected outputs is received then the next input is applied, otherwise the interaction with the IUT finishes. Since tests consider a single sequence of inputs, each intermediate input state of the sequence contains a single outgoing transition labeled by the next input and probability 1. Output states offer transitions with different outputs. Only one of the input states reached by these transitions offers a (single) transition; the interaction finishes in the rest of them.

Definition 8. A test $T = (S_I, S_O, I, O, \delta, s_0)$ is a PLTS such that for all $s \in S_I$ there is at most one transition $s \xrightarrow{i}_p s'$ (so, in this transition $p = 1$), and for all $s \in S_O$ there is at most one transition $s \xrightarrow{o} s'$ with a *continuation*, that is, $|\{s'' \mid \exists i \in I, o \in O, s''' \in S_O, p \in (0,1] : s \xrightarrow{o} s'' \xrightarrow{i}_p s'''\}| \leq 1$.   □

Let us note that, contrarily to other frameworks, tests are not provided with diagnostic capabilities on *their own*. In other words, tests do not have fail/success states. Since our framework is probabilistic, the requirements defined by specifications are given in probabilistic terms. Moreover, the absence of transitions labeled by specific outputs in specification states is considered in probabilistic terms as well: If there exists a state s, an input i, and an output o such that there do not exist p, s' with $s \xrightarrow{i/o}_p s'$ then we consider that the probability of producing o in the state s after receiving the input i is 0. As we will see in the next section, deciding whether the IUT conforms to the specification will also be done in probabilistic terms. In particular, we will consider whether it is *feasible* that the IUT *behaves* as if it were defined as the specification indicates. We will check this fact by means of a suitable *hypothesis contrast*.

Our testing methodology consists in testing the behavior of a system under the assumption that it is stimulated by a given user model. Thus, the sequences we use to stimulate it, that is, the tests, will be extracted from the behavior of the user model. Next we show how a test is constructed from a probabilistic trace of a user model. The input and output states of the test are identified with natural numbers. All the input states (but the first and last ones) are also endowed with an output action. In order to distinguish between input and output states we decorate them with and *, respectively.

Definition 9. Let $\rho = (i_1/o_1, i_2/o_2, \ldots, i_r/o_r)$ be a trace, I be a set of input actions such that $\{i_1, \ldots i_r\} \subseteq I$, and O be a set of output actions such that $\{o_1, \ldots, o_r\} \subseteq O$. We define the *associated test* to ρ, denoted by $\texttt{assoc}(\rho)$, as the test $(S_{IT}, S_{OT}, I, O, \delta_T, 0\ )$, where

- $S_{IT} = \{0\ , r\ \} \cup \{(j, o) \mid o \in O, 1 \leq j < r\}$ and $S_{OT} = \{j^* \mid 1 \leq j \leq r\}$.
- For all $1 \leq j < r, o \in O$: $(j, o_j) \xrightarrow{i_{j+1}}_1 (j+1)^*$, $j^* \xrightarrow{o} (j, o) \in \delta_T$. We also have $0 \xrightarrow{i_1}_1 0^*$, $r^* \xrightarrow{o_r} r \in \delta_T$.

Fig. 1. Normalization if composition of PFSMs and PLTSs

Let U be a PLTS. The *set of associated tests to U*, denoted by $\mathtt{assoc}(U)$, is the set of tests associated to its traces, that is $\{\mathtt{assoc}(\rho) \mid \rho \in \mathtt{tr}(U)\}$. $\qquad\square$

Next we define the composition of a PFSM (denoting either a specification or an IUT) with a PLTS (denoting either a user model or a test) in terms of its behavior, that is, in terms of traces and probabilistic traces. The set of traces is easily computed as the intersection of the traces produced by both components. In order to define the set of probabilistic traces, the ones provided by both components are considered. For a given input/output pair i/o, the probability of producing i will be taken from the corresponding transition of the PLTS, while the probability of producing o as a response to i will be given by a transition of the PFSM. Let us note that the states of a specification do not necessarily define outgoing transitions for all available inputs, that is, specifications are not necessarily *input-enabled*. So, a PFSM representing a specification could not provide a response for an input produced by a PLTS. Since the specification does not define any behavior in this case, we will assume that the PFSM is allowed to produce *any* behavior from this point on. The composition of a PLTS and a PFSM will be constructed to check whether the traces *defined* by the specification are correctly produced by the implementation (under the assumption that these machines are stimulated by the user model). Hence, undefined behaviors will not be considered relevant and will not provide any trace to the composition of the PLTS and the PFSM. In order to appropriately represent the probabilities of the relevant traces, their probabilities will be *normalized* if undefined behaviors appear. We illustrate this process in the following example.

Example 1. Let us suppose that a user model can produce the inputs i_1, i_2, and i_3 with probabilities $\frac{1}{2}$, $\frac{1}{4}$ and $\frac{1}{4}$, respectively (see Figure 1, left). At the same time, the corresponding specification provides outgoing transitions with inputs i_1 and i_2, but not with i_3 (see Figure 1, right). Since the specification does not define any reaction to i_3, the probabilities of taking inputs i_1 or i_2 in the composition of the specification and the user model are normalized to denote that i_3 is not considered. So, the probability of i_1 becomes $\frac{1/2}{3/4} = \frac{2}{3}$ while the probability of i_2 is $\frac{1/4}{3/4} = \frac{1}{3}$. $\qquad\square$

Definition 10. Let $M = (S_M, I, O, \delta_M, s_{0M})$ be a PFSM and let us consider a PLTS $U = (S_{IU}, S_{OU}, I, O, \delta_U, s_{0U})$ such that $s_{0M} \overset{\rho}{\Longrightarrow}_{\pi_1} s_1$ and $s_{0U} \overset{\rho}{\Longrightarrow}_{\pi_2} s_2$. We define:

– The sum of the probabilities of *continuing together after* ρ as

$$\text{cont}_{M\ U}(\rho) = \sum \left\{ p \ \middle| \ \begin{array}{l} \exists i \in I,\ o \in O,\ s_2' \in S_{OU},\ s_1' \in S_M,\ r \in (0,1]: \\ s_2 \xrightarrow{\ i\ }_p s_2' \ \wedge \ s_1 \xrightarrow{\ i/o\ }_r s_1' \end{array} \right\}$$

– The *normalization factor of* $M \parallel U$ *after* ρ as the sum of the previous probability plus the probability of U to stop after ρ, that is $\text{norm}_{M\ U}(\rho) = \text{cont}_{M\ U}(\rho) + \text{stop}_U(\rho)$.   □

Definition 11. Let $M = (S_M, I, O, \delta_M, s_{0M})$ be a PFSM and let us consider a PLTS $U = (S_{IU}, S_{OU}, I, O, \delta_U, s_{0U})$. The *set of traces* generated by the *composition* of M and U, denoted by $\text{tr}(M \parallel U)$, is defined as $\text{tr}(M) \cap \text{tr}(U)$. The *set of probabilistic traces* generated by the *composition* of M and U, denoted by $\text{pTr}(M \parallel U)$, is defined as the smallest set such that

– $(\epsilon, \epsilon) \in \text{pTr}(M \parallel U)$.
– If we have that $(\rho, \pi) \in \text{pTr}(M \parallel U)$, $s_{0M} \xRightarrow{\ \rho\ }_{\pi_1} s_1' \xrightarrow{\ i/o\ }_{p_1} s_1$, and $s_{0U} \xRightarrow{\ \rho\ }_{\pi_2} s_2' \xrightarrow{\ i\ }_{p_2} s_2'' \xrightarrow{\ o\ } s_2$, then $(\rho \cdot i/o, \pi \cdot \langle p \rangle) \in \text{pTr}(M \parallel U)$, where p is the product of p_1 and p_2 *normalized* with respect to the normalization factor of $M \parallel U$ after ρ, that is, $p = \frac{p_1 \, p_2}{\text{norm}_{M\parallel P}(\rho)}$.   □

Let us remark that the probabilistic behavior of the traces belonging to the composition of PFSMs and PLTSs is completely specified: The probabilities of inputs are provided by the PLTS while the probabilities of outputs are given by the PFSM. So, a random variable denoting the probability of *each* trace produced by the composition can be constructed. Moreover, the composition of the specification and the user model provides a source to *randomly* generate tests. In fact, tests are constructed by following a specific sequence of inputs and outputs of the user model. Hence, the random selection of tests can be represented by a random variable associating tests with the probability that the probabilistic trace guiding the test is taken in the composition of the specification and the user model.

Definition 12. Let $M = (S_M, I, O, \delta_M, s_{0M})$ be a PFSM and let us consider a PLTS $U = (S_{IU}, S_{OU}, I, O, \delta_U, s_{0U})$. We define the *traces random variable of the composition* of M and U as the function $\xi_{M\ U} : \text{tr}(M \parallel U) \longrightarrow (0,1]$ such that for all $(\rho, \pi) \in \text{pTr}(M \parallel U)$ we have $\xi_{M\ U}(\rho) = \prod \pi * (1 - \text{stop}_U(\rho))$.

Let $\mathcal{T} = \{T \mid \rho \in \text{tr}(M \parallel U) \ \wedge \ T = \text{assoc}(\rho)\}$. We define the *tests random variable of the composition of* M *and* U as the function $\zeta_{M\ U} : \mathcal{T} \longrightarrow (0,1]$ such that for all test $T \in \mathcal{T}$ we have $\zeta_{M\ U}(T) = p$ iff $(\rho, \pi) \in \text{pTr}(U)$, $(\rho, \pi') \in \text{pTr}(M)$, $T = \text{assoc}(\rho)$, and $p = \prod \pi * \prod \pi' * (1 - \text{stop}_U(\rho))$.   □

Let us note that the sum of the probabilities of all traces may be strictly less than 1. This is because random variables have to take into account some events that are not directly considered in the traces: The choice of a user to stop in a state. Next we identify some properties of our framework.

Proposition 1. Let S be a PFSM, U be a PLTS, $(\rho, \pi) \in \text{pTr}(U)$, and $T = \text{assoc}(\rho)$. The following properties hold:

- $\text{tr}(T) \subseteq \text{tr}(U)$ and $\text{tr}(S \parallel T) \subseteq \text{tr}(S \parallel U)$.
- if $(\rho, \pi') \in \text{pTr}(S \parallel T)$ then $(\rho, \pi \cdot \pi') \in \text{pTr}(S \parallel U)$.
- $\text{tr}(U) = \bigcup \{ \text{tr}(T) \mid T \in \text{assoc}(U) \}$.
- $\text{tr}(S \parallel U) = \bigcup \{ \text{tr}(S \parallel T) \mid T \in \text{assoc}(U) \}$. $\square$

4 Probabilistic Relations

In this section we introduce our probabilistic conformance relations. Following our user customized approach, they relate an IUT *and* a user model with a specification *and* the same user model. These three elements will be related if the probabilistic behavior shown by the IUT when stimulated by the user model appropriately follows the corresponding behavior of the specification. In particular, we will compare the *probabilistic traces* of the composition of the IUT and the user with those corresponding to the composition of the specification and the user. Let us remind that IUTs are input-enabled but specifications might not be so. So, the IUT could define probabilistic traces including sequences of inputs that are not defined in the specification. Since there are no specification requirements for them, these behaviors will be ignored by the relation. In order to do it, an appropriate subset of the traces of the composition of the IUT and the user must be taken. In the following relation, we require that the probabilities of the corresponding traces are *exactly* the same in both compositions. Later we will see another relation where, due to practical reasons, this requirement will be relaxed.

Definition 13. Let S, I be PFSMs, U be a PLTS, and s_{0S}, s_{0I}, and s_{0U} be their initial states, respectively. We define the *set of probabilistic traces generated by the implementation I and the user model U modulo the specification S*, denoted by $\text{pTr}(I \parallel U)_S$, as the smallest set such that:

- $(\epsilon, \epsilon) \in \text{pTr}(I \parallel U)_S$.
- If $(\rho, \pi) \in \text{pTr}(I \parallel U)_S$ and we have the following sequences of transitions:
  - $s_{0U} \overset{\rho}{\Longrightarrow}_{\pi_2} s'_1 \overset{i}{\longrightarrow}_{p_2} s''_1 \overset{o}{\longrightarrow} s_1$, and
  - $s_{0I} \overset{\rho}{\Longrightarrow}_{\pi_1} s'_2 \overset{i/o}{\longrightarrow}_{p_1} s_2$,
  - $s_{0S} \overset{\rho}{\Longrightarrow}_{\pi_3} s'_3 \overset{i/o}{\longrightarrow}_{p_3} s_3$,

  then $(\rho \cdot i/o, \pi \cdot \langle p \rangle) \in \text{pTr}(I \parallel U)_S$, where p is the product of p_1 and p_2 *normalized* with respect to the normalization factor of $S \parallel U$ after ρ, that is $p = \frac{p_1 \, p_2}{\text{norm}_{S \parallel U}(\rho)}$.

Let S, I be PFSMs and U be a PLTS. We say that I *conforms to S with respect to U*, denoted by $I \, \text{conf}_U \, S$, if $\text{pTr}(I \parallel U)_S = \text{pTr}(S \parallel U)$. $\square$

Although the previous relation properly defines our probabilistic requirements, it cannot be used in practice because we cannot *read* the probability attached to a transition in a black-box IUT. So, a more applicable version of the relation is required. Let us note that even though a single observation does not provide valuable information about the probability of an IUT trace, an *approximation* to this value can be calculated by interacting a high number of times with the IUT and analyzing its reactions. In particular, we can compare the empirical behavior of the IUT with the ideal behavior defined by the specification and check whether it is *feasible* that the IUT would have behaved like this if, internally, it were defined conforming to the specification. Depending on the empirical observations, this feasibility may be different. The feasibility degree of a set of samples with respect to its ideal probabilistic behavior (defined by a random variable) will be provided by a suitable contrast hypothesis. We will rewrite the previous relation in these terms. The new relation will be parameterized by two values: The samples collected by means of interactions with the IUT and a feasibility threshold. Then, by using an indirect approach, the new relation will impose the same probabilistic constraints as the relation defined before.

We must establish the way samples are collected. First, we generate the tests associated to the user model U and then we let these tests to interact with the IUT. Then, we must check if the obtained sample conforms to the random variable corresponding to the user and the specification, that is, $\xi_{S\ U}$, as introduced in Definition 12. This last point will be done via the hypothesis contrast. We will require that the feasibility of the hypothesis contrast reaches a required threshold. Before we present the new relation, we introduce the notion of *sampling*.

Definition 14. Let M be a PFSM and U be a PLTS. We say that a sequence $\langle \rho_1, \rho_2, \ldots, \rho_n \rangle$ is a *trace sample* of $M \parallel U$ if it is generated by $\xi_{M\ U}$. We say that a sequence $\langle T_1, T_2, \ldots T_n \rangle$ is a *test sample of* $M \parallel U$ if it is generated by $\zeta_{M\ U}$. Let $\langle T_1, T_2, \ldots T_n \rangle$ be a test sample of $M \parallel U$. We say that a sequence $\langle \rho_1, \rho_2, \ldots, \rho_n \rangle$ is a *trace-test sample* of $M \parallel U$ if for all $1 \leq i \leq n$ we have that ρ_i is the result of a probabilistic execution of $M \parallel T_i$. □

Next we introduce the new conformance relation defined in probabilistic terms. As before, we will ignore any implementation behavior involving sequences of inputs not considered by the specification. This will be done by removing them from the *trace-test sample* we use to compare the IUT and the specification. In the next definition, H_S represents the sequence of traces resulting after removing those traces from the original trace-test sample H.

Definition 15. Let S be a PFSM and $H = \langle \rho_1, \rho_2, \ldots, \rho_n \rangle$ be a sequence of traces. H_S denotes the sub-sequence $\langle \rho_{r1}, \rho_{r2}, \ldots, \rho_{rn} \rangle$ of H that contains all the probabilistic traces whose input sequences can be produced by S, that is, $\rho = (i_1/o_1, \ldots i_m/o_m) \in H_S$ iff $\rho \in H$ and there exist $o'_1, \ldots o'_n \in O$ such that $(i_1/o'_1, \ldots i_m/o'_m) \in \mathtt{tr}(S)$.

Let S and I be PFSMs, U be a PLTS, $H = \langle \rho_1, \rho_2, \ldots, \rho_n \rangle$ be a trace-test sample of $I \parallel U$, and $0 \leq \alpha \leq 1$. We write $S\ \mathtt{conf}_{(H,\alpha)}\ I$ if $\gamma(\xi_{S\ U}, H_S) \geq \alpha$. □

5   Upper Bound of Probability of Failure for a User

In this section we provide an alternative measure of the correctness of an IUT. Similarly to the conformance relation given in Definition 15, it will be calculated by using a sample collected from the interaction with the IUT. This measure is an *upper bound* of the probability that the user obtains from the IUT a trace whose probabilistic behavior is *wrong* with respect to the specification. That is, it provides an upper bound of the probability of finding an error in the IUT. Since this measure will be computed from a specific sample, it will also be parameterized by a feasibility degree α. We assess the measure as we sketched in the introduction: From a given interaction sample with the IUT, we consider the feasibility degree α that this sample was generated according to the probabilistic behavior defined by the specification. Next we consider the probability of producing a behavior that is *not* included in the sample. Since we can assume that the probabilities of the traces in the sample are correct with feasibility α, we can use these probabilities to compute the probability of producing any other trace by adding the probabilities of all traces in the sample and by considering the *complementary* probability. Then, we consider the worst case of these traces, that is, we suppose that the probabilistic behavior of all of them is *wrong*. Hence, we obtain an upper bound of the probability that the user interacts with the IUT and observes a trace whose propensity is not that given by the specification (with feasibility α). First, we define the *prefixes* of a probabilistic trace that will allow to structure samples in a suitable form.

Definition 16. Let $(\rho, \pi) = ((i_1/o_1, \ldots, i_n/o_n), \langle p_1, \ldots, p_n \rangle)$ be a probabilistic trace. We say that a probabilistic trace (ρ', π') is a *prefix* of (ρ, π) if either $(\rho', \pi') = (\epsilon, \epsilon)$ or $(\rho', \pi') = ((i_1/o_1, \ldots, i_j/o_j), \langle p_1, \ldots, p_j \rangle)$, for some $1 \leq j \leq n$. We denote by $\mathtt{prefix}(\rho, \pi)$ the set of all prefixes of the probabilistic trace (ρ, π). $\qquad\square$

Let us consider a finite set of probabilistic traces such that all their prefixes are also included in the set. In fact, if these traces are a sample produced by the IUT then we can represent our *knowledge* about the IUT by means of a *suitable machine* producing these traces. Since we assume that the IUT does not have non-observable non-determinism, if two observed samples share a common prefix then we can consider that the common parts of both traces traverse the same path of states in the IUT. This fact can be reflected in the machine we construct by making both traces to share the same states until they diverge. Besides, let us note that we cannot detect whether a *loop* of states is taken in the IUT during our interaction with it, since we consider that the IUT is a black box. So, the machine representing our knowledge about the IUT, extracted from a sample, will be a *tree*: All traces in the sample depart from the initial state and traces share paths while they traverse common prefixes.

Definition 17. Let tr be a prefix-closed set of probabilistic traces. We say that tr is a *probabilistic tree* if the following conditions hold:

- The nodes of the tree are labeled by probabilistic traces belonging to tr.
- The arcs between nodes are labeled by pairs $(i/o, p)$, where i is an input, o is an output, and p is a probability. There exists an arc between two nodes (ρ, π) and (ρ', π'), denoted by $(\rho, \pi) \xrightarrow{i/o}_p (\rho', \pi')$, if $(\rho', \pi') = (\rho \cdot (i/o), \pi \cdot \langle p \rangle)$.
- For each node (ρ, π), the probabilities of all the outgoing arcs is less than or equal to 1, that is, $L = \sum \{p \mid \exists i, o, p : (\rho, \pi) \xrightarrow{i/o}_p (\rho \cdot (i/o), \pi \cdot \langle p \rangle)\} \leq 1$. Hence, the probability of *stopping in* (ρ, π) is given by $\text{stop}_\pi(\rho) = 1 - L$.
- The probability of reaching a node (ρ, π) is equal to $\prod \pi$.   □

After a sample of traces H is extracted from the IUT, the $\text{conf}_{(H,\alpha)}$ relation given in Definition 15 allows to check whether the feasibility that H is produced by the specification is at least α. If this is the case then we can construct a set of *probabilistic traces* from H by attaching each trace in H with the probability given in the specification for that trace. The feasibility that the probabilities we attach are actually correct is equal to α. Then, the *probabilistic tree* representing this set shows, with feasibility α, the behavior of the IUT regarding the traces belonging to H. In order to compute the upper bound of the error probability, we will consider that any trace leaving this tree behaves *incorrectly*. We will identify these traces by considering a *higher* tree denoting all the traces that can be produced, not only those we observed in the sample. Then, the probability of producing any unobserved trace will be computed by considering the probability of performing a trace that leaves the lower tree. This probability is computed by adding the probabilities of all the traces reaching the border of the lower tree and performing an additional transition to leave it.

Definition 18. Let tr_1 and tr_2 be probabilistic trees such that $tr_1 \subseteq tr_2$ and tr_1 is finite. The probability of *reaching tr_2 from tr_1*, denoted by $\text{rch}(tr_1, tr_2)$, is defined as $\sum \{(1 - \text{stop}_\pi(\rho)) * \prod \pi \mid (\rho, \pi) \text{ maximal probabilistic trace in } tr_1\}$.   □

The *higher* tree tr_2 will be given by the set of probabilistic traces that are produced by the composition of the specification and the user model. This tree is used to compute the probability of leaving the lower tree. In particular, only the probabilities of transitions departing from *leaves* of the lower tree are considered. The following result shows how higher trees can be constructed.

Proposition 2. Let S be a PFSM and U be a PLTS. We have that the set of probabilistic traces $\text{pTr}(S \parallel U)$ is a probabilistic tree. Moreover, there exists an arc labeled by $(i/o, p)$ between two nodes (ρ, π) and $(\rho \cdot (i/o), \pi \cdot \langle p \rangle)$ iff we have the sequences $s_{0S} \overset{\rho}{\Longrightarrow}_{\pi_1} s_1 \xrightarrow{i/o}_{p_1} s_1'$ and $s_{0U} \overset{\rho}{\Longrightarrow}_{\pi_2} s_2 \xrightarrow{i}_{p_2} s_2'$, where p is the normalized product of p_1 and p_2 after ρ, that is, $p = \frac{p_1 \, p_2}{\text{norm}_{S \parallel U}(\rho)}$.   □

Next we show how the lower tree is created. A tree containing the traces of a given *sample* is constructed by considering both the sample and the composition of the specification and the user model. Let us note that, despite the sample being produced by the interaction with the implementation, the probabilities of traces will be taken from the specification. Let us also note that we will be able to do this if the sample passes the *hypothesis contrast* that compares it with the behavior of the specification. This hypothesis contrast is implicitly applied by the relation $\text{conf}_{(H,\alpha)}$.

Definition 19. Let S, I be PFSMs such that $S \text{ conf}_{(H,\alpha)} I$, U be a PLTS and $H = \langle \rho_1, \rho_2, \ldots, \rho_r \rangle$ be a trace-test sample of $I \parallel U$. The *probabilistic tree of H*, denoted by $\text{pTree}(H)$, is defined as $\bigcup_{\rho \in H, \ (\rho,\pi) \in \text{pTr}(S \ U)} \text{prefix}(\rho, \pi)$.   □

Due to the way probabilistic trees are constructed from implementations, specifications, and user models, the following result holds.

Proposition 3. Let S and I be PFSMs, U be a PLTS, and $H = \langle \rho_1, \rho_2, \ldots, \rho_n \rangle$ be a trace-test sample of $I \parallel U$ such that $S \text{ conf}_{(H,\alpha)} I$. We have that $\text{pTree}(H)$ is finite and $\text{pTree}(H) \subseteq \text{pTr}(S \parallel U)$.   □

Now we are provided with all the needed machinery to define the upper bound of the probability that a user interacting with the IUT observes a probabilistic behavior that does not conform to the specification.

Definition 20. Let S and I be PFSMs, U be a PLTS, and $H = \langle \rho_1, \rho_2, \ldots, \rho_n \rangle$ be a trace-test sample of $I \parallel U$ such that $S \text{ conf}_{(H,\alpha)} I$. The *upper bound of the probability that the user U observes a wrong probabilistic behavior in I with feasibility α*, denoted by $\text{wrong}(U, I, \alpha)$, is given by $\text{rch}(\text{pTree}(H), \text{pTr}(S \parallel U))$.   □

6   Conclusions and Future Work

In this paper we have presented a probabilistic testing methodology that allows to consider user models. On the one hand, by applying user models, we can focus on testing a specific critical behavior. On the other hand, since we explicitly consider the propensity of each non-deterministic choice of systems, we can study systems not only on the basis of what they do but also on how often they do it. Since actual probabilities cannot be extracted from a black-box system, the probabilistic behavior of implementations and specifications is compared by means of suitable hypothesis contrasts. In addition, the combination of user models and the probabilistic approach allows to compute a relevant measure that cannot be computed in other frameworks: For a given feasibility degree, we can provide an upper bound of the probability of finding an error in the IUT. After a finite test suite is applied to an IUT, this measure allows to assess how confident we are the IUT is correct. Moreover, it implicitly provides a method to evaluate the quality of a test suite to evaluate an IUT with respect to a specification: If a given test suite is passed and it provides a *lower* upper bound (or an upper bound with a *higher* feasibility) than another suite that is also passed, then the former suite is preferred.

As future work, we plan to extend our framework to deal with *symbolic* probabilities that allow to denote ranges of probabilities instead of fix probabilities [5]. Besides, we will also introduce *stochastic* temporal delays to denote the time consumed by actions, that is, temporal delays defined in probabilistic terms [9].

Acknowledgements. We would like to thank the anonymous referees of this paper for their suggestions and valuable comments.

References

1. D. Cazorla, F. Cuartero, V. Valero, F. Pelayo, and J. Pardo. Algebraic theory of probabilistic and non-deterministic processes. *Journal of Logic and Algebraic Programming*, 55(1–2):57–103, 2003.
2. I. Christoff. Testing equivalences and fully abstract models for probabilistic processes. In *CONCUR'90, LNCS 458*, pages 126–140. Springer, 1990.
3. R. Cleaveland, Z. Dayar, S.A. Smolka, and S. Yuen. Testing preorders for probabilistic processes. *Information and Computation*, 154(2):93–148, 1999.
4. M. Hennessy. *Algebraic Theory of Processes*. MIT Press, 1988.
5. N. López, M. Núñez, and I. Rodríguez. Specification, testing and implementation relations for symbolic-probabilistic systems. *Theoretical Computer Science*, 353(1–3):228–248, 2006.
6. R. de Nicola and M.C.B. Hennessy. Testing equivalences for processes. *Theoretical Computer Science*, 34:83–133, 1984.
7. M. Núñez. Algebraic theory of probabilistic processes. *Journal of Logic and Algebraic Programming*, 56(1–2):117–177, 2003.
8. M. Núñez and I. Rodríguez. Encoding **PAMR** into (timed) **EFSMs**. In *FORTE 2002, LNCS 2529*, pages 1–16. Springer, 2002.
9. M. Núñez and I. Rodríguez. Towards testing stochastic timed systems. In *FORTE 2003, LNCS 2767*, pages 335–350. Springer, 2003.
10. M. Núñez and D. de Frutos. Testing semantics for probabilistic LOTOS. In *Formal Description Techniques VIII*, pages 365–380. Chapman & Hall, 1995.
11. K. Sayre. Usage model-based automated testing of C++ templates. In *International Conference on Software Engineering. Proceedings of the first international workshop on Advances in model-based testing*, pages 1–5. ACM Press, 2005.
12. R. Segala. Testing probabilistic automata. In *CONCUR'96, LNCS 1119*, pages 299–314. Springer, 1996.
13. M. Stoelinga and F.W. Vaandrager. A testing scenario for probabilistic automata. In *ICALP 2003, LNCS 2719*, pages 464–477. Springer, 2003.
14. J. Tretmans. Test generation with inputs, outputs and repetitive quiescence. *Software – Concepts and Tools*, 17(3):103–120, 1996.
15. G.H. Walton, J.H. Poore, and C.J. Trammell. Statistical testing of software based on a usage model. *Software - Practice & Experience*, 25(1):97–108, 1995.
16. J.A. Whittaker and J.H. Poore. Markov analysis of software specifications. *ACM Transactions on Software Engineering and Methodology*, 2(1):93–106, 1993.

Generating Test Cases for Web Services Using Extended Finite State Machine

ChangSup Keum[1], Sungwon Kang[2], In-Young Ko[2],
Jongmoon Baik[2], and Young-Il Choi[1]

[1] BcN Research Division,
Electronics and Telecommunications Research Institute
{cskeum, yichoi}@etri.re.kr
[2] School of Engineering,
Information and Communications University
{kangsw, iko, jbaik}@icu.ac.kr

Abstract. Web services utilize a standard communication infrastructure such as XML and SOAP to communicate through the Internet. Even though Web services are becoming more and more widespread as an emerging technology, it is hard to test Web services because they are distributed applications with numerous aspects of runtime behavior that are different from typical applications. This paper presents a new approach to testing Web services based on EFSM (Extended Finite State Machine). WSDL (Web Services Description Language) file alone does not provide dynamic behavior information. This problem can be overcome by augmenting it with a behavior specification of the service. Rather than domain partitioning or perturbation techniques, we choose EFSM because Web services have control flow as well as data flow like communication protocols. By appending this formal model of EFSM to standard WSDL, we can generate a set of test cases which has a better test coverage than other methods. Moreover, a procedure for deriving an EFSM model from WSDL specification is provided to help a service provider augment the EFSM model describing dynamic behaviors of the Web service. To show the efficacy of our approach, we applied our approach to Parlay-X Web services. In this way, we can test Web services with greater confidence in potential fault detection.

1 Introduction

A Web service is any service available on the Internet that uses a standardized XML messaging system and is not tied to a operating system or programming language. In other words, Web service is a collection of components that are wrapped with SOAP (Simple Object Access Protocol) interfaces so they can exchange XML-based (Extensible Markup Language) messages [1]. Using Web Services, companies can integrate existing business applications into new and innovative business applications, publish them as services, discover and subscribe to other services, and exchange information [2].

Some testing techniques that are used to test software components are being extended to Web services. A few papers have presented testing techniques for Web services, but the dynamic discovery and invocation capabilities of Web services bring up many testing issues. Existing Web service testing methods try to take advantage of

M.Ü. Uyar, A.Y. Duale, and M.A. Fecko (Eds.): TestCom 2006, LNCS 3964, pp. 103–117, 2006.

syntactic aspects of Web service rather than semantic, dynamic, and behavioral information because standard WSDL is not capable of containing such information. Therefore, they focused on testing of single operations rather than testing sequences of operations. Furthermore, they heavily rely on the test engineers' experience.

In this paper, we propose a new approach to test Web services. This idea stems from similarities between communication protocol testing and stateful Web services testing. Web services can be either stateless or stateful. Stateful Web services have several operations which affect the service's state that are used by other operations. Operations in stateless Web service do not change the service's internal states. Each operation in Web services has a request and response message with parameters. It is hard to test such Web services because they are distributed applications with numerous runtime behaviors that are different from typical applications. Service consumers usually have to use black-box testing because specifications are available but design and implementation details of Web services are not available. The specification is written in WSDL (Web Services Description Language). Unfortunately, current WSDL does not contain sufficient information for a consumer to test the available Web services. Although a few technologies exist to verify syntactic aspects of the interactions, it is very difficult to find out whether Web services behave correctly with all possible messages.

Specifically, protocol testing and Web services testing both require to perform some message exchanges and to analyze the result. Furthermore, it is more important to test sequences of messages than to test of single message. Also these two testing methods are basically based on the black-box approach. In black-box testing, specification has a strong influence on testing. Stateful Web services have reactive characteristics similar to communication protocols; therefore specification languages for Web services are favored which precisely define the temporal ordering of interactions. FSM (Finite State Machine) model is often used for defining the temporal order of interaction. However, the FSM model is often too restrictive for defining all aspects of a Web service specification because a Web service has input and output messages with data parameters. In contrast with FSM, EFSM [3] includes additional variables, input and output events including parameters. It consists of transitions which are characterized by a so-called enabling predicate and a transition action. Therefore an extended FSM model seems to be a very promising model for describing Web services behaviors.

We utilize the EFSM model to test Web services. Since current WSDL does not contain sufficient information for a test engineer to test the available Web services, temporal ordering information is added to describe Web services behaviors. EFSM (Extended Finite State Machine) is well suited for describing Web services behavior because it has the control part of the specification represented by pure FSM model and the data part represented by the transition predicates and actions.

There are many benefits to constructing test cases on the basis of a formal model specification such as EFSM. The benefits arise from the ability to precisely describe and reason about potential faults. In particular, it means that test can be applied uniformly, with greater confidence in their fault detecting potential, and with the possibility of full automation. Using an EFSM formal specification for a Web service, we can generate test cases from the specification automatically if we are equipped with an appropriate tool set such as EFSM analyzer, test case generator, and monitor.

The remainder of the paper is organized as follows. After reviewing existing Web service testing methods in Section 2, we present a procedure from a WSDL specification to an EFSM model and introduce test case generation algorithm using EFSM in

Section 3. An application example is provided to show the efficiency of our method in Section 4. We conclude the paper with a discussion of future work in Section 5.

2 Related Works

In this section, we review various methods for test cases generation for Web services and discuss drawbacks of existing Web service testing.

Heckel and Mariani [4] generate test cases for Web services with individual rules by selecting "likely" inputs. Possible inputs are further restricted by the preconditions of the GT (graph transformation) rules [5]. This suggests the derivation of test cases using a domain-based strategy, known as partition testing [6]. The idea is to select test cases by dividing the input domain into subsets and choosing one or more elements from each domain [7]. The execution of an operation can alter parts of service's state that are used by other operations. GT rules specify state modifications at a conceptual level. By analyzing these rules we can understand dependencies and conflicts between operations without inspecting their actual implementation. In this method, data-flow testing technique is used to test the interaction among production rules if creation of nodes and edge is interpreted as "definition" and deletion as "use" [8]. Conceptually, each operation (rule) can add or remove nodes and edges to or from the conceptual state, and change the values of attributes. Authors expect sequences of operations, which include the creation of an entity and its subsequent uses are likely to expose (state-based) fault.

In short, this method applies existing domain-based testing (partitioning testing) to the GT rules to generate test cases which cover validation of both single operation and sequences. The major problem of this method is that the definition of GT rules does not contain the temporal aspects (control flow) of message interactions. This method only considers data-flow to generate test cases for sequences of operations. This means that [4] has no test criteria for control flow. Furthermore, splitting the input domain into subsets relies on the tester's experience. This could cause non-uniform and biased tests for Web services.

In the paper [9], data perturbation is used as main method for testing Web service components. The testing process operates by modifying request messages, retransmitting messages, and analyzing the response messages for correct behavior. To do this process, value data perturbation modifies values in SOAP messages in terms of the types of the data. Data value perturbation relies on ideas from boundary value testing [10]. Test cases are derived from default boundary values of XML schemas. Tests are created by replacing each value with each boundary value, in turn, for appropriate type.

Concisely, the authors present a new approach to testing Web services based on data perturbation. Data perturbation uses two methods to test Web services: data value perturbation and interaction perturbation. However, this approach relies strictly on syntactic information about the XML messages, does not use behavior information. They consider only the selection of appropriate input parameter values. The sequences of operations in Web service are not considered. They just focus on testing of single operation of Web service.

Li et al. [11] provide some techniques for various kinds of Web Services testing such as unit testing, functional testing, performance testing, Load/stress testing, security testing and authorization testing. They provide detailed information on the key aspects of Web service testing features related with performance, authorization, and security.

Furthermore, they designed an automatic testing tool including SOAP-based log analysis, script generator, recorder, and monitor. However, there is no detailed information on the method of test cases generations in their paper.

In the paper [12], the authors propose a method of extending WSDL to describe dependency information which is useful for Web service testing. They suggest several extensions such as input-output dependency, invocation sequences, hierarchical functional description, and concurrent sequence specification. Similar to [11], there is no test case generation method and experimental data using the extension.

In summary, the existing Web service testing methods try to take advantage of syntactic aspects of Web service rather than behavioral aspects of Web services because standard WSDL does not contain such information. Therefore, they focused on the test of single operations instead of sequences of operation. One of disadvantages using those methods is that they rely on test engineer's experience. This could lead to non-uniform and biased testing. All these problems can be solved by augmenting behavior information to WSDL file. The behavior information holds control and data dependencies of Web service operations because the information is represented as an EFSM formal model. Using the augmented EFSM model, we can generate test cases which cover control and data paths thoroughly. In the next section, we describe our approach in detail.

3   Test Cases Generation for Web Services Using EFSM

In this section, we describe our test generation approach for Web services in detail. In Section 3.1, we first give a procedure for deriving an EFSM model from a WSDL specification of a service and illustrate the procedure with a banking Web service example. Once an EFSM model is constructed, test cases can be generated easily using a well-known algorithm as described in Section 3.2.

3.1   Modeling Web Service with EFSM

A WSDL specification is used to describe how to access a Web service and what operations it can perform. However, a WSDL specification does not provide sufficient information for Web service test derivation because it only provides the interface for the service. An EFSM starts from an initial state and moves from one state to another through interactions with its environment. The EFSM model extends the FSM model with variables, statements and conditions. An EFSM is a 6-tuple $<S, s_0, I, O, T, V>$, where S is a non-empty set of states, s_0 is the initial state, I is an non-empty set of input interactions, O is a non-empty set of output interactions, T is a non-empty set of transitions, and V is a set of variables. Each element of T is a 5-tuple of the form: <source_state, dest_state, input, predicate, compute_block>, where "source state" and "dest state" are states in S corresponding to the starting state and the target state of t, respectively; "input" is either an input interaction from I or empty; "predicates" is a predicate expressed in terms of variables in V, the parameters of the input interaction and some constants, and "compute-block" is a computation block consisting of assignment and output statements. We will only consider deterministic EFSMs that are completely specified. In addition, the initial state is always reachable from any state with a given valid context.

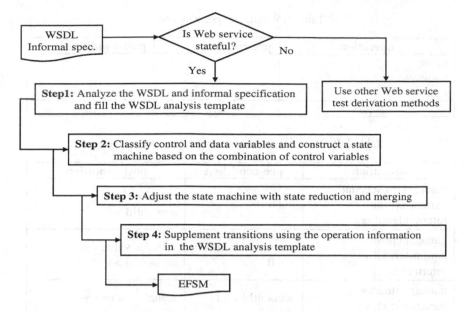

Fig. 1. Procedure for deriving an EFSM model from a WSDL description of a service

Figure 1 presents our procedure for deriving EFSM model from a WSDL specification. First of all, we have to decide whether the Web service to be modeled is stateful or not. A Stateful Web service in general can be modeled as an EFSM. Stateful Web service has several operations which change the service's internal state that are used by other operations. In that case, the operations may response with different output messages according to the internal state of Web service server. If the Web service is stateless, then we have to use other Web service testing methods such as [4] and [9]. Otherwise, we continue with Steps 1 through 4.

Step 1). We analyze the WSDL specification and the web service specification in informal language and fill the WSDL analysis template shown in Table 1. Each row of Table 1 describes an operation with its name, its parameter types and its return value type together with its pre-condition and post-condition for each operation in WSDL specification.

For example, Table 2 shows the WSDL analysis template filled out for a banking Web service. From WSDL description, we find out that the banking Web service provides four public operations, i.e. *openAccount*, *deposit*, *withdraw*, and *closeAccount*. The operation openAccount expects a single parameter *init* which means an initial deposit, and returns an account number *identifier*. The operation *closeAccount* expects a single parameter *id,* which means account number, and returns the result of operation such as *ResultOK* and *Error*. The operations *deposit* and *withdraw* expect two parameters *id* (identifier) and *v*(value), and return results such as *ResultOK* and *Error*. In Table 2, *value* holds the balance of the bank account created by openAccount operation and *accountId* means account number.

Step 2). To construct EFSM, it is necessary to classify variables in the pre-condition and post-condition of Table 2 into control variables and data variables. Then a

Table 1. WSDL analysis template

operation	pre-condition	post-condition
name: parameter: return:	...	...
...	...	...

Table 2. WSDL analysis template for banking Web service

operation	pre-condition	post-condition
name: openAccount parameter: init return: identifier	init > 0	value' = init accountId > 0
name : deposit parameter: id, v return : res	accountId = id v > 0	value' = value + v accountId > 0
name : withdraw parameter: id, v return : res	accountId = id value >= v	value' = value - v accountId > 0
name : closeAccount parameter: id return : res	accountId = id	accountId = 0 $\wedge$ value' = 0

Table 3. Classification of variables for banking Web service

operation	pre-condition		post-condition	
	control variable	data variable	control variable	data variable
name: openAccount parameter: init return: identifier	-	init	accountId value	init
name : deposit parameter: id, v return : res	accountId	v id	value	-
name : withdraw parameter: id, v return : res	accountId value	v id	value	-
name : CloseAccount parameter: id return : res	accountId	id	accountId value	-

combination of different values of the control variables makes a state of the EFSM under construction. For the banking Web service example, there are two control variables *accountId* and *value*. Table 3 presents the classification of variables for banking Web service.

Figure 2 shows an initial version of EFSM for the banking Web service. The states are constructed by combining possible value range of control variables. The variable *accountId and value* have two possible values: range 0 and greater than 0. If the control variables have value 0, it means that it is not initialized yet. When the variable *accountId* is initialized by *openAccout* operation, the variable has a value greater than 0 until it is closed by *closeAccout* operation. After initialization, the variable value keep a balance greater than 0 according to the operation *withdraw* and *deposit*. Therefore, we make four different states with combinations of the two control variables. Then we associate transitions with the appropriate operations by examining the pre-condition and post-condition of an operation.

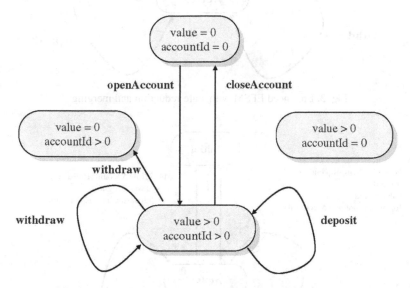

Fig. 2. EFSM construction with control variables

Step 3). It is desirable to reduce states in the initial version of EFSM model because first often the number of states would be otherwise huge and second there is a possibility that unreachable states may exist. For example, the state with *value* >0 and *accountId* = 0 is an unreachable state. Unreachable states should be deleted for the state reduction. Some states could be merged into one state according to test engineer's judgment. Figure 3 gives an enhanced EFSM obtained by removing an unreachable state and merging two states into a state named *Active*. For human readability, we assign a meaningful name to each state.

Step 4). To make a concrete transition in EFSM, operation information in the WSDL is used. An operation has input and output message. Input message is transformed into input event and output message is transformed into output event in the transition. Pre-condition is transformed into guard condition in the transition. Post-condition is transformed into actions in the transition. Figure 4 shows our final EFSM model derived from the WSDL specification for the banking Web service.

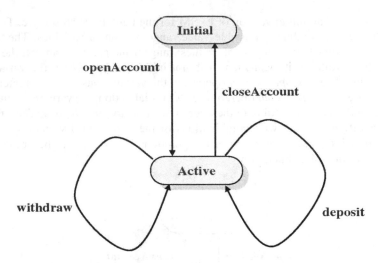

Fig. 3. Enhanced EFSM with state reduction and merging

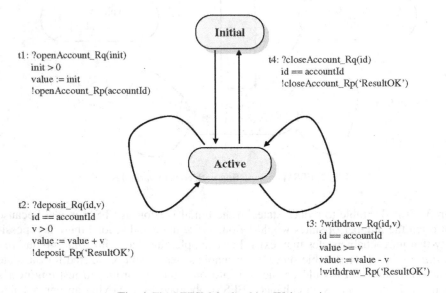

Fig. 4. Final EFSM for banking Web service

3.2 Test Cases Generation Algorithm Using EFSM

In the paper [3], the authors provide a comparison of single EFSM-based test generation methods. We choose Bourhfir's algorithm [13] as our test case generation method
for Web services because the algorithm considers both control and data flow with
better test coverage. The control flow criterion used is UIO (Unique Input Output)
sequence [14] and the data flow criterion is "all-definition-uses" criterion [15] where
all the paths in the specification containing a definition of a variable and its uses are

generated. Moreover, the algorithm uses a technique called cycle analysis to handle executability of test cases.

The detailed algorithm is described in Figure 5. For each state S in the EFSM, the algorithm generates all its executable preambles (a preamble is a path such that its first transition's initial state is the initial state of the system and its last transition's tail state is S) and all its postambles (a postamble is a path such that its first transition's start state is S and its last transition's tail state is the initial state). To generate the "all-definition-uses" paths, the algorithm generates all paths between each definition of a variable and each of its uses and verifies if these paths are executable, i.e., if all the predicates in the paths are true. After the handling executability problem, the algorithms removes the paths which is included in the already existing ones, completes the remaining paths (by adding postambles) and adds paths to cover the transitions which are not covered by the generated test cases.

Algorithm. *Extended FSM Test Generation*
Begin
    Generate the dataflow graph G form the EFSM specification
    Choose a value for each input parameter influencing the control flow
    Call *Executable-Du-Path-Generation(G)* procedure
    Remove the paths that are included in already existing ones
    Add a postamble to each du-path to form a complete path
    Make it executable for each complete path using cycle analysis
    Add paths to cover the uncovered transitions
    Generate its input/output sequence using symbolic evaluation
End.

Procedure *Executable-Du-Path-Generation(flowgraph G)*
  Begin
    Generate the shortest executable preamble for each transition
    For each transition T in G
      For each variable v which has an A-Use in T
        For each transition U which has a P-Use or a C-Use of v
          Find-All-Paths(T,U)
        EndFor
      EndFor
    EndFor
  End;

Fig. 5. Test case generation algorithm using EFSM

The following definitions that appeared in the paper [3] were used in the algorithm:

- A transition has an assignment-use (A-Use) of variable x, if x appears at the left-hand side of an assignment statement in the transition.
- When a variable x appears in the input list of a transition, the transition is said to have an input-use (I-Use) of variable x.
- A variable x is a definition (referred to as def), if x has an A-use or I-use.

- When a variable x appears in the predicate expression of a transition (Provided Clause), the transition has a predicate-use or P-Use of variable x.
- A transition is said to have a computational-use or C-use of variable x, if x occurs in an output primitive or an assignment statement at the right-hand side.
- A path $(t_1,t_2,\ldots,t_k,t_n)$ is said to a def-clear-path with respect to (w.r.t) a variable x if $t_2,\ldots,t_k$ do not contain defs of x.
- A path $(t_1,\ldots,t_k)$ is a Du-path (definition-uses) w.r.t a variable x, if $x \in def(t_1)$ and either $x \in c\text{-use}(t_k)$ or $x \in p\text{-use}(t_k)$, and $(t_1,\ldots,t_k)$ is a def-clear-path w.r.t x from t_1 to t_k.

In Table 4 shows a part of test cases and test sequences without input parameters for the EFSM in Figure 5.

Table 4. Test cases for the banking Web service

No	Test Cases	Input/Output Sequence
1	t1, t4	?openAccount_Rq!openAccount_Rp ?closeAccountRq !closeAccount_Rp
2	t1,t2,t4	?openAccount_Rq!openAccount_Rp ?deposit_Rq!deposit_Rp ?closeAccountRq !closeAccount_Rp
3	t1,t3,t4	?openAccount_Rq!openAccount_Rp ?withdraw_Rq!withdraw_Rp ?closeAccountRq !closeAccount_Rp
4	t1,t3,t2,t4	?openAccount_Rq!openAccount_Rp ?withdraw_Rq!withdraw_Rp ?deposit_Rq!deposit_Rp ?closeAccountRq !closeAccount_Rp
5	t1, t2, t3, t4	?openAcount_Rq!openAccount_Rp ?deposit_Rq!deposit_Rp ?withdraw_Rq!withdraw_Rp ?closeAccountRq !closeAccount_Rp

4 Application to Parlay-X Web Services

To show that our method can be effectively used for nontrivial real world problems, we applied it to Parlay-X Web services [16]. Parlay-X is a Web Services framework for telecommunications domain. The architecture of the framework in which Parlay-X Web services operate is shown in Figure 6. A Parlay-X Web service, *Third Party Call*, is used to create and manage a call initiated by an application. The overall scope of this Web service is to provide functions to application developers to create a call in a simple way. Using the *Third Party Call* Web service, application developers can invoke call handling functions without detailed telecommunication knowledge. The *Third Party Call* Web service provides four operations: *MakeCall, GetCallInformation, EndCall*, and *CancelCall*.

For comparison, we generated test cases for the *Third Party Call* Web service with three different methods, i.e. the method of Heckel et al [4], the method of Offtutt et al [9] and finally our method. For the method of Heckel et al [4], we defined a domain based on GT production rules. Eight production rules for the four operations were found. After that, we found attributes for each production rule. Test cases are generated by fixing a boundary value for at least one of them and randomly generating the other two values. In addition, we generated test cases using incorrect inputs for each rule. The sequences of operations are generated by analyzing dependencies and conflicts of operations. Finally, 36 test cases were generated using this method. For the method of Offtutt et al [9], 40 test cases were generated through the analysis of boundary values of message parameters.

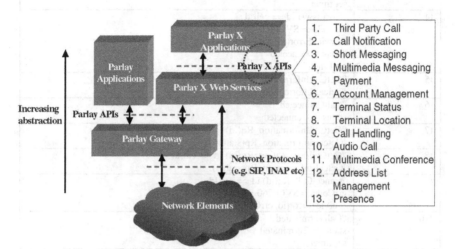

Fig. 6. Architecture of Parlay-X Web services

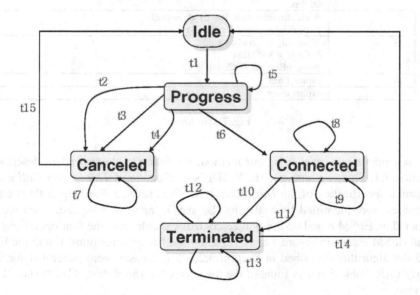

Fig. 7. EFSM model for the third party call Web service

Transition	Input/Output/Computation
t1	?MakeCall_Rq(cgNum,cdNum) callId := GenerateCallId() !MakeCall_Rp(callId) status := Initial
t2	?CancelCall_Rq(id) id == callId status := Canceled set timer
t3	?EndCall_Rq(id) id == callId status := Canceled set timer
t4	?NoAnswer id == callId errCode := SVC0001 !ServiceError(id, errCode) status := Canceled set timer
t5	?GetCallInformation_Rq(id) id == callId !GetcallInformation_Rp(status)
t6	?CallConnected status := Connected
t7	?GetCallInformation_Rq(id) id == callId !GetcallInformation_Rp(status)
t8	?GetCallInformation_Rq(id) id == callId !GetcallInformation_Rp(status)
t9	?CancelCall_Rq(id) id == callId errCode := SVC0260 !ServiceError(id, errCode)
t10	?CallTerminated status := Terminated set timer
t11	?EndCall_Rq(id) id == callId status := Terminated set timer
t12	?GetCallInformation_Rq(id) id == callId !GetcallInformation_Rp(status)
t13	?EndCall_Rq(id) id == callId errCode := SVC0261 !ServiceError(id, errCode)
t14	expire_timer
t15	expire_timer

Fig. 7. (*continued*)

To generate test cases using our method, we followed the procedure described in Section 3.1. First, we analyzed the WSDL specification of *Third Party Call* and the informal specification of the *Third Party Call* Web service. For Step 2, three control variables were identified by analyzing the WSDL analysis template. Then we constructed an EFSM based on these three control variables and the four operations. The final EFSM shown in Figure 7 has five states and fifteen transitions. Using the EFSM and the algorithm described in Section 3.2, 95 test cases were generated for *Third Party Call*. Table 5 shows some of the test cases for *Third Party Call* Parlay-X Web service.

Table 5. Test cases for Parlay-X Web service *Third Party Call*

No	Test cases
1	?MakeCall !CallId, ?GetCallInformation !CallStatus ?CallConnected ?CancelCall !ServiceError ?GetCallInformation !CallStatus ?CallTerminated ?TimeOut
2	?MakeCall !CallId ?CallConnected ?CancelCall !ServiceError ?CallTerminated ?TimeOut
3	?MakeCall !CallId ?GetCallInformation !CallStatus ?CallConnected ?CancelCall !ServiceError ?CallTerminated ?TimeOut
4	?MakeCall !CallId ?CallConnected ?GetCallInformation !CallStatus ?CancelCall !ServiceError ?CallTerminated ?TimeOut
5	?MakeCall !CallId, ?GetCallInformation !CallStatus ?CallConnected ?GetCallInformation !CallStatus ?CancelCall !ServiceError ?CallTerminated ?TimeOut

Table 6. Comparison of test criteria

	Data flow criterion	Control flow criterion
Method of Heckel et al [4]	all-definitions-uses	-
Method of Offtutt et al [9]	-	-
Our method	all-definitions-uses	UIO sequence

A test suite is a set of test cases and is said to satisfy a coverage criterion if for every entity defined by coverage criterion, there is a test case in the test suite that exercises the entity. Each method used in our experiment had its own test coverage criterion. The comparison of test coverage criterion for three methods is summarized in Table 6.

The method [9] had no test coverage criterion, but we could generate test cases easily through examining types of message parameters. There is a trade-off in choosing test coverage criteria. The program could be more thoroughly tested with the stronger criterion. However, usually the cost incurred by test cases generation and testing is negligible compared to the cost incurred by the presence of faults in programs.

Test cases and results of different methods are summarized in Table 7. As we expected, our method located more faults than the other methods even though it spent more time for executing a test case. Our method spent more time than other method because test cases generated using our method consist of the complex sequences of operations but almost all test cases generated using other method is made of a single operation. To show the efficacy of our method, the number of test cases and the accumulated number of faults detected are analyzed in Figure 8. As shown in Figure 8, our method detected many faults in the early phase of testing. Our methods detected many errors that occurred during executing complex sequences of operations. For example, the operation *GetCallInformation* worked well in the initial state and the progress state, but the operation caused an error when it executed in the connected state. The method [4] located some faults related with boundary value and incorrect input values in the case of testing for single operations. However, the sequences of operations derived from the method [4] were not effective for locating faults. Even if the method [4] expected the data flow coverage criterion "all-definitions-uses" for generated test cases, the generated test cases using relations of conflicts and casual dependencies between productions rules did not find out any faults which were located by our method. During testing using the method [9], it was difficult to find faults because faults rarely occurred when we executed single operations with different boundary values. Only two faults related with message parameter value with maximum length were founded.

Table 7. Test cases and results

	Method of Heckel et al [4]	Method of Offutt et al [9]	Our method
Number of test cases generated	36	40	95
Number of faults found	5	2	18
total execution time (sec.)	90	80	859
average execution time (sec.)	2.5	2	9

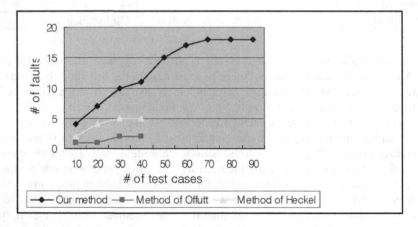

Fig. 8. Number of test cases and number of faults found

5 Conclusion

In this paper, we presented a new test cases generation method for Web services. The key idea is to augment a WSDL specification with an EFSM model that precisely describes the dynamic behavior of the service specified in the WSDL specification. Generally speaking, modeling an EFSM for a Web service is not a trivial task. To make this task easy and systematic, we suggested a procedure to derive an EFSM model from WSDL description of a service.

In summary, the main contributions of this paper are as follows: First, this paper introduces a new Web service testing method that augments WSDL specification with an EFSM formal model and applies a formal technique to Web service test generation. Second, using the EFSM based approach, we can generate a set of test cases with a very high test coverage which covers both control flow and data flow. Third, we applied our method to an industry level example and showed the efficacy of our method in terms of test coverage and fault detection.

One of drawbacks of our approach is the overhead to generate test cases based on an EFSM. Even if we suggest a procedure to derive an EFSM model from a WSDL specification, it may require additional jobs besides Figure 1 to complete a fully described EFSM in case of very complicated WSDL files. The algorithm described in Section 3.3 is also a heavy-weight algorithm. Without any automatic tool for generating test cases using EFSM, it is a very tedious task to generate test cases manually.

In this paper, we focused on testing of a Web service with single EFSM derived from a WSDL specification. For future work, we plan to extend our method to treat more complex situations such as test cases generation for compositions of Web services.

References

1. E. Cerami, Web Services Essentials, O'Reilly, 2002.
2. D. Booth, H. Haas, F. McCabe, E. Newcomer, M. Champion, C. Ferris, and D. Orchard, Web Services Architecture. W3C working group note, W3C, 2004.
3. C. Bourhfir, E.Aboulhamid, F.Khendek, and R.Dssouli, "Test cases selection from SDL specifications," Computer Networks 35(6), pp.693-708, 2001.
4. R. Heckel and L. Mariani, "Automatic Conformance Testing of Web Services," FASE 2005, LNCS 3442, pp. 34 – 48, 2005.
5. P. Baldan, B.Konig, and I.Sturmer, "Generating test cases for code generators by unfolding graph transformation systems," Proc. 2nd Intl. Conference on Graph Transformation, Rome, Italy, 2004.
6. L. White and E. Cohen, "A domain strategy for computer program testing." IEEE Transactions on Software Engineering 6, pp. 247–257, 1980.
7. E. Weyuker and B. Jeng, "Analyzing partition testing strategies," IEEE Transactions on Software Engineering 17, pp. 703–711, 1991.
8. S. Rapps, and E. Wejuker, "Data flow analysis techniques for program test data selection," 6th Intl. Conference on Software Engineering. pp. 272–278, 1982.
9. J. Offutt and W. Xu, "Generating Test Cases for Web Services Using Data Perturbation," ACM SIGSOFT SEN, 2004.
10. B. Beizer, Software Testing Techniques, Van Nostrand Reinhold, Inc, New York NY, 2nd edition, 1990.
11. Y. Li, M. Li, and J. Yu, "Web Service Testing, the Methodology, and the Implementation of the Automation-Testing Tool," GCC2003, LNCS 3032, pp.940-947, 2004.
12. W.T.Tsai, R. Paul, Y. Wang, C. Fan, and D. Wang, "Extending WSDL to Facilitate Web Services Testing," HASE 2002, 2002.
13. C. Bourhfir, R. Dssouli, E.Aboulhamid, and N.Rico, "Automatic executable test case generation for EFSM specified protocols," IWTCS'97, pp.75-90, 1997.
14. K.Sabnani and A.Dahbura, "A new Technique for Generating Protocol Tests," ACM Comput. Commun. 15(4), 1985.
15. Weyuker, E.J. and Rapps, S., "Selecting Software Test Data using Data Flow Information", IEEE Transactions on Software Engineering, April, 1985.
16. Parlay X Working Group, Parlay-X White Paper, http://www.parlay.org, 2002.

Towards the Testing of Composed Web Services in 3rd Generation Networks

Abdelghani Benharref, Rachida Dssouli, Roch Glitho, and Mohamed Adel Serhani

Concordia University,
1455 de Maisonneuve West Bd, Montreal, Quebec,
H3G 1M8, Canada
{abdel, m_serhan}@ece.concordia.ca,
{dssouli, glitho}ciise.concordia.ca

Abstract. With the proliferation of web services in business and as the number of web services is increasing, it is anticipated that a single web service will become insufficient to handle multitude, heterogeneous, and complex functions. Hence, web service composition will be used to create new value added services with a wide range of functionalities. Management of a composed web service is a complex issue compared to the management of a non-composed (basic) web service. In this paper, we propose a multi-observer architecture for detecting and locating faults in composed web services. It makes use of a network of observers that cooperate together to observe a composed web service. An observation strategy based on a set of heuristics is presented to reduce the number of web services to be observed. Observers are developed as mobile agent observers to help reducing the load introduced by the observation. Algorithms for fault detection, notification, and collaboration between observers are described. Finally, the architecture is illustrated through a case study for observing a composed teleconferencing web services in a 3G network. Different components of the architecture are developed. The network load introduced by the observation is measured and the fault detection capabilities of the architecture are discussed.

1 Introduction

Web services offer a set of mechanisms for program-to-program interactions over the Internet [1]. They make use of a multitude of emerging standard protocols, such as Simple Object Access Protocol (SOAP), Web Services Description Language (WSDL), and Universal Description, Discovery and Integration (UDDI).

Managing web services is critical because they are being deployed actually in heterogeneous environments and used in a wide range of applications especially in 3G networks. In 3G networks, they are being used for engineering Value Added Services (VAS). VAS are telecommunication standards that add value to those services already available on the network. Another use is digital imaging where their use is being standardized [2]. Their use in telecommunications networks is being standardized by the Open Mobile Alliance (OMA) [3].

Testing web services in open environments with multitude of participants is a hot issue. Testing can be active or passive. In active testing, fault detection is usually

M.Ü. Uyar, A.Y. Duale, and M.A. Fecko (Eds.): TestCom 2006, LNCS 3964, pp. 118–133, 2006.
© IFIP International Federation for Information Processing 2006

based on test cases that are applied to the web service under test. Passive testing, known also as *passive observation,* is based on traces collection and traces analysis.

A new kind of web services is known as "composed web services". A composed web service is any web service that makes use of a set of available web services to provide a different, more complex, service. Web services composition is generating considerable interest in recent years ([4], [5], [6]). It has a considerable potential of reducing development time and effort for new applications by reusing already available web services. The composed web service is also known as the final web service and a service participating in a composition as a basic web service.

Currently, there are standards or languages that help building composed web services such as: WSFL [7], DAML-S [8], and BPEL [9]. These languages make the web services composition process easier by providing concepts to represent partners and orchestrate their interactions. BPEL, which represents the merging of IBM's WSFL and Microsoft's XLANG, is gaining a lot of interest and is positioned to become the primer standard for Web service composition. This is the main reason for which BPEL is used in our work and then will be considered in the remaining parts of this paper.

Observation of composed web services is more complex than observation of basic web services. For instance, a fault occurring in the composed web service can originate in one of the basic web services and propagate to another basic web service. Furthermore, some faults may occur due to the composition itself, these faults are known as feature interaction.

Tracking a fault into its originating web service, will require the passive observation of all, or a subset of, the basic web services. As in distributed systems [10], this observation requires a network of observers rather than a single observer.

In this paper, we propose a novel architecture for online fault management of composed web services by observing their basic web services as well as the composed web service. The architecture is rooted in passive observation. The observers are model-based and are designed and implemented as mobile agents. The architecture makes available a web service observer that can be invoked and mobile observers that are sent back following an invocation.

The remainder sections of the paper are organized as follows: section 2 presents briefly web services composition and involved technologies followed by related works on management of composed web services. Section 3 and 4 discuss respectively the requirements and the fault model of the new architecture for observation of composed web services. Section 5 introduces different components of the multi-observer based architecture. It also discusses the limitations of the architecture in terms of necessary resources and network load. In section 6, we illustrate the observation procedures through a case study where a conferencing composed web service is observed. Finally, we provide a conclusion that summarizes the paper and discusses items for future work.

2   Related Work

Management of composed web services is a key issue for their success. Nowadays, this management is vendor-dependent and too much coupled to the application servers on which the composed web services are deployed. Few companies provide limited management features embedded within their platforms. The BPEL process manager [11], ActiveBPEL [12] and Process eXecution Engine [13] provide web-based consoles to deploy/undeploy services and manage their instances. These tools can only be used by the service provider. They manage the composed web services as if it was a basic web service, that is, without taking into consideration the management of basic web services participating in the composition. Moreover, since the tools managing basic web services are also vendor-dependent, exchange of information between different tools managing different entities is not straightforward.

Most of research activities on management of composed web services are actually on non-functional aspects of composed web services such as Quality of Service (QoS) ([14], [15], [16], [17]). For functional aspects, the authors in [18] propose the publication of some testing scripts in the registries. These scripts can be used by entities to test the correctness of desired web services. This approach requires active testers and not transparent to concerned web services.

The web services-based architecture for management of web services presented in [19] is limited to the observation of basic (non-composed) web services. It does not offer mechanisms to observe composed web services. The observation starts by invoking the web service observer. The latter generates a mobile agent and sends it to the hosting platform. The mobile observer checks all the traffic between the client and the observed web service and reports misbehaviors.

In this paper, we extend this architecture to observe composed web services while respecting its initial properties of transparency and availability. The observation is transparent since it does not invoke the web service for the sake of testing. The architecture is also available to all involved parties including the web service provider and the requestor since the architecture is based on web services.

3 Requirements

As stated above, observation of composed web services is based on the observation of the final web service and the participating basic web services. A set of information/resources is required for the sake of this observation. First of all, the web service observer must have access to the choreography document describing this composition. Another issue to solve toward making this observation possible is how to get the exact locations of the participating web services. Once this list of locations is known, models of the web services in this list (FSM or FSM annotated with timing properties) and WSDL documents should also be handed to the observers before the observation.

Discussions of possible mechanisms to satisfy these requirements are presented in section 5.4.

4 Fault Model

Fault detection is based on the information contained within the available resources. The models of these resources can be grouped in two groups: statefull (FSM/Annotated FSM, BPEL) and stateless (WSDL). The observers use this information to detect the following classes of faults:

From BPEL:

- **Ordering Faults (OF):** This fault occurs when the order of invocations of different participating web services is not respected. The "activities" section of the BPEL document describes the rules and order in which participating web services must be invoked. It is in fact a violation of an orchestration scenario that is a global property. This fault only can be detected by a global observer.

From FSM/FSM with timing annotations:

- **Input Fault (IF):** An input fault occurs if a requestor invokes an operation unavailable from the actual state of the web service. Unlike OF, this is a local property.
- **Timing Constraints Fault (TCF):** When monitoring the response time of web services, observers can measure the response time of web services and compare it to the threshold described in the model.

From WSDL:

- **Input Type Fault (ITF):** An input type fault is observed when a method is invoked with a wrong number and/or wrong types of parameters with regards to its signature published in the WSDL.
- **Output Type Fault (OTF):** This fault occurs if the type of the returned result is different from the type expected in the WSDL document.

5 Multi-observer Architecture

In this section, we will present different functional entities of the architecture and their interactions. We will then detail required steps for observation, starting from the invocation of the web service observer and ending with the result of the observation. Some steps of this procedure will smoothly change depending on the participation of different involved web services as will be presented latter.

5.1 Overall Architecture

The multi-observer architecture is illustrated in Figure 1 where the global observer and the local observers cooperate for fault management of composed web services. Local observers check exchanged messages observed at their points of observation and send these messages to the global observer. Whenever a local observer detects a fault, it informs the global observer then location and isolation procedures take place. Due to the position of observation points, fault location at this level is restricted to the location of a faulty web service, not the exact component within this web service.

In the case where a basic web service is itself a composed web service, the same architecture applies for its observation. That is, the web services participating in its composition will be observed also. This gives the architecture a tree structure and makes it very flexible in the observation of composed web services.

One of the design keys to be studied is the number of observers. In some cases, observing all participating web services in a composition is nothing but costly and useless. A full observation can dump the observers and the network with redundant

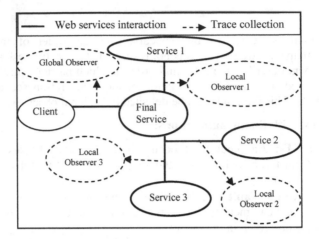

Fig. 1. Multi-observer architecture

information. Observing, for example, just the main web services that represent the core of the composition can be enough from a fault detection point of view. Potential suggestions and hints to select the web services to observe are discussed in section 5.3.

For each web service in this list, a mobile agent will passively observe its behavior. Two problems to solve: who hosts the mobile agent and how to provide it with exchanged messages? These problems are implementation-related issues and will be discussed in section 6.3.

5.2 Procedure

Observation is performed in two main steps. The first step consists of the configuration of the observation components, and the second step is fault management.

Observation is initiated by invocation of the web service observer. This is done by the entity willing to observe, which can be the provider of the composed web service, its client, one of the basic web services, or a mandated third party. After a successful invocation, the web service observer generates a set of mobile agents and sends them to the location(s) specified during invocation. Once the mobile agents reach their target locations, one of them becomes the global observer; other mobile agents are local observers. The local observers must inform the global observer of their locations and information about particular basic web services they are observing. At that point, all the components of the architecture are ready to start observation at the time specified during the invocation.

After deployment and configuration of all the observers, they start traces analysis at the time specified during the invocation. This observation will end at the time specified also during invocation. Whenever misbehavior is observed, local observers report to the global observer who reports to the web service observer.

5.3 Optimization

In this section, we discuss a set of suggestions and potential criteria that can be considered to build the list of web services to be observed.

For the selection of web services to observe, an important criterion is the number of interactions between the final web service and a basic web service. If a web service has few published interfaces and is invoked few times while others are invoked very often, observing the latter web services can be more appropriate. Another criterion is the complexity of a basic web service, from its FSM model: more a model of a web service is complex (number of states, number of transitions, etc.), more is the necessity for its observation.

Statistics on previous detected faults is another criterion. If faults occurred in a web service a certain number of times, a periodic observation of this web service can be a wise decision.

Selection of web services to observe might be implied by preferences of the final web service provider. These preferences depend on the importance a basic web service is playing in the composition or the tolerance of the final web service to some specific faults generated by some specific basic web services.

5.4 Participation of Web Services in Their Self Observation

The information required for observation (section 3) can be gathered through participation of involved web services providers: the provider of the composed web service, the providers of basic web services or from both. We designate these types of participation, respectively, as *final web service provider's participation*, *basic web services providers' participation* or *hybrid participation*.

Final Web Service Provider Participation. In this participation, the final web service provider supplies all the required information and resources necessary for the observation. This includes the BPEL description, WSDL documents, models of the web services (basic and final), and the list of nodes to host the mobile agents observers.

This kind of participation is completely transparent to basic web services and their providers. Additionally, due to the cloning nature of mobile agents, the web service observer sends only one mobile agent to the final web service provider's side instead of a separate mobile agent for each web service to be observed. This mobile agent will clone itself once it gets into its location. Doing so reduces significantly the traffic generated by moving mobile agents. If n is the number of web services to be observed, the complexity of the introduced load decreases from $\Theta(n)$ to $\Theta(1)$.

The load that will be introduced by the cooperation of observers to detect and locate a fault is limited to in-site load, that is, within the provider's domain since all observers are located there. The complexity of this load can be considered as $\Theta(1)$. Synchronization of observers is also easier than if observers were scattered between many sites.

The major weakness of the participation of one side is that all information, resources and observation activities will be within one web service provider.

Basic Web Services Providers' Participation. Unlike the centralized participation, the basic web services' participation requires the participation of the providers of all web services that have to be observed, including the final web service. Each web service provider supplies the WSDL document and the model of its web service and hosts the associated mobile observer. In addition, the final web service provides the BPEL document.

The network load is the major weakness of this type of participation. First, a mobile agent is generated and sent to each web service in the list of web services to be observed. The complexity of the load here is $\Theta(n)$. The cooperation of the observers introduces also another $\Theta(n)$ network load since observers are in different locations.

Hybrid Participation. The hybrid participation is a compromise between the two kinds of participation presented above. The participation is neither completely distributed nor centered. The final web service provider supplies a portion of the required information and resources while a subset of the list of web services to be observed supplies the remaining portions.

This can be a possible alternative when the final web service can not provide all the information and resources and only a subset of basic web services' providers are willing to participate in the observation. Those basic web services providers' who accept to participate in the observation will supply the information related to their web services and host the associated mobile observers. The final web service provider's furnishes information for other basic web services.

The configuration of the hybrid participation ranges between the centralized configuration and the distributed information, depending on how many basic web services providers' are participating in the observation and how much. Thus, the complexity of the load generated by moving the mobile observers ranges from $\Theta(1)$ to $\Theta(n)$ and for the cooperation of observers from in-site load to $\Theta(n)$. In the average, these complexities are around $\Theta(\log n)$.

In the following section, algorithms implemented by observers are presented and discussed.

5.5 Algorithms

Passive observation is performed in two steps: 1) passive homing and 2) fault detection [20]. The homing procedure is required to bring the observer to the same state as the observed web service. It is needed if the observation starts while interaction between entities has already started. When the observation starts at the same time as the interaction between observed entities, the homing sequence is empty.

For fault detection, every observed event in traces (request or response) is checked against the expected behavior in the corresponding model. We must note here that there are some cases where the observer can not decide if a response is expected or not. This is due to the fact that when the observation starts, it may miss some previous requests and the homing procedure might not give indication on requests not yet served. This is mainly the case for asynchronous invocations.

Each time a fault is detected by a local observer, a notification is sent to the global observer. Notifications must be purged before their correlation. This is done through two methods: purgeFinalNotification implemented by the global observer and purgeLocalNotification implemented by local observers. The main purging role is the ability of a receiver (client, final web service or basic web service) to detect a faulty received request or response. When a local observer detects an output fault, it notifies the global observer. It waits then for the reaction of the invoked web service. If the response of the latter contains a fault indication (in the SOAP message), the local observer informs the global observer. Otherwise, it sends a second notification to the global observer requesting fault location.

A faulty output generated by a web service will be detected by its associated observer. It will also be detected as an input fault by the observer of the receiving web service. Both observers will generate fault notification. The two notifications must be correlated since they refer to the same fault.

After receiving a notification from a local observer, the global observer associates it, if possible, to a previous fault or notification and updates the fault records accordingly. It waits then for a second notification for a specific period of time before starting the correlation. This starts by checking the fault records for previously detected fault. If the same fault has been detected before, the list of suspected web services is updated with the faulty web service(s) in the fault record. The list of suspects is then augmented by all basic web services invoked before the notification. This list is derived from the "activities" section of the BPEL document. Traces observed by the local observers of the web services in this list are checked to find the faulty web service. This process is repeated until a faulty web service is identified, remaining web services are not observed, or no decision can be made due to a lack of information on behaviors.

In the next section, we illustrate the applicability of the architecture through a motivating example of a composed web service for conferencing. The detailed requirements and steps for observation are depicted all along this example.

6 Case Study

In this section, we present our experiments using the multi-observer architecture to observe a composed web service. We introduce the context of utilization of the composed web service and its participating basic web services. We show a situation where the observation of basic web services gives more insights for fault identification. We present implementations of different components of the architecture, and discuss results with analysis.

6.1 Context

For the end of year meetings, a general manager has to meet with managers from different departments (Sales and R&D for example). Managers are located in different locations and due to their time tables cannot meet in a single meeting room. A practical option is to perform these meetings in a series of teleconferences. Only mangers are concerned and only those of them that are in their offices can join a conference. This is implied by security issues since confidential information will be exchanged during the meetings and communication between different locations is secured (VPN for example). At the end of each meeting, meetings' reports must be printed and distributed among all participating managers.

The manager decides to use a "Conferencing Web Service" (CWS), a composed web service, who performs all of the required tasks. In fact, it allows creation of a conference, add and remove users depending on their locations and profiles. At the end of each meeting, the CWS submits the produced reports for printing. Once printed and finalized, the paper version is distributed to appropriate locations.

6.2 Web Services

To perform the tasks presented above, the CWS is a composition of the following basic web services:

- Presence WS: this web service contains information on users' profiles (name, address, location, status, position, availability).
- Sensors: this web service detects the physical presence of users.
- Call Control: this web service creates and manages a multiparty conference (initiates the conference, adds/removes users, and ends conferences).
- Printing: at some points during the conferences or later on, managers may want to print documents (meeting reports ...). The printing web service will print these documents and keeps them for shipping.
- Shipping: documents printed during and after the conference should be distributed among users located in different locations. The CWS informs the shipping web service of the location of the documents to be shipped and their final destinations.

Figure 2 shows the composed CWS and its interactions with the basic web services.

Fig. 2. Composed/composing web services

6.3 Implementation Issues

All web services, including the web service observer, are implemented in BEA WebLogic. In fact, CWS is implemented in BEA even if it has a BPEL description. This is due to some limitations of the BPEL language and the available (non commercial) application servers. Implementing the CWS in BEA does not affect the observation process since the latter deals only with the exchanged SOAP messages which are independent from the adopted platform.

To host the mobile observers, a mobile agent platform should be available. In this case study, we use JADE [21], an open source platform easy to configure and deploy. All nodes willing to host mobile observers must download and configure the JADE libraries. The configuration of jade consists of adding the path of different libraries to the "path" system environment variable.

For trace collection, in this case study, we make use of the SOAP Handlers available within the BEA platform. A SOAP Handler, a special java class, intercepts a request or a response to/from a web service before it gets to the core web service or the client respectively, and can perform operations on it. In our case, the SOAP handler sends each event (request or response) in a UDP Datagram to the concerned mobile observer. The date of occurrence of the event is also sent in this datagram so that the observer can compute the response time. To be able to detect lost UDP datagrams, a sequence number field is used. When a mobile observer detects a lost Datagram (wrong sequence number), it suspends the fault detection and re-perform the homing procedure. It restarts the fault detection once this procedure is achieved correctly. Since the behavior of SOAP handlers within all observed web services is similar, a unique generic SOAP Handler is developed and then distributed to all providers.

6.4 Single Observation

When using the single-observer architecture initially presented in [19], the observer will check only the traffic between the manager and the CWS. Figure 3 shows the overall configuration and the information (traces) available to the observer where it is not aware of the interactions (request/response pairs) between CWS and basic web services. By doing so, if the CWS fails to provide the requested service or if the QoS degrades, the observer cannot designate the faulty web service. For example, if the "Sensors" web service (basic WS) fails to check the actual physical location of a manager, the CWS can not create a conference. From the observer's point of view (and then the manager's point of view), the CWS failed to create the conference. No more indication on the failure is available. Figure 4 shows a typical observation scenario from invocation of the observer (WSO) to the delivery of the verdict of observation. In this scenario, traces are collected through a participation of the web service's provider.

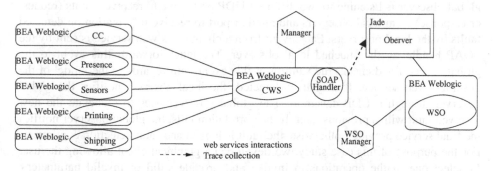

Fig. 3. Single-observer configuration

As will be illustrated in the following subsections, the multi-observer architecture gives more information in case of misbehaviors. This capability is made possible by using a network of observers rather than a single observer. Whenever an abnormal event occurs, cooperation between observers is initiated to track the faulty web service.

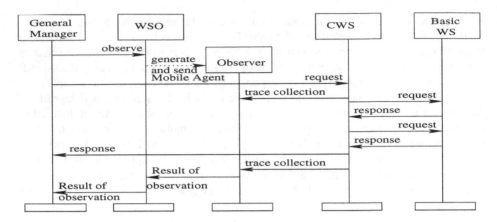

Fig. 4. Single-observer scenario

6.5 Multi-observer Observation Procedure

The general manager is highly concerned about the environment in which meetings will be carried out using CWS. He decides to make use of the passive observer available as a web service (WSO) to observe the behavior of the CWS. In addition to the observation of the CWS, the manager needs to assure that all the steps are performed according to the agreed on contract and QoS. All the providers accept to participate in the observation. The provider of the CWS will host all the mobile observers. It will also provide the BPEL and WSDLs documents, and the FSM models of each of the basic web services.

Once deployed and configured, mobile observers start by performing the homing procedure. When this procedure is carried out correctly, fault detection starts. Each local observer is listening to a UDP port to receive events from SOAP handlers. The global observer is listening to two different UDP ports: one to receive events (request or response) from local observers and another port to receive information on detected faults by the local observers. Each event from a client to its web service is sent by the SOAP handler to the attached local observer. The latter forwards this event to the global observer and checks the validity of this event with regards to the model of the observed web service. If a fault is detected, the local observer notifies the global observer through a UDP datagram. The global observer tries to associate the new received fault with a previous fault. If the correlation fails, the global observer notifies the final service provider, otherwise, the fault is logged and fault detection continues.

For the purpose of this case study, we developed a graphical client allowing the user to select one of the operations to invoke and provide valid or invalid parameters. Figure 5 shows the overall configuration of interacting web services, mobile observers and communication between these entities.

The observation procedure of CWS is performed following the steps illustrated in Figure 6. To keep the figure simple, just one web service handler and one web service client are depicted in the figure.

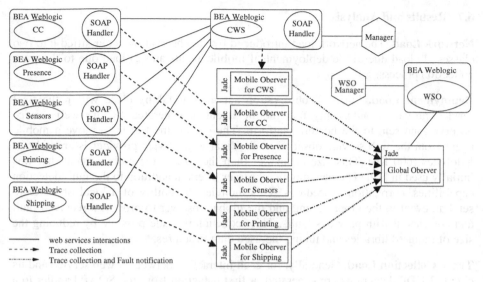

Fig. 5. Multi-observer configuration

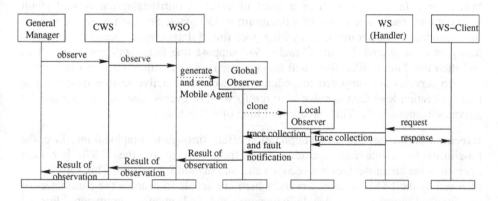

Fig. 6. Multi-observer scenario

6.6 Optimization

The main web service in the composition of the CWS is the Call Control web service. For this reason, we decide to observe it. Moreover, security of communication during conferences is of prime importance. As requested by the general manager, only managers that are in their offices should participate in a conference. So, the observation of the Presence and the Sensors web services is required. The Printing and Shipping web services are the only web services that deal with documents, so we decide to observe only one of them, the Printing web service. We assume that in case of a misbehavior during printing and shipping procedures, if the fault is not detected at the Printing web service by its attached observer, the fault is then within the Shipping web service.

6.7 Results and Analysis

Network Load. The network load introduced by the observation is classified into two classes: 1) load due to the deployment of mobile agents and 2) load due to the trace collection process.

Deployment Load. Since all observation activities is taking place within the final service provider's side, only one mobile agent is generated by the web service observer and sent to the hosting platform. The size of the traffic to move a mobile agent from the web service observer to the final web service provider is around 600 Kilobytes (600 Kb). This size is smaller than the size of the mobile agent that was initially used in [19]. The new mobile observer offers in addition to the fault detection capabilities, correlation procedures. This includes the ability of a local observer to send an event to the global event, and the global observer to process a received fault and correlate it with previous faults. This reduction is made possible by reducing the size of required libraries and tuning the used data structures.

Trace Collection Load. Generally, for each interaction between a web service and its client, 2 UDP datagrams are generated: a first datagram from the SOAP handler to a local observer, and a second datagram from this local observer to the global observer. Whenever a fault is detected in a local observer, a third datagram is sent (fault notification). The average size of a datagram is 150 bytes. So, each response/request pair introduces 4 datagrams if everything goes fine, 5 datagrams if one of the events is faulty, or 6 datagrams if both are faulty. We suppose that faults will not occur often, and then few fault notifications will be generated. This assumption is realistic since all web services are supposed to undergo an acceptable active testing process. The trace collection load then is reduced to the forward of events, that is, 4 datagrams for a request/response pair. This represents a load of 600 bytes.

Executed Scenarios. The client application offers, through its graphical interface, the possibility to invoke any operation from those offered by the CWS. For each operation, the client decides between a valid and invalid invocation. This selection is imposed by the FSM-based observers, which are unable to process the parameters of the invoked operation to decide between valid and invalid parameters. For all operations, the web service should return the output "true" if the operation is valid and "false" if the operation is invalid, otherwise a fault occurred.

To illustrate the detection capabilities of our architecture, we injected faults to the web services and or in the network and monitored the behaviour of the observers. Most of the injected faults have been detected by the observers. The global observer was also able to link related notifications that are originated by the same faulty event. From the BPEL document, the global observer builds the list of partners and the order in which they are invoked. Correlation is based on this information and the event sent within the fault notification message.

A fault that cannot be detected occurs when the last event in a communication between a web service and its client is lost. As discussed before, traces are sent as UDP packets. To be able to detect lost packets and recover the observation, a sequence number attribute is used. An observer detects a lost packet if the sequence number of the following received packet is different than expected. When a lost packet carries the last event in a communication, observers will not be able to detect this incident since no future packets will arrive. Table 1 shows brief descriptions of some of the executed scenarios and the reactions of observers (both local and global) to the fault.

Table 1. Some of the executed scenarios

Target web service	Fault description	Comments
CWS	Submit a printDocument request before creating a conference	Fault detected by local and global observer
Call Control	Add a user before creating a conference	Fault detected by local and global observer
Shipping	A trace collection event (shipDocument response) from a handler to the local observer is lost (Figure 7.a)	Neither the local observer nor the global observer will detect the fault.
Shipping	A trace collection event (shipDocument response) or a fault notification from a local observer to the global observer is lost (Figure 7.b)	The global observer will not be able to detect the fault or process the notification (correlation)

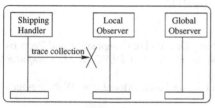

a. trace event does not reach the local observer

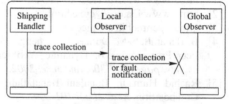

b. trace event or fault notification does not reach the Global Observer

Fig. 7. Scenarios of non-detected faults

7 Conclusion and Future Work

As web services, both basic and composed, are rapidly emerging as a new concept for business-to-business interactions, their management becomes a critical requirement for their success. Management of composed web services is more complex than the management of basic web services. This complexity is implied by the fact that a composed web service aggregates a set of basic web services to provide a different, more complex service. In fact, in addition to the management of the composed web service in its own, management of basic web services must be performed accordingly and all management entities should share management information.

In passive observation, the single observation of a composed web service does not give insights on the behaviors of the basic web services. Many events observed between a final web service and its client can not be studied and explained without information on the exchanged events between the final web service and its basic web services. Thus, observation of all basic web services or at least a subset of these web services is needed.

In this paper, we presented a multi-observer architecture for the observation of composed web services. The architecture proposes to observe the final web service and a set of basic web services. Heuristics to select the basic web services to be observed are also discussed. To reduce the network load generated by the observation, the architecture considers mobile agent observers. We discussed also the network load in terms of mathematical complexity for each type of participation of web services: final web service provider's participation, basic web services providers' participation or hybrid participation.

As a proof of concept, we developed a set of basic web services and a composed web service for conferencing management. We also evaluated the network load introduced by the observation and the fault detection capabilities of different observers.

Future work includes the consideration of an Extended Finite State Machine based observers. This is a main issue in web services interactions where data flow is important and fundamental.

References

[1] http://www.w3.org/TR/2004/NOTE-ws-arch-20040211/
[2] http://www.i3a.org/i_cpxe.html
[3] http://openmobilealliance.org
[4] B. Benatallah, M. Dumas, Q. Z. heng, and A. Ngu. Declarative Composition and Peer-to-Peer Provisioning of Dynamic Web services. *In Proc. of ICDE'02, IEEE Computer society, pages 297-308, and Jose, 2002.*
[5] Rachid Hamadi, Boualem Benatallah. A Petri Net-based Model for Web services Composition. *ADC 2003: 191-200.*
[6] S. Narayanan, and McIlraith, S.Simulation, verification and automated composition of web services. *In Proceedings of the World Wide Web Conference, 2002.*
[7] F. Leymann, Web service flow language (WSFL) 1.0. Available online at http://www-4.ibm.com/software/ solutions/webservices/pdf/WSFL.pdf 2001.
[8] A. Ankolekar, M. Burstein, J.R. Hobbs, O. Lassila, D. McDermott, D. Martin, S.A. McIlraith, S. Narayanan, M. Paolucci, T. Payne, and K. Sycara, "DAML-S: Web Service Description for the Semantic Web," *Proc. First Int'l Semantic Web Conf. (ISWC 02), 2002.*
[9] BPEL4WS Version 1.1 specification, May 2003 ftp://www6.software.ibm.com/software/ developer/library/ws-bpel.pdf
[10] S. Ghosh and A. Mathur. Issues in testing distributed component-based systems. 1st ICSE Workshop on Testing Distributed Component-Based Systems. May 1999.
[11] www.oracle.com
[12] http://www.activebpel.org
[13] http://www.fivesight.com/pxe.shtml
[14] A. Mani and A. Nagarajan, "Understanding quality of service for web services", January 2002. IBM paper: http://www-106.ibm.com/developerworks/library/ws-quality.html

[15] M.A. Serhani, R.Dssouli, A. Hafid, H. Sahraoui "A QoS broker based architecture for efficient web services selection" IEEE international conference on web services, July 2005, Orlando Florida, USA.

[16] Hongan Chen; Tao Yu; Kwei-Jay Lin, "QCWS: an implementation of QoS-capable multimedia web services", Proceedings of the Fifth International Symposium on multimedia software engineering, 2003.

[17] M.A. Serhani, R.Dssouli, H. Sahraoui, A. Benharef, E. Badidi "QoS Integration in Value Added Web Services" In second international conference on Innovations in Information Technology (IIT05) Dubai, U.A.E, 26-28 September 2005.

[18] Tsai, W.T.; Chen, Y.; Paul, R.; Liao, N.; Huang, H.; "Cooperative and Group Testing in Verification of Dynamic Composite Web Services" Computer Software and Applications Conference, 2004. Proceedings of the 28th Annual International, Volume 2, 2004 Page(s):170 - 173 vol.2

[19] A Benharref, R. Glitho and R. Dssouli, Mobile Agents for Testing Web Services in Next Generation Networks, 2nd International Workshop on Mobility Aware Technologies and Applications, (MATA 2005), Montreal , Canada, October 2005

[20] D. Lee et al. Passive Testing and Applications to Network Management. *Proceedings of IEEE International Conference on Network Protocols*, pages 113-122, October 1997.

[21] http://jade.tilab.com

Application of Two Test Generation Tools to an Industrial Case Study

Ana Cavalli[1], Stéphane Maag[1], Wissam Mallouli[1],
Mikael Marche[2], and Yves-Marie Quemener[2]

[1] Institut National des Télécommunications GET-INT,
Evry, France
{wissam.mallouli, ana.cavalli, stephane.maag}@int-evry.fr
[2] France Télécom R&D Division,
Lannion, France
{mikael.marche, yvesmarie.quemener}@francetelecom.com

Abstract. Many tools for test generation already exist and are used in industry; others are under development or improvement to allow faster generation and more effective tests. Comparing testing tools permits to acquire in-depth knowledge of the characteristics of each tool and to discover its strong points and limitations. Thus, the analysis of different automatic test generation tools provides a precise idea on the appropriate tool to be used to attain the expected results. This paper describes the application of two test generation tools to an industrial case study: a reverse directory telephone service similar to deployed services of this category developed by France Telecom. The tools used, for the automatic test generation, are a commercial tool *TestComposer* and *SIRIUS*, a tool developed by INT team. France Telecom R&D division provided the test campaign designed manually by a France Telecom service expert used to define the test objectives. The goal of this paper is to present the experimental results of tools application, to compare their performances and analyze some issues related to test execution.

Keywords: Case study, telephonic service, extended finite state machine, conformance testing, service testing, automatic test generation, formal specification, test generation tools.

1 Introduction

In the telecommunication field, the complexity and the variety of the implemented systems, as well as the high degree of reliability required for their global functioning, justify the care provided to the design of the best possible tests. Moreover, it is significant to automate these steps with an aim of reducing the time and the development cost and especially of increasing the reliability of the offered products. Manual tests are expensive in terms of time, and are less reliable. Thus methods of automatic test generation are proposed. The tools for test generation are varied and closely related to the language (formal or not) in which the system, protocol or service specification is written. Nevertheless,

M.Ü. Uyar, A.Y. Duale, and M.A. Fecko (Eds.): TestCom 2006, LNCS 3964, pp. 134–148, 2006.
© IFIP International Federation for Information Processing 2006

even if automatic test generation [5, 12] is seen as one of the most profitable application of formal methods, there are still very few commercial tools capable of automatically generating tests from a formal description and to execute them on the real system. The main reason for this is the difficulty of incorporating algorithms, to the methods, sufficiently powerful to make them scalable, i.e., applicable to real systems.

Our contribution: In this paper we present an industrial experience that consists of the application of two test generation tools to an industrial case study, a reverse directory service proposed by France Telecom, to perform test experiments and to compare their performances. One of the tools is a commercial one, *TestComposer*, while the other, *SIRIUS*, is a prototype developed in an academic environment. Both tools automatically generate test sequences from test objectives. These latter have been selected taking into account the testing campaign designed by a service expert of France Telecom. The comparison between these tools can help us to take a faster choice when modeling a system to validate it. It is indeed important for the validation phase to know as soon as possible which tool will be used. This comparison can be based on several criteria according to the technical resources we have. Among these criteria we can quote: test generation time, length of test sequences, complexity of the objectives, etc. Other aspects are also treated to clarify the strong points and the limits of each tool.

Different types of tests exist to ensure the reliability of a tested product. The tests presented in this paper are conformance tests [4, 2], which consists of testing that an implementation conforms its specification. It must be noted that at the beginning standardization activities related to conformance concerned the strict field of communication protocols. Later many researchers proposed [8] the extension of the applicability of conformance testing methods and its techniques to cover all the fields where it is possible to specify a system interface or an operation in a formalism close to that used for protocols (automata, process algebra, etc.). As this was the case for services specification, it was proposed to apply these methods to services, in particular, to those provided to a network user. However, this extension has several consequences on the test generation: it is necessary to redefine the test methodology according to the characteristics of the services to be tested; it is necessary to extend the capacity of expression of the test description languages; it is necessary to define strategies for test selection adapted to the various types of systems. The test of a transport protocol OSI is not performed in the same way as a telephone service (i.e. we do not seek the same errors). In this paper, we propose to apply and compare methods used for protocol conformance testing to the test of services.

The rest of the paper is organized as follows. Section 2 gives an outline of the test methodology and test assumptions. In Section 3 the tools applied to the case study are presented. Section 4 presents the reverse directory service specification, the test objectives to be checked on the service and provides an outline of the test generation methods used by each tool. Section 5 presents the results and compares the performance of each tool. Finally, Section 6 concludes the paper.

2   Basics

2.1   Service Definition

Before introducing our test methodology for services, it is necessary to provide the service definition, the concept of service in telecommunication being indeed very important. However, there is no unique definition of a service. In this paper, we used the definition that considers a service as a product offered by an operator to a customer in order to satisfy his needs.

2.2   Testing Methodology

The service conformance test methodology is based on the one presented in the standard ISO9646 [2] and it is divided into four aspects:

1. The definition of test architecture: in general, services are hosted on servers where direct access is difficult for the tester. The service access is done by users and consequently, the points of control and observation (PCO) are placed on the user side in order to initialize the transactions, to inject the events (valid, inappropriate or invalid) and to recover the results. Points of observation (PO) can be attached to certain strategic points to observe that data are well transcribed during the transfer between the various entities (customer, server, Proxy, gateway...). Vocal services such as the reverse directory studied here add a difficulty to test execution: the interaction is made over the phone network, either by DTMF[1] (phone keys) or by voice which are difficult to handle automatically (see Section 5.6).

2. The description of service behavior using a formal specification language. The description shows the behavior of the service by taking in consideration the test architecture and includes the actions of the various entities that intervene in the correct operation of the service. The formal specification languages used for the description of the reverse directory service are SDL [6] and IF[2] .

3. The characterization of the tests to be carried out and the test generation. It consists of a selection of tests according to certain preset criteria and the tests generation following a given procedure. As the size of the specifications for the services is rather considerable, the use of traditional methods that produce a global reachability graph from which tests are generated is not possible. Therefore, it is necessary to use methods which are based on a partial generation of the accessibility graph by applying different algorithms. These methods are based on the definition of test objectives to guide the generation of the tests. In this work, test selection has been based on test objectives that represent relevant aspects of the reverse directory service behavior. Different test generation algorithms have been used by the tools *TestComposer* and *SIRIUS*. Many of these algorithms can be embedded in

---

[1] Dual Tone Multi-Frequency.
[2] http://www-verimag.imag.fr/ãsync/IF/

several graph searching strategies like depth-first search (DFS), breath-first search (BFS) and BDFS which is a random combination of the two methods. In Section 5.2, we will explain the weight of these strategies in automatic test generation.

4. Test execution. The generated tests are executed according to the defined test architecture. Verdicts are established according to a conformance relation on each PO and PCO. All verdicts are directed to a central tester that deduces the final verdict. For this case study, test execution has been performed by France Telecom internal tools, which enable to send DTMF or vocal requests and to give a verdict from the vocal answers of the service.

After the description of the test methodology for services, we present in the following section the two tools used for the automatic test generation which are the commercial tool *TestComposer* and *SIRIUS*, a tool developed by INT team.

3 Tools Presentation

3.1 TestComposer

TestComposer [11] is a test generator tool commercialized by Telelogic. This tool corresponds to a major revision of the ObjectGeode TTCgeN tool which is a test generator produced by Verilog. *TestComposer* is based on two complementary prototypes: Tveda and TGV.

Tveda is a tool developed by the research laboratory of France Telecom CNET (currently France Telecom R&D division). The main features of Tveda are the automatic generation of test purposes, its heuristic approach to the use of reach-ability analysis, and the wealth of pragmatic customizations that have been included to cater to the needs of many different applications within CNET. The first versions of Tveda only accepted the Estelle language as an input language. However, an extension that allows the use of SDL language as an input language was introduced in the version 3, which is integrated with TGV in ObjectGeode. And a new semantic approach based on the analysis of the reachability graph produced by the Véda simulator was also applied in this version. The Tveda tool allows the generation of test sequences by using the heuristics related to explo-ration of the reachability graph. Its main advantage is that it calculates automat-ically the test objectives, which is not the case with other existing approaches.

TGV (Test Generation using Verification Technology) is an automatic pro-totype for the generation of conformance tests developed by IRISA/INRIA and Verimag within the framework of the PAMPA[3] project. It allows generating test cases using an on-the-fly exploration of the state space based on test objectives. The on-the-fly exploration permits to carry out the reachability graph creation and the checking of a property at the same time. The advantage of that method is that it just keeps the required part of the graph necessary for checking the properties. Moreover, this technique can give a partial result even if it stops with

---

[3] http://www.irisa.fr/pampa/

a memory fault. Using on-the-fly algorithms makes the execution of the principal algorithm activates the execution of each intermediate stage. TGV needs the
specification and the test objective as input and generates the test case in an
Aldebaran automaton format [9]. This format can be then translated into the
TTCN language [13, 10].

3.2 SIRIUS

SIRIUS belongs to a set of powerful test tools (TestGen-SDL [3, 7]) developed by
the INT team. The major advantage of this tool is that it avoids the exhaustive
generation of the reachability graph of a system specified by building only partial
graphs, allowing solving combinative explosion constraints. *SIRIUS* is based
on one of the flagship algorithms of TestGen-SDL, namely Hit-or-Jump [7].
In particular it allows the automatic generation of test sequences for the test
in context. This is very significant mainly for the features integration in an
operating platform. In addition to the generation of test sequences from test
objectives, *SIRIUS* also offers many other functionalities. Indeed, it also allows:

- The automata minimization according to the bisimulation.
- The detection of sink nodes.
- The transformation of a possibly partial reachability graph in an Aldebaran
  automaton.
- The parsing and lexing in the purpose of debugging all the output files of
  ObjectGEODE.

4   The Reverse Directory Case Study

4.1 Vocal Services Presentation

The development of vocal services at France Telecom R&D division uses many
distinct competencies. There are specialists in the techniques of voice recognition
and text-to-speech systems which provide reusable components. The use and the
adaptation of these components are themselves rather technical: it is necessary
to build models of the sentences being able to be pronounced by the users and
to check the good pronunciation by the text-to-speech system of precise information.

However, the main activity in the development of a vocal service consists in
the conception of the dialogue between the human being and the service. An
automaton should be defined, which, reacting to the input pronounced by the
user as determined by the voice recognition, will provide output information
via the text-to-speech system. This dialogue is often specified by a document in
natural language.

In order to formalize the process of the development and the validation of
vocal services, France Telecom R&D division developed a design tool based on
TauG2[4]. This tool makes it possible to design the dialogue of the service in

---

[4] http://www.telelogic.com/

the form of automata, and to produce a formal specification of it in IF. This specification can be used for doing simulations and automatic test generation.

At first approximation, the dialogue to be specified can be seen like a reactive system composed of a single automaton communicating by synchronous messages. This approximation corresponds to a rather simple problem looking to the state of the art. However, the following points complicate this first approximation.

- The vocal services often need to have access to data provided by the information system. That makes it necessary to model the vocal service in several distinct components.
- The assumption of synchronism and atomicity of the vocal messages is not entirely correct. It is necessary to be able to take into account phenomena such as the duration of pronunciation of the statements, the fact that the user can interrupt them or that this interruption can be blocked by the vocal service (barge-in can be allowed or not).

As a consequence, each dialogue is formalized by a distinct automaton, and the execution of the service corresponds to the parallel execution of each one, activated or not in a given state.

Each automaton composing the service can react to inputs of types DTMF or speech of the user. However, it is not important to model the voice recognition and its possible errors in the behavior of the vocal service. It is enough to consider an abstraction of this one starting from events in input of the service, and to consider an event corresponding to a failure of the voice recognition.

4.2 A Reverse Directory Specification

France Telecom team has written a specification of a reverse directory vocal service similar to real services of this category deployed over the network, using the IF language. It is an intermediate language developed by VERIMAG[5] to describe timed communicating systems. Many tools were also developed allowing the translation from languages like SDL, LOTOS, UML to IF and the generation of code or labeled transition systems in order to formally verify some properties.

INT team has used the following tools: *TestComposer* from ObjectGeode and *SIRIUS* developed at the INT. These two tools are based on a well adapted language for the formal specification of interactive systems: SDL (Specification Language Description). As the original specification was written in IF, we carried out a translation from IF towards SDL. For that aim, we used the respective relations between the semantics and syntax of SDL and IF. Moreover, Object-GEODE tool gives the possibility to draw the specification using a graphical mode which facilitates the comprehension of the service. Figure 1 produced by ObjectGEODE tool presents two states of Dialog_NumSpec process of the reverse directory specification.

After the redaction of the specification, the ObjectGeode tool makes syntactic and semantic verifications of the specification. The syntactic analysis ensures

---

[5] http://www-verimag.imag.fr/

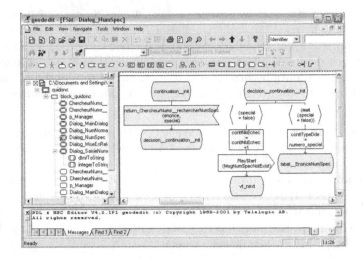

Fig. 1. Reverse directory specification using ObjectGEODE tool

that the specification complies with the syntactic rules of SDL in order to have a correct specification, whereas the semantic verification ensures the consistency of the specification. This step is carried out not only by this static analysis but also by an automatic exhaustive exploration of the specification. This is performed by testing all possible ways of system execution, with a certain number of rules and the cases of violations such as deadlocks, loops etc. During the verification, the main analyzed properties are:

- Safety (absence of deadlock, unspecified reception, blocking cycles, etc). Deadlock takes place when a state of the system, reachable from the initial state can not trigger a transition anymore.
- Promptness (livelock), indeed, a state is known as alive if it can be reached starting from all the states of the global system.

4.3   Determination of Test Objectives

France Telecom provided the test campaign designed manually by a France Telecom service expert. This document was the first reference we used to define our test objectives. In this document, the validation of the reverse directory service was described and organized in several phases related on one hand to the service conformance, and on the other hand to interoperability with the components used for voice synthesis and recognition. 130 test cases that were informally described cover the whole interactions of the service with the user, which includes:
- Unitary checking of each service's functionality; - Checking of system behaviors related to eventual user inactivity; - And checking of behaviors activated at the time of voice recognition failures. The unit checking makes it possible to validate that all the service functionalities conform to the specification. The other

verifications allow the validation of the service ergonomics in the case of defective use, related either to the user or to the service.

Using this test campaign, we have chosen to test the most relevant service reactions and functionalities. Therefore, 17 test objectives were obtained which represent more than 95% of the specification states and transitions. These test objectives can be classified in four categories: - Errors when dialing a number: we suppose that the user makes mistakes while trying to dial the telephone number of whom he wants to know the name. - Functionalities offered after dialing a normal number: Three functionalities are offered to the user. Only some reactions of the system are described. - Functionalities offered after dialing a special number: Some reactions of the system are described. - Errors due to user inactivity: we suppose that the user stops typing on the keys for a significant period. Messages for assistance will try to guide him for a good use of the service.

The three first categories describe functional tests of the service. It represents the details of what have been already called in France Telecom test campaign "the unitary checking if each service's functionality". The last category allows exactly, like in the France Telecom document, the system behaviors verification related to eventual user inactivity. We did not take into account in our test objectives the system behaviors related to voice recognition failures because this service feature was not specified.

Following are the test objectives we have defined in this case study:

N	Objectives to be tested: To test the correct reaction of the system, following this case:
1	The user does not press on any key after dialing the 3288. She then receives messages for assistance before receiving three messages to ask her to hang up followed by the stop of the service and the disconnection of the user.
2	After dialing the 3288, the user presses on the star key as it is required by the welcome message. Then, if she does not press on any key, an assistance message is transmitted to her explaining the functionalities of the service.
3	Error when dialing a number: the user presses on # key without having entered before a number. Sharp is a key which indicates the end of dialing.
4	Error when dialing a number: the user gives a telephone number that doesn't start with 0.
5	Error when dialing a number: the user gives a phone number which contains less than 10 digits.
6	Error when dialing a number: the user gives a telephone number which contains more than 10 digits.
7	Error when dialing a number: After having begun dialing, the user cancels his input number and dials * to give another number.
8	The user gives a good normal number (a normal number is a number starting with 0 and containing 10 digits). This number does not exist in the telephone directory of the reverse directory, an informative message is then transmitted to the user.

Fig. 2. Test Objectives for the case study

N	Objectives to be tested: To test the correct reaction of the system, following this case:
9	The user gives a normal number, receives the name of the corresponding person and remains inactive; this stops the service.
10	The user gives a normal number, listens to the name of the corresponding person and then dials 1 to listen to the spelling of this name.
11	The user gives a normal number, receives the name of the corresponding person and presses # that stops the service.
12	The user gives a normal number, listens to the name of the corresponding person and then presses 3 to contact this person. The telephone of the called rings without answer.
13	The user gives a normal number, listens to the name of the corresponding person and presses 3 to contact this person. The connection succeeds.
14	The user gives a normal number, listens to the name of the corresponding person and then dials 2 to listen to the address of the correspondent.
15	The user gives a good special number. This number does not exist in the telephone directory of the reverse directory; an informative message is then transmitted to the user.
16	The user gives a good special number that exists in the telephone directory. Information concerning this special number is transmitted to the user.
17	The user gives a good special number that exists in the telephone directory. Information concerning this special number is transmitted to the user. Some variables' values are transmitted to the system to allow generation of statistics.

Fig. 2. (*continued*)

4.4   Generation with TestComposer

According to [11], *TestComposer* is based on two complementary prototypes Tveda, which makes it possible to calculate automatically the test objectives and TGV, which allows, starting from calculated objectives or specific objectives, to calculate the corresponding tests. In our case, we give specific test objectives using a "test condition" formalism. The defined or calculated test objectives may be applied on a global SDL specification representing the system under test (the case of the reverse directory). We are also able to apply them on a part of this specification to test a component in a context. From the test objectives, test cases are produced and stored in a database. Then, the test sequence built from the obtained test cases is written in a TTCN file. Experimental results are presented in the Section 5.

4.5   Generation with SIRIUS

As mentioned above, *SIRIUS* is based on Hit-or-Jump, an algorithm especially used for components testing to perform test sequences generation through the specification. This research is guided by objectives which are illustrated by predicates on transitions. The research in the partial reachability graphs is performed in depth, width or both at the same time, and is restricted by a limited depth. In order to initialize the generation of test sequences, several parameters are necessary. Four main files must be developed. The first is the specification of the service (component to be tested), the second allows to initialize certain variables if necessary, the third one mentions the stop conditions (i.e. test objectives) and finally the last one allows the expert to guide the system at the beginning of the simulation, this file is called preamble. This latter is very important; it allows reducing in a consequent way the length of the sequence and the duration of its generation. As for *TestComposer*, we present our analysis and experimental results from the *SIRIUS* use in the following section.

5 Experimental Results

This section presents the results of the experimentation performed by both tools. Different aspects related to test coverage and execution of the tests are also discussed.

5.1 The Proposed Tools Are Scalable

The first objective of this paper is the application of two test generation tools to an industrial case study. The reverse directory service is an example of significant size as illustrated by the following table that provides some metrics of the specification, both in IF and SDL. These numbers are representative of real examples that have been treated at INT and France Telecom (MAP-GSM protocol, TCP/IP, SSCOP). It must be noted that the specification in SDL of the service has 4603 lines and the specification in IF has 1213 lines. The difference is due to the fact that SDL specification contains many lines of comments describing graphical aspects of the service specification (cf Figure 1). The SDL specification has been used by both tools to generate the tests.

	Blocks	Processes	States	Transitions	Signals	Channels	Number of lines
IF	—	8	77	160	31	15	1213
SDL	1	8	77	160	31	15	4603

Fig. 3. Metrics of IF and SDL reverse directory service specification

5.2 Test Objectives Generation

The results of test generation for the seventeen objectives are recapitulated in the following table. The tests are performed in the order in Section 4.3. For each test, the length of the sequence (number of transitions), the test generation time and the length of the preamble to be provided to guide the simulator to lead the test objectives are given.

The results established in this table are obtained after a BFS (breath-first search) exploration of the reachability graph. This choice is due to the specificity

	TestComposer			SIRIUS		
	Nb of transitions	Generation duration	Preamble length	Nb of transitions	Generation duration	Preamble lenght
Test 1	8	0mn 0s	0	7	0mn 0s	0
Test 2	9	0mn 0s	0	5	0mn 0s	0
Test 3	22	0mn 2s	0	7	0mn 12s	0
Test 4	64	0mn 33s	35	15	3mn 36s	35
Test 5	11	0mn 0s	0	5	0mn 0s	0
Test 6	13	0mn 0s	0	5	0mn 1s	0
Test 7	13	0mn 0s	0	5	0mn 0s	0
Test 8	87	0mn 18s	50	28	1mn 40s	50
Test 9	83	0mn 34s	45	24	3mn 47s	45
Test 10	84	3mn 47s	45	25	2mn 33s	50
Test 11	82	0mn 24s	45	25	1mn 52s	45
Test 12	85	0mn 9s	50	24	0mn 36s	50
Test 13	89	0mn 10s	50	25	0mn 51s	50
Test 14	83	0mn 38s	45	24	3mn 59s	45
Test 15	47	0mn 32s	8	14	3mn 34s	8
Test 16	48	4mn 34s	8	14	3mn 36s	13
Test 17	49	0mn 12s	14	16	1mn 02s	14

Fig. 4. Some automatic test generation results

of the service which has to take into account after each transition all the possible inputs injected by the user, to analyze them and generate the right output. The test objectives we defined are reachable via quite short sequences (almost 50 transitions) and do not need a DFS or BDFS exploration that tries to search in depth of the reachability graph.

5.3   Performance Analysis and Discussions

According to the Figure 4, we can elaborate on performancewise comparison between *TestComposer* and *SIRIUS*. First, we can easily notice that the test generation duration using *SIRIUS* is larger than using *TestComposer*. This fact constitutes a positive point for *TestComposer*. But if we refer to another criterion which is the length of test sequences, *SIRIUS* becomes more efficient. Actually, the length of tests sequences constitutes a very significant comparative data since test execution duration depends on it, and mainly when this execution is manual. This is often the case for vocal services where the test automatization is difficult because of the peculiarities of DTMF or vocal interaction (see Section 5.6). This length is on average three times shorter for *SIRIUS*. Indeed, *SIRIUS* has an advantage compared to *TestComposer* since it allows the automatic elimination of silent transitions. With this operation, the length of the test sequence becomes shorter and more comprehensible for the person carrying out the test. This fact constitutes one of the strongest points of *SIRIUS*. The Figure 5 recapitulate the length of the test sequences for the seventeen predefined objectives.

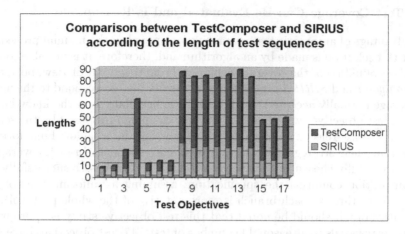

Fig. 5. Graphics comparison according to the length of test sequences

The automatic test generation is not always possible especially if the system to be tested presents a complex reachability graph. It is sometimes necessary to guide the simulator by providing a preamble. This preamble comprises the first transitions to be followed in order to begin the research of the test objective. The shortness of the preamble constitutes a strong point of the used tool. In our application, *TestComposer* shows superiority compared to *SIRIUS*. Indeed, we may in certain cases (test 10 for example) automatically generate tests using *TestComposer* giving a 45 transition length preamble, whereas a generation with *SIRIUS* requires 5 more transitions.

Another significant comparison criterion between various tools of simulation and test generation is the memory use. With ObjectGEODE, the memory consumption for the reverse directory telephone service is about 35.0 MB; this consumption is 10 times lower with *SIRIUS* (3.1 MB). This is easy to understand, first because *SIRIUS* only builds a part of the reachability graph for the test generation and second because ObjectGEODE has a graphic interface and several other functionalities which is not the case of *SIRIUS*.

	TestComposer	SIRIUS
Generation duration	+ Fast generation	-
Length of sequence	-	+ Sequence 3 times shorter
Memory used	-	+ Space 10 times smaller
Length of Preamble	+ Shorter preamble	-

Fig. 6. Summary table

According to this summary table, the user can make his choice based on his personal criteria and the characteristics of his application.

5.4   Test Coverage Can Be Evaluated and Is Reasonable

The advantage of automatic test generation compared to the manual procedures is that test selection is made by an algorithm and, therefore, it is possible to have a precise evaluation of the coverage achieved. From this point of view, both tools *TestComposer* and *SIRIUS* produce similar results that correspond to the notion of coverage normally accepted. Both tools have been based on the know-how of experts: test objectives were based in the test campaign proposed by the service expert of France Telecom. By placing this know-how in an algorithmic form we realize that their strategy for selecting tests correspond to branch coverage of a subset carefully chosen in the specification. If the test passes successfully, the implementation conforms the specification; assuming a uniformity hypothesis (passing once through each branch is representative of the whole protocol).

Furthermore, it should be noted that this test objective study is appropriate, since it corresponds to a reasonable number of tests. 17 test objectives have been selected, which cover more than 95% of the specification states and transitions.

5.5   Tests Are Really Usable

The produced tests are really usable, since they have the same format as (and are of comparable size to) test suites developed manually. Both tools, *TestComposer* and *SIRIUS*, generated tests that are composed of a preamble (shortest path between the initial state and the starting state for the transition to be tested), followed by the transition being tested. The structure and length of the tests produced by both tools correspond therefore to the usual standards. As a final remark, it is interesting to note that tests can be produced in TTCN [13] and MSC [1] notation facilitating the portability of the tests.

5.6   Automatic Test Execution

If the phase of automatic tests generation is a problem that can be adequately treated by test generation tools, because of the nature of reactive system of the specification, the phase of automatic tests execution encounters many difficulties.

The first difficulty, central with the problem, is related to the heterogeneity of the platforms executing the services. This heterogeneity, characterized by the lack of a standardized API to access to the platform, implies that a generic solution for the tests automation must be based on an emulation of the interaction between the human and the service.

This consists in emulating the input of tests starting from text to speech or recorded sound files, and to observe/control the outputs by using techniques of voice recognition. However, even if it is technically rather easy to associate each tests input with sound files, the recognition of the output is not good enough.

Concretely, France Telecom R&D division developed a test automaton based on the voice recognition. This automaton ensures the validity of the PASS and FAIL verdicts, but produced many false positive characterized by verdict UN-CONCLUSIVE. Typically, a verdict is FAIL if the service produced an output instead of a silence, and UNCONCLUSIVE if the output is well envisaged, but not recognized by the voice recognition.

Taking into account these difficulties, many tests must be carried out manually at the time of the validation of the service. However, the manual execution of tests is expensive in resources, because it implies the mobilization of people and time. In order to optimize the resources as well as possible, a great interest is related in advance to the quality of the generated tests. Those tests must cover in a minimum of occurrences the functionalities of the service, and must especially comprise a minimum preamble to validate the test objective.

In this context, the quality of the preambles to the generated tests is a discriminating element at the time of the selection of a tool for generating tests.

6   Conclusion

In this paper, we attempt to compare the performances of two test generation tools, *TestComposer* and *SIRIUS*, by applying them to a real case study provided by France Telecom, a reverse directory telephone service . These tools perform automatic test generation based on test objectives. The test objectives used for the experiments were provided by the France Telecom test plan. Results of the experimentation show that these test objectives covered quasi completely the tests provided by the test plan, showing the interest of the use of formal testing methods. Performance analysis shows that even if test generation time could be no so important for industrials (they are looking for pertinent and correct test sequences), this criteria could be important from an academic point of view in order to compare the algorithm performances. In addition, experiment show that the length of test sequences is an important criterion to evaluate the test sequences. In particular, for this case study this element was imperative because the tests were executed manually by France Telecom. Tests were executed manually because the reverse directory is a vocal service and it was very difficult to automate test execution and mainly to automate voice recognition that remains non deterministic. Finally, tools comparison allows us to improve our understanding of the strong points and limitations of each tool.

It must be mentioned also, that this study could be extended to other test generation tools. France Telecom's authors are ready to give access to the specification and the original test plan. This could provide a realistic problem to designers of automatic test generation tools.

References

1. *ITU-T Rec. Z. 120 Message Sequence Charts, (MSC)*. Geneva, 1996.
2. ISO/IEC 9646-1. *Information Technology - Open Systems Interconnection - Conformance testing methodology and framework Part 1: General Concepts*.
3. R. Anido and al. Engendrer des tests pour un vrai protocole grâce à des techniques éprouvées de vérifications. In Proceeding of CFIP96/Cinquième Colloque Francophone sur l Ingénierie des Protocoles, editor, *In ENSIAS*, pages 499–513, Rabat, Maroc, octobre 1996.

4. A.V.Aho, A.T.Dahbura, D. Lee, and M.U.Uyar. An optimization technique for protocol conformance test generation based on uio sequences and rural chinese postman tours. In *IEEE transactions on Communications, 39(3), pages 1604-1615.*

5. C. Bourhfir, R. Dssouli, E. Aboulhamid, and N. Rico. Automatic executable test case generation for EFSM specified protocols. In Chapman & Hall, editor, *IWTCS97*, pages 75–90, 1997.

6. A. Cavalli and D. Hogrefe. Testing and validation of SDL systems : Tutorial. In *SDL'95 forum*, 1995.

7. Ana Cavalli, David Lee, Christian Rinderknecht, and Fatiha Zaïdi. Hit-or-Jump: An Algorithm for Embedded Testing with Applications to IN Services. In Jianping Wu, Samuel T. Chanson, and Qiang Gao, editors, *Formal Methods for Protocol Engineering And Distributed Systems*, pages 41–56, Beijing, China, october 1999.

8. M. Clatin, R. Groz, M. Phalippou, and R. Thummel. Two approaches linking test generation with verification techniques. In A. Cavalli and S. Budkowski, editors, *Protocol Test Systems VII.* Chapman & Hall, 1996.

9. J.-C. Fernandez, H. Garavel, A. Kerbat, L. Mounier R. Mateescu, and M. Sighireanu. Cadp : A Protocol Validation and Verification Toolbox. In Rajeev Alur and Thomas A. Henzinger, editors, *The 8th Conference on Computer-Aided Verification, CAV'96*, New Jersey, USA, August 1996. Springer Verlag.

10. G. Rethy I. Schieferdecker A. Wiles J. Grabowski, D. Hogrefe and Colin Willcock. An introduction to the testing and test control notation (ttcn-3). In *Computer Networks 42(3)*, pages 375–403, 2003.

11. A. Kerbrat, T. Jeron, and R. Groz. Automated test generation from SDL specifications. In R. Dssouli, G.V. Bochman, and Y. Lahav, editors, *SDL'99.* Elsiever Science, 1999.

12. J. Tretmans and A. Belinfante. Automatic testing with formal methods. In *Proceedings of the 7th European International Conference on Software Testing, EuroSTAR'99*, November 1999.

13. ETSI. TTCN-3. *TTCN-3 – Core Language.*

Performance Analysis of Concurrent PCOs
in TTCN-3

Máté J. Csorba[1], Sándor Palugyai[1], Sarolta Dibuz[1], and Gyula Csopaki[2]

[1] Ericsson Hungary Ltd., Test Competence Center,
H-1117 Budapest, Irinyi J. u. 4-20, Hungary
Tel.: (36) 1-437 7489; Fax: (36) 1-437 7576
[2] Department of Telecommunications and Media Informatics,
Budapest University of Technology and Economics,
H-1117 Budapest, Magyar tudósok körútja 2, Hungary
{Mate.Csorba, Sandor.Palugyai, Sarolta.Dibuz}@ericsson.com
Csopaki@tmit.bme.hu

Abstract. This paper deals with a study and a mathematical model of concurrent Points of Control and Observation (PCOs) realized in Testing and Test Control Notation version 3 (TTCN-3). We study test scenarios that are gaining importance as TTCN-3 is emerging as a notation suitable for conducting load tests too. We investigate communication between parallel test components (PTCs) and analyze race conditions between the queues underlying the implemented PCOs. This way, we build an analytic model to investigate behavior of PCOs under stress conditions and to assess possible latencies messages in a TTCN-3 based load test system might suffer. We present a discrete-time Quasi Birth-Death process to predict performance indices of test components and we propose to use the results to avoid indefinite postponement in the communication of PTCs. Also, we aim to use the model for calculating traffic intensity limits under which it is feasible to use TTCN-3 for load testing. Furthermore, we present the output of the model together with an example load test scenario that is vulnerable to that types of latencies.

1 Introduction

The subject of our investigation is the standardized test specification language the Testing and Test control Notation version 3 (TTCN-3) [1]. TTCN-3 is widely used in different areas of testing including different fields of telecom and datacom and is gaining acceptance even in automotive systems testing. The language itself has a variety of applications including testing various systems for interoperability, robustness and conformance.

Recently, there has been a sore need to assess the capabilities of TTCN-3 as a testing solution not only for conformance and interoperability testing but for performance evaluation of telecommunication systems as well. Since TTCN-3 is a rather high level specification language, concerns have arisen regarding its applicability in load tests that require a significant amount of processing power, e.g. a high number of packets per second generated. On the other hand, tests and numerous pioneer projects show us that

M.Ü. Uyar, A.Y. Duale, and M.A. Fecko (Eds.): TestCom 2006, LNCS 3964, pp. 149–160, 2006.

the language is capable of working at the edges of what the underlying hardware is capable of. Besides, emerging real-time extensions to the original notation also exist [2], [3], [4]. So, usage of TTCN-3 in load tests should not imply a bottleneck if tests are designed carefully.

Our work examines the event processing capabilities of test components that exchange messages via communication ports competing with each other for system resources. By analyzing race conditions between the queues underlying the implemented PCOs, we build an analytic model to investigate behavior of PCOs under stress conditions and to assess possible latencies messages in a TTCN-3 based load test system might suffer. A discrete-time Quasi Birth-Death process is presented to predict performance indices of test components. We aim to use the model for predicting traffic intensity limits under which it is feasible to use TTCN-3 for load testing purposes. The remainder of this paper is organized as follows.

In Section 2, we introduce briefly a few basic issues in TTCN-3, the operation of Points of Control and Observation (PCOs), and in Section 3, we present the mathematical formalism used in this paper. Section 4 presents a parametrical model of test components using multiple PCOs. In Section 5, analytic results of the model for an example test component are detailed, together with an *ns-2* [5] simulation that was used mainly for validating and fine tuning of the model. Finally, in Section 6, concluding remarks are given and future work is detailed.

2   Concurrent PCOs and Alternative Behavior in TTCN-3

This section outlines the basic structures in TTCN-3 we aim to model, together with examples of load test scenarios that might be vulnerable to certain types of latencies during execution. Moreover, we point out the performance indices we evaluate to predict the behavior of TTCN-3 test components.

In TTCN-3, configuration of the test system can be set dynamically. This means that a configuration may consist of several components participating in the test [6], either as a component that communicates directly with the System Under Test (SUT), or as a component having only registration or internal purposes, meaning that it communicates only with parts of the test system itself. Within every configuration there is only one designated component, the Main Test Component (MTC) that is created automatically. Other components can be created dynamically and are called Parallel Test Components (PTCs). PTCs do not have any hierarchical order among them. PTCs communicate with each other, and with the MTC also, via test ports. Similarly, communication towards the SUT is established via test ports too, as in Figure 1.

Each test port that connects either two test components (internal port) or a test component and the interface towards the SUT is modeled as a FIFO queue for the incoming/outgoing messages. Each component can access messages in its correspondent queue one by one. Properties of the FIFO queues assigned to a test port are dependent on the actual implementation of the TTCN-3 compiler. The queues can be infinite in principle, as long as the system memory lasts, but might overflow indeed. More importantly, in a load test system response time must be considerably short. This means that it is inexpedient to implement a virtually infinite buffer for a PCO and forget about message

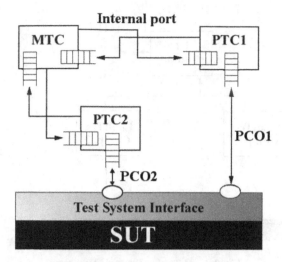

Fig. 1. Test Components and Test Ports

loss at all. Although a sufficiently long buffer might eliminate message loss, response time increases significantly at the same time. Accordingly, we investigate message loss in relatively short queues.

The actual behavior of a test case is defined by dynamic behavioral statements in a test component that communicates over certain test ports. Usually, sequences of statements can be expressed as a tree of execution paths, that is alternatives. In TTCN-3, the *alt* statement is used to handle events possible in a particular state of the test component. These events include reception of messages, timer events and termination of other PTCs.

The *alt* statement uses a so-called snapshot logic [7]. This means, that before evaluating the actual alternatives in the *alt* a snapshot of the test component containing any information that is relevant (e.g. status of the test ports involved in the *alt*, running timers) is taken. Branches of the *alt* might have a Boolean guard expression assigned to them that is evaluated before the branch is examined. The guard expression might be based on the snapshot as well.

Different types of branches exist in an *alt* statement (e.g. timeout, receive). Each time a receiving branch is found during execution a matching operation is done first. In case the incoming message, that is first in the corresponding PCO's FIFO, matches the criteria the *alt* branch will be executed and the message will be removed from the top of the queue. Otherwise, the execution continues and the next branch will be examined. However, execution does not stop after a snapshot was taken, so the state of the test component and the queues assigned to it might change in between. However, these events do not change the actual snapshot, until the *alt* statement is not executed again.

So, generally the two most significant factors we consider, while evaluating the performance of a test component are the matching mechanism and the queuing at the test ports. See for example the following scenario (Figure 2).

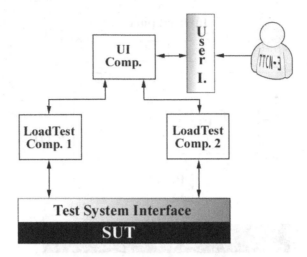

Fig. 2. Load Testing Example Scenario

In this simple example we have two load test components that connect to the SUT via two different PCOs and handle the actual protocol behavior. One test component is receiving commands from the user via a user interface (UI). Besides, this component can supply the user with additional data regarding the test case execution by querying the load test components periodically. Let us say that in this setup LoadTest Comp.1 stimulates the SUT, which is for example an ATM switching center, and due to this stimulation the SUT forwards a high number of calls towards LoadTest Comp.2. In this case, LoadTest Comp.2 should be able to handle and examine a very high amount of calls per second coming from the switch. The actual race condition results from the different functionalities of LoadTest Comp.2. Firstly, it receives a high amount of incoming calls from the SUT and secondly it must answer status request messages, coming from UI. Comp. on an internal port periodically. Although, status messages might be relatively infrequent compared to the messages participating in the load test, they also need to be handled by LoadTest Comp.2. most probably in the same *alt* structure. This setup leads to a race condition between the two separate FIFO queues assigned to the two test ports of the component. Namely, in the test component branches of the *alt* referring to the PCO towards the SUT receive significantly more hits in a unit of time than branches referring to the internal port do, this way increasing the risk of an indefinite postponement of the status messages.

In our modeling approach, we build an analytic model of test components that use alternatives to handle internal messages and messages coming from the SUT. We describe concurrent queues underlying the test ports of a component with a stochastic process and calculate the steady-state solution. After solving the analytical model, we predict the probability that one of the queues contain a message that is postponed indefinitely, because of the fact that race conditions arise between the queues. The probability of an indefinite postponement is calculated as a function of arrival intensities at the corresponding queues and of other parameters relevant to the implementation of the actual test component.

3 Discrete-Time Quasi Birth-Death Processes

Our method uses a mathematical model that can be evaluated and performance indices of the described components can be derived by existing solver techniques. The mathematical formalism behind our evaluation method can be identified as discrete-time Quasi Birth-Death processes (QBDs) [8]. A simpler M/G/n type solution would not allow us to use finite queues and to calculate state distributions [9].

The mathematical analysis in turn, is based on matrix-geometric solution techniques and matrix analytic methods [10]. In order to become acquainted with QBD processes let us consider processes $N(t)$ and $J(t)$, where { $N(t), J(t)$ } is a DTMC (Discrete-Time Markov Chain). The two processes have the following meaning: $N(t)$ is the level process and $J(t)$ is the phase process. { $N(t), J(t)$ } is a QBD if the transitions between levels are restricted to one level up or down, or inside the actual level. The structure of the transition probability matrix P of a simple QBD is the following:

$$P = \begin{bmatrix} A_1^* & A_0 & 0 & 0 & \cdots \\ A_2 & A_1 & A_0 & 0 & \cdots \\ 0 & A_2 & A_1 & A_0 & 0 \\ & 0 & A_2 & A_1 & A_0 \\ & & & \ddots & \ddots & \ddots \end{bmatrix} \tag{1}$$

Matrix A_0 describes the arrivals (transitions one level up), matrix A_1 describes transitions inside each level and matrix A_2 describes departures (transitions one level down). Matrix A_1^* is an irregular transition matrix at level 0. The row sum of P is equal to 1 (discrete-time model). The tangible meaning of levels and phases can be seen in Figure 3.

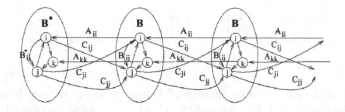

Fig. 3. Logical representation of a QBD process

The two-dimensional property of a QBD is utilized by our method significantly. On one hand, we map the sequential branch examination and matching mechanisms of the *alt* structures in a given test case we model, into the internal phases inside each level of the QBD. On the other hand, we use the levels to describe queuing at the test port.

After the system's behavior is described by the QBD model, with the aid of matrix analytic methods the following properties of the model can be calculated: steady state solution, queue length in steady state, phase distribution, mean value of time to transition back to level zero (meaning that the buffer is empty). As a result, a variety of performance indices can be derived from the QBD model, such as packet loss ratios, delays, and available throughput under stress conditions.

4   A QBD Model for PCOs

For the performance evaluation of PTCs that use alternatives to handle messages we build the following novel QBD model. For the sake of simplicity consider a test component using two test ports, each of them with a separate FIFO queue (e.g. in Section 2). The model will macroscopically look like a simple QBD that is infinite in one dimension similar to the example in Figure 3. So, in the first dimension the model is an infinite QBD. This dimension represents the PCO that drives the system into a race condition, call it PCO1. This might be the PCO serving a significant load coming inwards from the SUT. But at a lower level another embedded QBD is to be found, representing the PCO (call it PCO2) that is suppressed by the heavily loaded PCO1. Levels of the QBD describe queue sizes in this case, accordingly the second dimension of the model is in direct connection with the size of the queue assigned to PCO2. This dimension will be finite with a parametrical size, in order to assure matrix-analytic solutions to work [11], [12] and to allow the model to predict infinite postponement in the PCO, which is underprivileged because of the race conditions between the concurrent PCOs (Figure 4).

Example 1 (Example alt structure).

```
alt {
/*Group 1*/
[] PCO1.receive(template1) { Statements... }
[] PCO1.receive(template2) { Statements... }
[] PCO1.receive      { // Trash unmatched messages
                        repeat }
/*End of Group 1*/
/*Group 2*/
[] PCO2.receive(template3) { Statements... }
[] PCO2.receive(template4) { Statements... }
[] PCO2.receive      { // Trash unmatched messages
                        repeat }
/*End of Group 2*/
}
```

The vertical dimension, that is the number of distinct groups of states in Figure 4 is restricted by the size of the FIFO queue of PCO2, which can be set as a model parameter. Also, relatively short buffers are considered for the investigated PCO in order to keep response time, that is crucial in a load test system, under a considerably low level. If we look at this model as a simple QBD, transitions upwards and downwards are described by matrices C and A respectively. Whereas, matrix B describes internal phase transitions. Irregularities, denoted by *, occur at the first level, where there is no possibility of a transition downwards. Each group of phases (within ellipses) corresponds to a group of *alt* branches that use the same PCO. Group to group (vertical) transitions describe queuing and service of messages at PCO2. For example, see the excerpt in *Example 1*.

In this example branches that refer to the same PCO are grouped together. Two groups are formed that correspond to the separate groups of phases in Figure 4. In

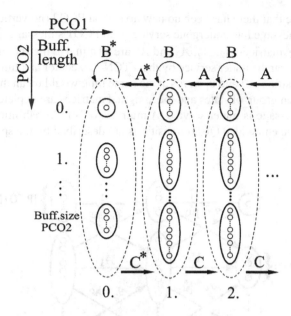

Fig. 4. Two dimensional model for two concurrent PCOs

effect, only branches of PCO2 are active in the first horizontal level of the model as horizontal levels represent the queue of PCO1. Being at the first level means the buffer of PCO1 is empty, so in this case queueing at the FIFO queue of PCO2 is considered only, in addition to the matching mechanism. Similarly, when the queue of PCO2 is empty but PCO1 is active the model moves on the horizontal axis across the first groups of phases representing queueing at the FIFO of PCO1 and the matching mechanism on the correspondent group of branches. Of course, the other phases describe mixed states, when both queues contain messages.

Description of the transitions are realized by matrices A, B, C, A^*, B^* and C^* that consist of several submatrices, for example matrix A and its irregular counterpart is shown here (2).

$$\mathbf{A} = \begin{bmatrix} \widetilde{\widetilde{A_1}} & \widetilde{A_0} & \underline{0} & \underline{0} & \cdots \\ \underline{A_2} & A_1 & A_0 & \underline{0} \\ \underline{0} & \underline{A_2} & A_1 & A_0 & \underline{0} \\ & & \ddots & \ddots & \ddots \\ & & & \underline{A_2} & \widehat{A_1} \end{bmatrix} \quad ; \mathbf{A}^* = \begin{bmatrix} \widetilde{\widetilde{A_1^*}} & \widetilde{A_0^*} & \underline{0} & \underline{0} & \cdots \\ \underline{A_2^*} & A_1^* & A_0^* & \underline{0} \\ \underline{0} & \underline{A_2^*} & A_1^* & A_0^* & \underline{0} \\ & & & \ddots & \ddots & \ddots \\ & & & & \underline{A_2^*} & \widehat{A_1^*} \end{bmatrix} \quad (2)$$

Submatrices A_0 and A_0^* describe a step forward (that is downwards vertically) in the embedded QBD. This means, a new message has arrived into the queue of PCO2 (step downwards) while a message has been served from the queue of PCO1 (one step left). In a similar manner, matrices A_1, A_1^*, $\widetilde{A_1}$, $\widehat{A_1^*}$, $\widetilde{A_1}$, and $\widehat{A_1^*}$ contain transition probabilities confine to the case that there has been no new arrival on PCO2 (no vertical movement), but exactly one message has undergone service from PCO1's message queue (one step left). Furthermore, matrices A_2, A_2^*, $\widetilde{A_2}$ and $\widetilde{A_2^*}$ are all equal to the zero matrix, because only one message can be under service by the TTCN-3 executor at a time, naturally the meaning of a transition left and upwards at the same time would mean that. In Figure 5, transitions between group of states and the assigned matrices are depicted, restricted to the case that a message is under service (A matrices only). Probabilities of the actual phase (states of the embedded QBD) transitions are described by the specific elements of the matrices.

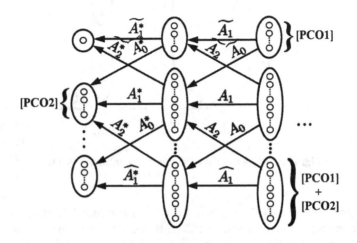

Fig. 5. Possible transitions of the model while a message is under service in PCO1

Important cornerstones of the matrices that build up the generator matrix of the model are the matching probabilities of branches the *alt* contains. Consider these values as predicted hit rate counters for each branch stored in vectors (p), one for each PCO, e.g. in case of two PCOs we apply p' for PCO1 and p'' for PCO2. Naturally, length of the *alt* structures (groups of different PCOs) is equal to the length of the hit rate vectors and determines the size of the submatrices among others in Equation 2. Elements of the vectors (e.g. p_i) contain the probability that a message under service does not match the template at branch number i. In this case, it is checked against the template at the next branch.

A further important input parameter of the model is the time slot (T) needed for one elementary action in the TTCN-3 executor, e.g. the average time needed for the examination of one *alt* branch, the average time in which a template matching operation is finished. Samples of messages traversing the PCOs the test component needs to deal with are taken in discrete T time intervals (DTMC).

Furthermore, we describe the arrival intensities of the PCOs involved with a λ assigned to each of them. λ values are relative to the selected elementary T, e.g. $\lambda_1 = 1000$ and $T = 10^{-6}$ (1 μsec) means that one message arrives approximately every millisecond ($\lambda_1 \cdot T = 10^{-3}$) on PCO1. Accordingly, we use four different variables to include arrival intensities into the model. D_{00} denotes no arrival at a time slot. D_{10} and D_{01} indicate that one message has arrived on PCO1 or on PCO2 respectively. The meaning of D_{11} is that a message has arrived on both PCOs at the same time concurrently, but this is disallowed by the model, since the TTCN-3 executor handles only one message at a time. This leads to the following scheme (3).

$$D_{00} = 1 - (\lambda_1 + \lambda_2) \cdot T; D_{01} = \lambda_2 \cdot T; D_{10} = \lambda_1 \cdot T; D_{11} = 0; \qquad (3)$$

The transition submatrices are constructed using the D matrices and the matching probability vectors until every state transition of the state space is covered. For the sake of clarity we present only one submatrix as it is defined (4). Matrix $\widehat{A_1}$ has elements only in the first column and only until the s_1th row, where s_1 is the number of branches referring to PCO1 (that is the size of p'). $(D_{00} + D_{01})$ represents the probability that either no message has arrived or exactly one arrival happened on PCO2.

$$\widehat{A_1} = \begin{bmatrix} (1 - p'_1) \cdot (D_{00} + D_{01}) & \cdots & 0 \\ \vdots & & \vdots \\ (1 - p'_{s_1}) \cdot (D_{00} + D_{01}) & \cdots & 0 \\ \vdots & & \vdots \\ 0 & & 0 \end{bmatrix} \qquad (4)$$

When the matrices are built up we then calculate the ratio of time spent at the nth and $n+1th$ level before returning to level n, in matrix R. R is calculated using a logarithmic reduction algorithm described in [8]. The complete distribution of states in the QBD model is then calculated using R (5).

$$P_1 = P_0 \cdot \underline{R}; P_2 = P_1 \cdot \underline{R}; \ldots P_n = P_0 \cdot \underline{R}^n \qquad (5)$$

Where vector P_i contains distribution of the phases at level i and it is not to be confused with p' or with p'' that contain match probabilities, which are architectural constants of the system. After the state distribution is available we can derive performance indices of the system. In this case, we calculate the probability that a message suffers indefinite postponement while in the buffer of PCO2, because PCO1 is under a significantly heavier load. This means that we sum state probabilities at every (horizontal) level, but only those at the vertically last group of states (subsets l and m). This gives us the following equation, where matrix I is the identity matrix. Practically, we need state distributions only at the first two levels and matrix R for the evaluation.

$$Pr(PCO2loss) = \sum_i P_{oi} + \sum_{j=1} (\sum_k P_{jk}) = \ldots \underline{P_{0_l}} + \sum_{j=1} (P_{j_m}) s$$

$$= \underline{P_{0_l}} + \sum_{j=1} (\underline{P_{1_m}} \cdot \underline{R}^{j-1}) = \underline{P_{0_l}} + \underline{P_{1_m}} \cdot \sum_{k=0} (\underline{R}^k)$$

$$= \underline{P_{0_l}} + \underline{P_{1_m}} \cdot (\underline{I} - \underline{R})^{-1} \qquad (6)$$

This way, after obtaining steady state distributions of the QBD process, we can predict the probability that messages are postponed in the queue of the PCO that is in lack of system resources because another PCO of the same test component used them up. In the stochastic model this means that we have a new demand incoming in the queue of that PCO that is already at the last level possible, so further messages cannot be received or they will be lost. This analysis allows us to evaluate load test components implemented differentially by setting the parameters of the model accordingly. Also, analysis can be set up to examine a test component with different load requirements as the arrival intensities are parametrical.

5   Model Results

After the transition matrices are defined according to the architectural constants, the matching probability vectors and the arrival intensities for each PCO, the model has been built up. Together with the arrival intensities the elementary time slot T has to be set also. Besides, one more important input parameter exists, the size of the queue at the PCO we investigate for possible latencies.

Accordingly, considering the simple example in Section 4 we can get the following results on the probability that a message suffers indefinite postponement in the queue of a PCO. In this example the test component uses two separate PCOs. Say PCO1 is a load generator with considerably high amount of packets generated, while PCO2

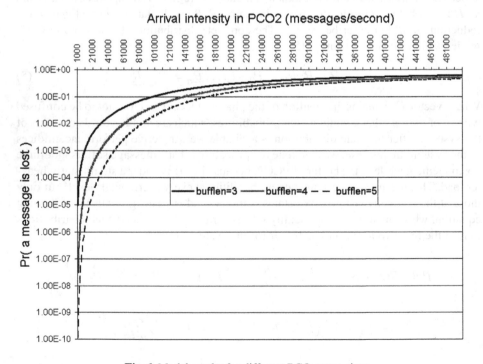

Fig. 6. Model results for different PCO queue sizes

Table 1. Message loss probability, model and simulation results

Arrival intensity (λ)	Messages per second	Buffer size	Model result	Simulated value
0.08	80000	5	0.00831	0.01252
0.10	100000	4	0.04333	0.04333
0.11	110000	3	0.12333	0.59024
0.12	120000	4	0.06971	0.07333
0.14	140000	4	0.10102	0.14333
0.19	190000	5	0.12718	0.13333

handles lower priority operations, such as communicating with the user, or receiving maintenance messages from the main test component. In the analytical model (Figure 6), message loss is calculated according to (6). Different sizes of queues underlying the analyzed PCO are represented by each curve. In Figure 6 and in Table 1, the arrival intensity in PCO1 is considered to be constant, while the arrival rate in PCO2, that is the port we investigate, is variable from 0 to 0.5. An arrival rate ($\lambda \cdot T$) of 0.5 means, in case we have a $T = 1$ μsecond, that approximately 500 messages arrive each millisecond to PCO2.

Simulations in *ns-2*, that has been used for validating the model, show similar behavior. The simulated scenario consists of two PCOs implemented as FIFOs with variable buffer length. Among others, each simulated PCO has a variable length delay loop sequence to simulate the matching mechanism too. Table 1 shows simulated results together with the model results for the same parameters. On one hand we use this simulation to develop the analytic model for more precise results, on the other hand an analytic model is necessary to analyze more complex test scenarios with more than two PCOs involved.

6   Conclusions

Generally, load testing is not an easy task, neither with TTCN-3, nor with any other test environment [13], [14]. Tests that work somehow are not sufficient, but load tests must also work well. Careful test design and test optimization is a must in these kinds of tests, meaning that tests have to be designed to work efficiently on the execution platform that accommodates them. Efficiently, that is e.g. with low CPU load, even using stock hardware elements. To evaluate a load test component and to verify it meets the identified requirements, the complete path traversed by the corresponding messages has to be taken into consideration.

Our analytical model is designed to evaluate load testing TTCN-3 parallel test components to satisfy user requirements using discrete-time QBDs. In order to be able to design test components that simulate behavior of real nodes in a telecom test network and that are even interchangeable with a real node in terms of performance, modeling of the components performance is nearly inevitable.

This modeling approach makes it possible to evaluate existing test components and also to give a feedback to designers of load test scenarios. Also, the model can aid distribution of load test traffic among the test components participating in test. The aim of successful traffic mix composition is to simulate behavior of real nodes, or even to produce an equivalent counterpart of the real node in TTCN-3.

Our future work on this topic includes extending the model to be capable of describing PTCs containing more than two concurrent PCOs at a time. Besides, further simulations and measurements of real components are ongoing and observations are gathered to improve our model beyond the current level.

References

1. ETSI ES 201 873-1 (V3.1.1): "Methods for Testing and Specification (MTS); The Testing and Test Control Notation version 3; Part 1: TTCN-3 Core Language", 2005.
2. H. Neukirchen, Z. Ru Dai, J. Grabowski. Communication Patterns for Expressing Real-Time Requirements Using MSC and their Application to Testing. R. Groz, R. M. Hierons (Eds.) TestCom 2004, LNCS 2978, pp. 144-159, 2004.
3. Z. Ru Dai, J. Grabowski, H. Neukirchen. Timed TTCN-3 Based Graphical Real-Time Test Specification. D. Hogrefe and A. Wiles (Eds.): TestCom 2003, LNCS 2644, pp. 110-127, 2003.
4. Z. Ru Dai, J. Grabowski, H. Neukirchen. Timed TTCN-3 - A Real-time Extension for TTCN-3. I. Schieferdecker and H. König and A. Wolisz (Eds.) Proceedings of the IFIP 14th International Conference on Testing Communicating Systems - TestCom 2002, pp. 407-424, 2002.
5. S. McCanne, S. Floyd. ns Network Simulator. http://www.isi.edu/nsnam/ns/
6. I. Schieferdecker, T. Vassiliou-Gioles. Realizing Distributed TTCN-3 Test Systems with TCI. D. Hogrefe and A. Wiles (Eds.) TestCom 2003, LNCS 2644, pp. 95-109, 2003.
7. ETSI ES 201 873-4 (V3.1.1): "Methods for Testing and Specification (MTS); The Testing and Test Control Notation version 3; Part 4: TTCN-3 Operational Semantics", 2005.
8. G. Latouche, V. Ramaswami. Introduction to Matrix Analytic Methods in Stochastic Modeling. by the American Statistical Association and the Society for Industrial and Applied Mathematics, 1999, pp. 83-99, pp. 221-237.
9. S. Palugyai, M. J. Csorba. Performance Modeling of Rule-Based Architectures as Discrete-Time Quasi Birth-Death Processes. Proceedings of the 14th IEEE Workshop on Local and Metropolitan Area Networks, 2005, Chania, Greece.
10. M. F. Neuts. Matrix-Geometric Solutions in Stochastic Models. Johns Hopkins University Press, 1981, pp. 81-107, pp. 63-70, pp. 112-114.
11. T. Osogami, A. Wierman, M. Harchol-Balter, and A. Scheller-Wolf. A recursive analysis technique for multi-dimensionally infinite Markov chains. Performance Evaluation Review, 32(2):3-5 (2004).
12. T. Osogami. Analysis of a QBD Process that Depends on Background QBD Processes. Technical Report CMU-CS-04-163 (2004).
13. G. Rößler, T. Steinert. Traffic Generator for Testing PABX and Call Center Performance. I. Schieferdecker and H. König and A. Wolisz (Eds.) Proceedings of the IFIP 14th International Conference on Testing Communicating Systems - TestCom 2002, pp. 139-151, 2002.
14. I. Acharya, H. Kumar Singh. Testing of 3G 1xEV-DV Stack - A Case Study. D. Hogrefe and A. Wiles (Eds.): TestCom 2003, LNCS 2644, pp. 20-32, 2003.

Use of TTCN-3 for Software Module Testing

Andreas Johan Nyberg

Nokia Research Center, P.O. Box 407,
FIN-00045 NOKIA GROUP
Andreas.J.Nyberg@nokia.com

Abstract. Efficient testing of software modules remains a challenging task for complex software implementations. TTCN-3 has so far been applied mainly in the telecom domain but not yet in a larger extent to software module testing. This paper describes a multi purpose TTCN-3 test system solution primarily targeted for concurrent software and testing of software modules in isolation. Apart from a test system solution, an approach for type mappings from C to TTCN-3 is discussed, followed by an example of how test cases could be implemented and how the discussed test system be utilized for a simple software module.

Keywords: Software testing, concurrent software, mock objects, TTCN-3.

1 Introduction

Software module testing can be a complex task for non-trivial code, especially in concurrent programming. Testing modules of concurrent software in isolation has a number of requirements on the test framework used. The test system and test case implementations must be able to handle concurrent behavior and support the use of emulated components for module isolation. Testing of concurrent software has a number of additional challenges[1],[2] compared to sequential software. These include shared critical areas, synchronization, deadlocks, livelocks[3] and non-deterministic behavior. When testing modules in isolation missing parts of the implementation have to be emulated e.g., components and services.

This paper describes a multi purpose test system targeted for testing of complex software implementations with concurrency aspects and requirements to test modules in isolation. Several testing concepts must be handled by the test system which are treated as requirements: The test system has to provide (1) *test control and test management*, (2) *test suite trace consistency*, (3) handle *non propagating errors* from earlier executed test cases and (4) support free choice of *system under test (SUT) execution environment*. In addition, (5) *language independency* for the implementation under test (IUT) is supported that comes for free with the choice of the test language used in this paper, TTCN-3.

For well known CUnit[5],and JUnit[6] frameworks, software module testing using the native language of the tested software module as test language has its advantages. Among them are easy data type creation and instantiation, conditional tests, and already existing knowledge of the language to write test cases in. Involving another

M.Ü. Uyar, A.Y. Duale, and M.A. Fecko (Eds.): TestCom 2006, LNCS 3964, pp. 161–176, 2006.

language for test case implementations in the testing process creates an overhead in form of learning another language. Nevertheless using TTCN-3, a standardized test language for test suite implementations and test systems, has some advantages and might even opens up new possibilities for software testing.

TTCN-3 has evolved from the telecom testing industry and is a test implementation language that looks very much like an ordinary programming language. It is currently the best established standardized testing language on the market. TTCN-3 has been used for testing in telecom related areas, e.g., protocols such as IPv6[7], SIP and WiMax. What these test systems have in common is that they are fully asynchronous and heavily rely on the message-based communication mechanism part of the TTCN-3 language. TTCN-3 supports however in addition also procedure-based communication. The procedure-based paradigm can be applied to cases where there is a clear distinction between a caller and replier of procedures. This makes TTCN-3 suitable for calling remote procedures in a SUT, in the same way as Sun/ONC RPC[RFC 1831] works.

TTCN-3 provides the concept of test components which makes it possible to implement concurrent behavior in a natural synchronous procedure-based manner. TTCN-3 has a built in extension mechanism that enables the import of notations in to a test system. Examples of these extension mechanisms can be found in the standardization work involving the IDL and XML language integrations [13]. As TTCN-3 is independent of the implementation language of the tested code the language has to be mapped and imported to TTCN-3 through the extension mechanism. The mapping has to cover up to the extent of functions, function parameter and return value types. No semantics need to be mapped, only what is publicly visible in interfaces. This paper covers a subset of the C to TTCN-3 mapping which will be a future part of the TTCN-3 standards. The C language has language specific features like pointers allowing for definition of complex types that have not been covered in other TTCN-3 language mappings before.

2 Background

The targeted problems in this paper are to exemplify the possibility to create multi purpose software testing test system based on the TTCN-3 standards. Difficult testing areas such as concurrent software and module isolation with emulated components will be addressed.

Two ways of testing concurrent software is through multiple execution and deterministic execution of test cases[2],[8]. Test case implementation based on methodologies for testing concurrent software such as above and in addition mutation testing[9] and reachability testing[8] will only be addressed in the context of applicability by using TTCN-3. The test system approach in this paper makes it possible to perform multiple and deterministic execution of test cases utilizing testing methodologies that are suitable for concurrent software testing by using the execution semantics of TTCN-3.

Testing software modules in isolation is a difficult task even for non-trivial code. One approach is to use mock objects[4] which are dummy implementations that emulate real implementations to replace functionality in the testing domain of the tested software module. There are many good reasons to use mock objects[10] such as non-deterministic behavior, test object setup and call back functionality. This approach can be utilized efficiently by using TTCN-3 not only for code level mock objects but also for emulating behavior of external services something that will be exemplified in this paper.

TTCN-3 telecom protocol test systems are already in product use, e.g.[7], in addition established testers also for OMG CORBA interfaces are available. Test systems for OMG CORBA implementations based on IDL interfaces require that the interface definitions are mapped to TTCN-3[11]. The test system approach in this paper requires that the language of the tested software module is mapped to TTCN-3. To be more precise the interface of the module under test has to be represented in the TTCN-3 test suite as is with parameters and type mappings.

3 TTCN-3 Test Systems

Suitable test architectures for testing software modules can be approached from several different angles. Architectures and approaches can differ depending on the stage in the development process or the kind of testing that is looked for. The approach described in this section resembles a unit testing platform which is applicable and targeted for unit testing and higher level testing such as functional testing and system testing. The test language and the test system design allow for efficient test suite implementations by using a mixture of message-based and procedure-based communication. For example, a module can be tested through its available interface at function level and at the same time it can be tested by using message-based communication with external interfaces and mock objects, all from within the same test suite. With multiple test capabilities software faults can be detected at several levels of testing: early on in development using unit testing-like approaches (with high code coverage), integration testing for interface and integration faults as defined in[12], and in system and conformance testing for error detection at a pure black box level.

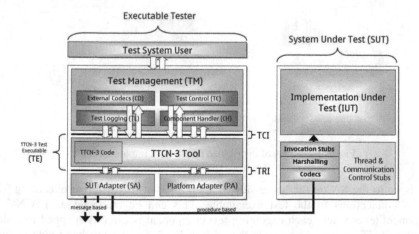

Fig. 1. Overview of a TTCN-3 test system and SUT

Using TTCN-3 for test suite implementations requires a TTCN-3 test system which contains more than TTCN-3 code[16]. The left part in Figure 1 shows the composition of a TTCN-3 test system as defined in the TTCN-3 standards[13]. There are several different entities involved in the test system which communicate over standardized interfaces named TTCN-3 Control Interface (TCI) and TTCN-3 Runtime Interface (TRI). The most central entity, the TTCN-3 Executable (TE) handles the execution of the TTCN-3 code, either as compiled code or interpreted at runtime. The user interacts through the Test Management (TM) entity which also provides functionality for encoding and decoding of data values. The lower level containing the SUT Adapter (SA) and Platform Adapter (PA) interfaces as the names implies relate to the interaction towards the SUT and towards the test system operating system.

The interaction between the SUT and the SUT Adapter is a design choice for the test system. In the illustrated case a separated Executable Test Suite (ETS) and SUT has been chosen. This is defined as the distributed test method in ITU-T X.290[14].

3.1 IUT and the Test Harness

An IUT cannot be tested as it is. There needs to be a test harness present in the SUT. The IUT together with test harness composes the SUT. As the test system that is discussed in this paper is distributed, the test harness has to be able to handle, communication between the test system and SUT, marshalling, and invocation of the functions in the module under test. Marshalling builds valid C values to use in function invocations. The distributed approach automatically puts requirements on the mapping between C and TTCN-3 in order to make sure that values really can be built from the mapped C value instances in the TTCN-3 test suite. Implementation of the SUT test harness depends on the function interface, programming language of the IUT and available tools for stub generation.

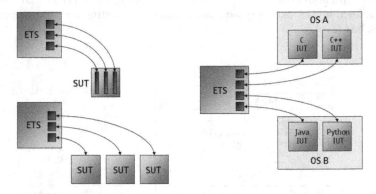

Fig. 2. Examples of distributed test system implementations

It is up to the SA on the tester side to keep track of the different executing SUTs and its connections to the test executable. This can be managed at TTCN-3 test component level where each executing test component is directly mapped to a single executing SUT or even to a single thread in a SUT. On the left hand side in Fig. 2

software components or SUT threads which run in parallel are mapped to individual TTCN-3 test components. The right hand side in Fig.2 shows an example of the implementation language independency that TTCN-3 can provide.

3.2 TTCN-3 Concepts and Usability for Software Testing

Using TTCN-3 for software testing provides several very useful features. Apart from what has been mentioned already in form of standardized test system interfaces and independency of the language of the IUT there are other very testing specific advantages.

TTCN-3 has very well defined execution semantics that guarantees that a test case execution of a TTCN-3 test system will be the same with any TTCN-3 tool. There is also an internal matching mechanism, that when combined with the TTCN-3 language construct template, allows for the writing of very compact test cases. Using a template construct which represents a value instance, all or selected parts of the value instance can be targeted for matching i.e. when matching two structured types against each other.

Concurrency is something that can be handled very gracefully in TTCN-3 by using test components and a test case verdict that can be set from any test component without any need for component synchronization. The TTCN-3 test components also allow performing of tests on several modules written in different languages, with the same test system - even from within the same test case. Combining the procedural and message-based aspects with the concept of TTCN-3 test components that can be executed in parallel with individual behavior and complex test case scenarios can be implemented in a clear and concise manner.

3.3 Distributed Test System Main Concepts

Distributing the test system as in Fig. 1 by separating the SUT from the ETS provides capabilities that solve some of the test system requirements stated in the introduction. These capabilities come with a number of benefits that outweigh the drawbacks of a distributed test method, e.g., its communication latency:

- *Test control and test management*
A potential problem when running large test suites is the possibility to recover and continue a test suite execution in case of a crash, deadlock or livelock in the IUT. With the ETS running in isolation from the IUT the SUT can be restarted by the ETS if needed, e.g. after a crash in the IUT or between each test case.

- *Test suite trace consistency*
Logs and test suite verdicts and summaries can be kept intact and will not be corrupted. The ETS does not crash from a crash in the IUT and all test execution logs can be finalized gracefully without the risk of loosing valuable test execution information.

- *Non propagating errors*
Most software testing literature defines that the first step in the unit testing procedure shall be to test the smallest possible unit, e.g. a function, to make sure errors do not propagate into further development. With the separated SUT and ETS the SUT can be

restarted for every test case. This usually results in re-initialization of memory through which the propagation of errors could propagate.

- *Language independency*

Most software unit testing tools are written in the same language as is used for writing the tested software. Using a TTCN-3 based solution the test cases will not be written in the same language as the IUT. In Fig. 2 the implementation language of the IUT is independent of the test suite language, which adds to the goal of a multi purpose test system.

- *SUT execution environment*

A distributed test system approach allows a free choice in execution environment with regards to operating system and platform of the SUT. The only requirement is that the target system must allow communication between the SUT Adapter in the test system and the SUT test harness.

Test system distribution has advantages, but also has a few drawbacks that have to be considered. Communication overhead is one issue while invocation latency and marshalling are others. The latency has been measured in a comparison between two almost identical test systems. One test system had the IUT directly attached to the SUT adapter and thus execute in the same process as the ETS. The other test system is the one described in this paper. The additional latency of having the SUT separated from the ETS was approximately three to one. Invocation, marshalling and communication latency are not alone the reasons for causing the performance drawback. There is also additional routing functionality that has to be handled in the SUT adaptation due to the possibility to have many separately executing SUTs or SUT threads. There are a number of possible solutions which can be applied to the distributed test system case for overcoming obstacles such as latency and memory access overhead. One is through the use of inter-process communication (IPC) over UNIX sockets utilizing shared memory between processes running at the same machine which will reduce the time spent on pointer handling in the test cases.

Another drawback is that a model for memory access has to be defined before writing test code for software which is implemented in programming languages that utilize pointers or object references. Such references and pointers only reside in the memory space of the SUT. However, a memory address or object reference has no real meaning inside the test system which is executing in its own separate process. This lack of pointer transparency requires that memory access is explicitly defined in TTCN-3 test cases through the language mappings.

3.4   Testing Multithreaded and Concurrent Software Modules

One of the biggest challenges when testing concurrent software is to make sure that it will be tested thoroughly when the software includes critical areas and shared data. The order of execution of concurrent statements in the different threads or processes must be considered. This non-deterministic behavior can cause execution sequences to be different from execution to execution even with the same input data. This adds synchronization complexity to test case implementations. Synchronization between test components can be solved by connecting test component ports to each other through the TTCN-3 operation *connect* which allows for message communication internal to the TTCN-3 executable. Deterministic execution sequences can then be synchronized utilizing this internal communication. Functional testing of individual functions in isolation does not constitute enough assessment because none of the functions under test will then deal with concurrency. What is needed is a test system that can simultaneously trigger several functions to access critical sections in the IUT at the same time. This problem can be solved by assigning one TTCN-3 test component per thread or software component in the SUT. Here, each test component in the test system creates its own connection to the SUT who in turn spawns a new working thread for every connection and thereby establishes a coupling between the test component and, e.g. a POSIX thread, in a C test harness.

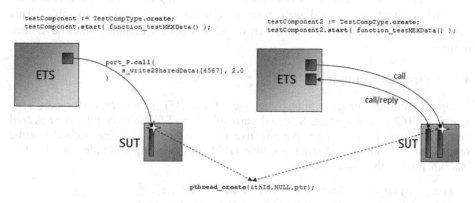

Fig. 3. TTCN-3 test components spawn working threads in the SUT

All procedure-based calls between a test component and a working thread in the SUT are transmitted via such an established connection. The connection which is not visible from within the test case is a feature of the test system design. The spawned working thread is part of the test harness in the SUT and is utilized for simulating concurrent behavior in the IUT.

3.5 Usage of Mock Objects

When a software module is to be tested as an isolated entity, the test system has to emulate any calls that a particular module makes to other functions or methods, that are external to the tested module. This clearly moves away from pure black box testing, as knowledge and understanding of internal functional details about the module is needed. To properly isolate but still test the module in its intended environment, we need to create test configurations with test components (mock objects) that can act as the external, collaborating components.

By using parallel test components of TTCN-3, we can easily simulate the external collaborating code modules and their behavior by allowing the test components to make or expect calls to and from the tested module. The strength of this approach is that it allows control of the behavior of these external parts in order to simulate correct or incorrect external behavior and then analyze if the tested module is able to handle problematic situations. This means that it will be possible to test error handling code as well.

4 C to TTCN-3 Language Mappings

Specifying test cases in TTCN-3 requires that the tested functions of the software module to test can be called from the TTCN-3 test suite. This is possible provided that a mapping between the language of the software to test and TTCN-3 exists. Utilizing the procedure-based paradigm in TTCN-3 with the concept of signatures a mapping can be defined in a clear and evident way. This applies to all the tested functions, functions in mock objects and to the types used as parameters and return values. This section gives an overview of a subset of the C to TTCN-3 mappings that have been defined.

4.1 A Practical Example

Mapping from the programming language of the IUT is an important part of a working TTCN-3 test system. Selected parts of a C to TTCN-3 mapping is presented and exemplified in this paper. We will use a simple reentrant thread safe C module representing an automated teller machine (ATM) as ongoing example. The module has functions with critical sections.

```
int login   ( int userId, int code );
int balance ( int userId, int* balance );
int withdraw( int userId, int amount, int* balance );
int logout  ();
```

The ATM server example requires the existence of a component that handles user verification. This component in the example is replaced by a mock object to make it possible to test the ATM module in isolation. Below is the sole function of the mock object which is called from the IUT.

```
int verifyUser( int userId, int code );
```

The C module of the ATM can be tested in many ways, primarily through unit and functional testing of functions followed by testing of concurrent scenarios.

The ATM example above is a typical example where a number of problematic scenarios easily can be identified, e.g. read/write locks, starvation and partial failures. Before exemplifying possible test cases for the ATM module the necessary mapping from the language of the IUT (C) to TTCN-3 has to be defined.

4.2 Mapping of Functions

TTCN-3 signatures may specify parameter lists and return values where the direction of the signature parameters is also defined. The interface of the ATM module and mock object function are mapped and exemplified in Table 1.

Table 1. TTCN-3 signatures mapped from a set of the functions in the example

```
001  /* ATM interface */
002  signature s_login( in CInt p_userId,
003                     in CInt p_code )
004     return CInt;
005  signature s_balance( in CInt p_userId,
006                       in CIntPtr p_balance )
007     return CInt;
008  signature s_withdraw( in CInt p_userId,
009                        in CInt p_amount,
010                        in CIntPtr p_balance )
011     return CInt;
012  signature s_logout  () return CInt;
013
014  /* mock object interface */
015  signature s_verifyUser( in CInt p_userId,
016                          in CInt p_code ) return CInt;
```

All parameters are mapped with the direction set to *in* parameters as if passed by value. This also applies to the case of pointer type parameters since a pointer address cannot change during a function call.

4.3 Mapping of Pointers

Mapping of primitive and built in data types in C is rather straightforward. But pointers need some more attention. A C pointer can been thought of as a four or eight byte aligned value[15] and can therefore be represented as a TTCN-3 integer. The contents pointed at by a pointer on the other hand calls for a more constructive mapping than a simple string representing the memory address of the contents. One possibility is to represent the pointer value by the data type it is pointing at in the test suite. However, in the case of more complex types, e.g. cyclic data structures, this approach fails. Another issue is that a pointer to e.g. an integer may not always reference to a single integer. It can just as well point to the first element of an array of integers of an unknown size. In general, a pointer must always refer to a data array with one or more elements representing the memory space it is pointing at. Therefore the C pointer's *content* has to be mapped to the TTCN-3 list type record of. The pointer itself is still mapped to an integer but when the contents are retrieved they are represented by an ordered list of the pointed at data type. One benefit of this mapping is that pointer arithmetic can be done easily in a test case implementation, e.g. increment and decrement by direct indexing. This mapping is not sensitive regarding the type of the pointer, complex and abstract data types are mapped in the same way as for simple types.

Further functionality in the mapping is needed when working with pointers. A check for null pointers can be enough for a simple test case. What is required are means for allocating memory, assigning data and retrieving data. This functionality is handled through type specific signatures as exemplified in Table 2. The test case writer has to be aware of this mapping due to the nature of the transparent memory access that has to be provided.

Table 2. Mapping of pointer to integer

```
c
001   int* balance

TTCN-3
001   type integer CInt;
002   type integer CIntPtr;
003   type record of CInt CIntArr;
004
005   signature s_MallocCInt( in CInt size ) return CIntPtr;
006   signature s_FreeCInt( in CIntPtr ptr );
007
008   signature s_SetCInt( in CIntPtr ptr, in CIntArr data );
009   signature s_GetCInt( in CIntPtr ptr, in CInt size )
010       return CIntArr;
```

For data assignment and retrieval two signatures are sufficient, where the number of elements to retrieve or assign has to be given. At all times a pointer's content is represented by the ordered list type, i.e. the record of type. Dealing with a pointer to a single element is treated as a list containing only one element. For C pointers two TTCN-3 signatures for every type representing the malloc and free calls are enough for complete memory handling.

5 Test Case Implementations

Implementing unit test cases in TTCN-3 is done in a conditional test based manner where return values and pointer parameters are matched in conditional statements individually through implicit templates or through defined templates. If parameters and return values are not enough for evaluation, the test system has to be extended with additional functionality for testing outside a TTCN-3 signature definition, e.g., file streams and connections. Test cases can implement testing functionality resembling the assertion macros and functions of, e.g. the well known testing frameworks CUnit[5] and JUnit[6]. Table 3 shows fragments of a simple unit test case illustrated in Fig. 4. It tests for correct behavior when verifying a user against a mock object that acts as the missing user verification component. The test case in Table 3 tests the case where one user 'known' to the ATM system tries to log in with an incorrect password. It also checks that the login function under test realizes that the password is not correct from the user verification reply. On lines 10 to 11 in Table 3 two test components are created, one for the mock object and one for the function to test. The behavior of the components is implemented in the functions that are passed as arguments when the components are started on lines 13 and 14.

The user verification should fail and the failure shall be detected by the function under test and then later in the test case the test case verdict is set to pass. This is exemplified in Table 5 where the function login in the IUT is called on line 9 and the alternative replies are evaluated on lines 12 to 21.

Table 3. The test case body where two components are created and started

```TTCN-3
001  testcase TC_UserVerification_Invalid_001()
002       runs on IprPBComp
003       system IpRouterPBTSI {
004
005      // Start two components here. One for the login and one for
006      // the mock object.
007      var IprPBComp testComponent;
008      var IprPBComp mockObject;
009
010      testComponent := IprPBComp.create;
011      mockObject    := IprPBComp.create;
012
013      mockObject.start( f_verifyUser_mockObject( 0 ) );
014      testComponent.start( f_login_invalid( 1 ) );
015
016      f_waitForComponentsToFinish( c_COMP_TIMEOUT );
017  }
```

The component representing the mock object for the user always returns a failed reply. The behavior is defined in the function in Table 4 where a call from the SUT is accepted (line 10) and the *rejected user* reply is returned (line 11).

Table 4. Code fragments from the mock object function *f_verifyUser_mockObject*

```TTCN-3
001  function f_verifyUser_mockObject( in integer p_compNo )
002       runs on IprPBComp {
003
004      f_initPBTcp( localIp,
005                   localPort+p_compNo,
006                   sutIp,
007                   sutPort );
008
009      // Accept any user and always return invalid.
010      pt_pb.getcall( s_verifyUser:{?,?} );
011      pt_pb.reply( s_verifyUser:{-,-} value c_REJECTED_USER );
012
013      unmap( self:pt_pb, system:pt_pb );
014  }
```

For concurrent software modules the behavior of the threads executing on the IUT has to be controlled by a dedicated TTCN-3 test component. Each test component has to establish a new connection to the SUT which at the point of the accepted connection spawns a working thread in the test harness. This is exemplified in Fig. 3 and in the right part of Fig. 4 which illustrates a test case with mutually exclusive data and multiple accessing threads.

The purpose of the test case defined in the function in Table 6 is to determine if the ATM module can handle (not crash) and recover from a function call that purposely introduces an invalid parameter (null pointer), Table 7. Recovering means in this case

Table 5. The test purpose functionality that verifies that the function under test can detect a failed login

```
TTCN-3
001  function f_login_invalid( in integer p_compNo )
002    runs on IprPBComp {
003
004    f_initPBTcp( localIp,
005                 localPort+p_compNo,
006                 sutIp,
007                 sutPort );
008
009    pt_pb.call(
010      s_login:{c_USERID_A, c_BADSECRETCODE_A}, c_TIMEOUT_SEC ) {
011
012      [] pt_pb.getreply( s_login:? value c_REJECTED_USER ) {
013        setverdict( pass );
014      }
015      [] pt_pb.getreply( s_login:? ) {
016        setverdict( fail );
017      }
018      [] pt_pb.catch( timeout ) {
019        setverdict( fail );
020        stop;
021      }
022    }
023
024    unmap( self:pt_pb, system:pt_pb );
025  }
```

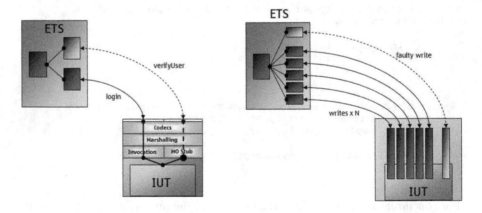

Fig. 4. Test cases visualized

that an invalid parameter shall not halt the ATM module by e.g. keeping mutexes locked. The test case creates a set of components that all but one executes the behavior in Table 6.

In Table 6 the behavior of the correctly executing test components can be seen. A pointer is created (lines 13 to 25) and is used through out the test case execution. A number of withdrawal requests are performed in a loop (lines 28 to 45) to keep the server busy and to determine if it can recover when another test component introduces an error. All components include detection of timeouts which will indicate that the IUT has crashed or got locked by a deadlock or livelock (Table 6 line 14 and Table 7 line 25).

Table 6. The test behavior of the test components that correctly calls the withdraw function

```
TTCN-3
001    function f_write_valid( in integer p_compNo )
002      runs on IprPBComp {
003
004      f_initPBTcp( localIp,
005                   localPort+p_compNo,
006                   sutIp,
007                   sutPort );
008
009      // Create the integer pointer where to store the account
010      // balance.
011      var CIntPtr v_balance;
012
013      pt_pb.call( s_MallocCInt:{ 1 }, c_TIMEOUT_SEC ) {
014        [] pt_pb.getreply( s_MallocCInt:? value c_NULLPTR ) {
015            setverdict( fail );
016            stop;
017        }
018        [] pt_pb.getreply( s_MallocCInt:? ) -> value v_balance {
019            setverdict( pass );
020        }
021        [] pt_pb.catch( timeout ) {
022            setverdict( fail );
023            stop;
024        }
025      }
026
027      // Continuous withdrawals, no checking.
028      for ( i:=0; i<c_LOOPS; i:=i+1 ) {
029
030        pt_pb.call(
031            s_withdraw:{ c_USERID_A, 20/*€*/, v_balance },
032            c_TIMEOUT_SEC ) {
033
034            [] pt_pb.getreply( s_withdraw:? value c_WITHDRAW_OK ) {
035                setverdict( pass );
036            }
037            [] pt_pb.getreply(s_withdraw:? value c_WITHDRAW_ERROR) {
038                setverdict( fail );
039                stop;
040            }
041            [] pt_pb.catch( timeout ) {
042                setverdict( fail );
043                stop;
044            }
045        }
046      }
047      unmap( self:pt_pb, system:pt_pb );
048    }
```

The function in Table 7 is executed by the component that introduces a possible problem by passing a null pointer to the function withdraw (line 15). The correct behavior of the IUT shall at this point be that the null pointer is detected, any locked mutexes are released and an error is returned. Test case verdict according to how the IUT behaves is set on lines 18 to 28.

Table 7. Code fragments of the test component that introduces an error by passing a null pointer argument

```
TTCN-3
001   function f_write_invalid( in integer p_compNo )
002     runs on IprPBComp {
003
004     f_initPBTcp( localIp,
005                  localPort+p_compNo,
006                  sutIp,
007                  sutPort );
008
009     // Introduce the error after a few withdrawals have been done
010     // already.
011     f_sleep( 1.0 );
012
013     // Withdrawal with NULL pointer.
014     pt_pb.call(
015       s_withdraw:{ c_USERID_B, 20/*€*/, c_NULLPTR },
016       c_TIMEOUT_SEC ) {
017
018       [] pt_pb.getreply( s_withdraw:? value c_WITHDRAW_OK ) {
019         setverdict( fail );
020       }
021       [] pt_pb.getreply( s_withdraw:? value c_WITHDRAW_ERROR ) {
022         setverdict( pass );
023         stop;
024       }
025       [] pt_pb.catch( timeout ) {
026         setverdict( fail );
027         stop;
028       }
029     }
030     unmap( self:pt_pb, system:pt_pb );
031   }
```

The TTCN-3 verdict type is special in a way that its value can never be degraded, meaning that if a verdict is set to fail it cannot later be set to pass. Any test component can set the test case verdict eliminating the need for verdict synchronization after the test components have executed.

6 Conclusions

This paper presented an approach to implement a TTCN-3 test platform for software testing. The capabilities of TTCN-3 enable efficient ways of testing software modules in a standardized way with standardized test system interfaces. One language can be used for all test cases independent of the language of the IUT so there is no need to learn test tool proprietary languages. Utilizing standardized interfaces for modular test platforms make it possible to efficiently set up new test systems and at the same time eliminate possible overlapping test system development work.

Setting up a distributed test system provides a number of advantages that cannot be achieved with a tighter test system, e.g. the problem of a shared memory where the tested IUT can crash the process responsible for executing and evaluating test suites, and target platform independence.

The core TTCN-3 core language allows for creation of multi component test cases and easy means to set up synchronization between the components, making it possible to approach testing problematic domains as concurrency and non-deterministic behavior.

As the procedure-based paradigm is only a part of the TTCN-3 language a combination with the message-based parts can be used for test suites that interact with the SUT through different kinds of interfaces. This can lead to multi purpose test suites with a lot of useful capabilities such as unit testing with mock objects and possibilities to control the SUT with function calls while performing protocol testing. Multi purpose functionality that can be hard to achieve based on test area targeted frameworks e.g. unit testing and protocol testing frameworks.

Further case studies will be done to evaluate the usefulness of TTCN-3 based software module testing in several areas such as ease of use, test suite reusability, concurrent testing efficiency and implementation of synchronized deterministic test cases. Also its applicability to testing of distributed operating systems and language independent SUTs will be further explored. We believe that TTCN-3 has enough advantages to make it as a powerful alternative and language for future software module testing test systems. One test system and one test language offer the ability to perform multi purpose software testing capabilities written in any language.

Currently a lot of mapping work has still to be done manually. However, this can based on the language mappings be generated by tools, including TTCN-3 signatures and types, marshalling, invocation and mock object parts for the SUT based on the definition of the software module to test. The requirement is that a specified mapping exists or can be made from the languages of the IUT to TTCN-3, something that has proven to be a challenge especially for C and C++.

Acknowledgements

A lot of useful ideas for a C mapping has come from the long discussions with Matti Kärkki and Pekka Pulkkinen who are the originators of some the C++ to TTCN-3 mapping work done. Special thanks also to the TTCN-3 people at the Nokia Research Center, Thomas Deiß and Stephan Schulz for feedback and review of this paper, and also to Federico Engler, Sami Heinonen, Martti Söderlund, Stephan Tobies and Colin Willcock.

References

1. Itoh, E. Furukawa, Z. Ushijima, K. A prototype of a concurrent behavior monitoring tool for testing of concurrent programs. IEEE, 1996.
2. Tai, K, C. Testing of Concurrent Software, Computer Software and Applications Conference, 1989. COMPSAC 89, Proceedings of the 13th Annual International 20-22 Sept.

3. Tai, K. Definitions and detection of deadlock, livelock, and starvation in concurrent programs. Proceedings 1994 International Conference Parallel Processing. 1994.
4. Mackinnon, T. Freeman, S. Craig, P. Endo-Testing: Unit Testing with Mock Objects, Proceedings XP2000.
5. CUnit 2005: CUnit (2005) Retrieved November 9, 2005, from CUnit Web site: http://cunit.sourceforge.net/.
6. JUnit 2005: JUnit (2005) Retrieved November 9, 2005, from JUnit Web site: http://junit.org/index.htm.
7. Moseley, S. Randall, S. Wiles, A. Schulz, S. IPv6 Test Specifications from ETSI. Global Ipv6 Summit, Barcelona, June 2005.
8. Hwang, G. Tai, K. Huang, T. Reachability Testing: an approach to testing concurrent software. Software Engineering Conference, 1994. Proceedings. 1994 First Asia-Pacific 7-9 Dec.
9. Carver, R. Mutation-based testing of concurrent programs. Test Conference, 1993. Proceedings. International 17-21 Oct. 1993.
10. Thomas, D. Hunt, A. Mock Objects. Software, IEEE Volume 19, Issue 3, May-June 2002.
11. Ebner, M. (2001). A Mapping of OMG IDL to TTCN-3. University of Lübeck, Germany.
12. Beizer, B. Software Testing Techniques, 2nd ed, p41-54. Van Nostrand Reinhold, 1990.
13. ETSI ES 201 873 "Methods for Testing and Specification (MTS); The Testing and Test Control Notation version 3"; V3.0.0, Sophia Antipolis, March 2005.
14. InformationTechnology, OSI conformance testing methodology and framework. ISO/IEC, 1994-1997. International Telecommunication Union recommendation X.290.
15. ISO/IEC 9899:1999: "Programming languages - C". New York, NY, USA (1999-12)
16. Willcock, C. Deiß, T. Tobies, S. Keil, S. Engler, F. Schulz, S. (2005). TTCN-3 Test Systems in Practice: An Introduction to TTCN-3. England: John Wiley and Sons Ltd.

Distributed Load Tests with TTCN-3

George Din[1], Sorin Tolea[1], and Ina Schieferdecker[1,2]

[1] Fraunhofer FOKUS, MOTION, Kaiserin-Augusta-Allee 31,
10589 Berlin, Germany
[2] Technical University Berlin, Faculty IV,
Straße des 17. Juni 135, 10623 Berlin, Germany
{din, tolea, schieferdecker}@fokus.fraunhofer.de

Abstract. The design of TTCN-3 focused on extensions to address testing needs of modern telecom and datacom technologies and widen the applicability to many kinds of tests including performance tests. One of the most important features of TTCN-3 is the platform independence which allows testers to concentrate on the test specification while the complexity of the underlying platform (i.e., operating system, hardware configuration, etc.) is left behind the scene. As far as the test distribution is concerned, TTCN-3 provides the necessary language elements for distributed tests. This is however supported in a transparent fashion so that the same test may run either locally or distributed. The distributed execution of a test enables the execution of test components belonging to one test configuration on different computers (the test nodes), sharing thus a bigger amount of computational resources. Test distribution is a research challenge when it comes to the problem of how to distribute the test components efficiently on the test nodes. Specifically for load testing – a particular kind of performance test – we investigate strategies to distribute tests on heterogeneous hardware in order to use the hardware resources of the test nodes efficiently.

1 Introduction

Performance testing is a qualitative and quantitative evaluation of a System under Test (SUT) under realistic conditions to identify problems for scalability or usability aspects under heavy load and to collect measurements as success/fail rate, response times or round-trip delay. Although performance testing is often used in different ways, performance testing usually determines how fast a system reacts or how much load a system can handle. The literature distinguishes [13]: load, robustness, stress, or volume testing. *Load testing* simulates various loads and activities that a system is expected to encounter during production time. The typical outcome of a load test is the level of the load the system can handle but also measurements like fail rate, delays under load etc. Load testing helps detecting problems of the SUT (like abnormal delays, availability or scalability issues, or failover) when the number of emulated users is increased. A load test defines real life like volumes of transactions to test system stability, limits or thresholds. Typically, a number of emulated users interacting with the SUT have to be created and managed while the functional behaviour of their communication with the SUT has to be observed and validated.

M.Ü. Uyar, A.Y. Duale, and M.A. Fecko (Eds.): TestCom 2006, LNCS 3964, pp. 177–196, 2006.

Scalability testing is a special kind of load testing, where the system is put under increasing load. *Robustness testing* is load testing over extended periods to validate an applications stability and reliability. *Stress testing* is the simulation of activities that are more "stressful" than the application is expected to encounter when delivered to real users. Stress tests measure various performance parameters of the application under "stressful" conditions. Examples of stress tests are: *spike testing* (short burst of extreme load), *extreme load testing* (load test with huge number of users), *hammer testing* (continuous sending of requests). *Volume testing* is the kind of performance test we run in order to find which volume of load an application under test can handle.

TTCN-3 (Testing and Test Control Notation) [2] enables systematic, specification-based testing for various kinds of tests including functional, inter-operability, integration, load, robustness, volume and stress testing. It allows an easy and efficient description of complex distributed test behaviours in terms of sequences, alternatives, and loops of stimuli and responses. The test system can use a number of test components to perform test procedures in parallel. The task of describing the dynamic and concurrent configuration is easy to perform since it is developed at a platform independent level. The advantage of this approach is that the distribution configuration is abstract and it does not depend on a particular test environment. The same (potentially distributed) test specification can be executed on different hardware environments and various distribution setups. For example, a test case which creates a number of N test components can be distributed on 5 hosts, but can run also on 10, 2 or just 1 host.

The test workload definition belongs to a performance test plan. It is a description of the test actions against a tested system and should reflect how users typically utilize that system. Overloading an SUT with a huge number of requests tells us how robust the system is, but this kind of test does not reflect normal performance requirements and gives no information about the behaviour of the system in daily scenarios. The workload definition should describe performance tests according to real world scenarios taking into account social, statistical, and probabilistic criteria.

Test distribution is a technique to realize the load as required by the workload definition on several test nodes. Only one test node might not be enough to emulate a big number of users. A distributed test case may consist of two or more parts that interact with each other, but each part is being processed on a different test node.

We concentrate our study on developing and executing distributed load tests with TTCN-3. TTCN-3 offers the required flexibility in specifying load tests. Quite a number of language concepts help to design complex workloads in an intuitive way. However, the real distribution and deployment of the executable tests is out of consideration of TTCN-3. Therefore, an additional layer of specifications is needed to describe the real test configuration on a real target network of test nodes (being potentially only one test node).

This paper discusses in Section 2 related work, and presents in Section 3 foundations like load test specification, test component distribution and factors which influence the distribution. Next, in section 4 the architecture for load tests execution is presented. Section 5 presents the distribution algorithms and discusses their characteristics. Section 6 presents an example. The paper is concluded by a summary.

2 Related Work

Related work on applying TTCN to performance testing targets either the specification of distributed tests with TTCN (TTCN-3 or previous versions of it) or concerns the test execution and test distribution over several test nodes.

The first experiments with TTCN applied to performance testing were done with the version 2 of TTCN language. PerfTTCN [5] is an extension of TTCN-2 with notions of time, measurements and performance. In [7] SDL specifications are used to generate tests for distributed test architectures. This paper discusses also concepts related to distributed and concurrent testing. TimedTTCN-3 [9] is a real-time extension for TTCN-3 that supports the test and measurement of real-time requirements. This paper introduces concepts like absolute time, definition of synchronization requirements for test components and provides possibilities to specify online and offline evaluation procedures for real-time requirements. Most of these ideas can be also reused in performance testing with TTCN-3. The work in [4] presents a number of patterns in specification of distributed tests. It uses also TTCN-3 to specify distributed tests and discusses different facets of distributed testing. The test architecture topic is discussed also in [8] where a generic test architecture is presented. As far as the execution of distributed tests is concerned, [6] introduces TCI (Test Communication Interfaces) and discusses the possibility to use TCI to realize distributed test execution environments.

Our paper uses principles of these works and analyses in particular, how test components for load test scenarios can be efficiently specified, distributed and executed.

3 Foundations

Test distribution with TTCN-3 implies, on first hand, the use of the language elements to describe distributed load tests and, on the second hand, the development of an execution environment capable to distribute tests. In this section, we investigate first the TTCN-3 language capabilities to specify load tests along small examples. In additions to this, we look into possible patterns the tester may use for specifying load tests and analyse which are the constraints with respect to the distribution strategies for those patterns.

3.1 Load Test Specification with TTCN-3

TTCN-3 offers various concepts to design load tests such as test components to emulate SUT users/clients, ports to handle connections to the SUT, send/receive or call/reply statements to communicate with the SUT, timers to measure the responsiveness. These concepts are introduced along with small examples of how they are useful in load testing.

component is the structural element which is used to define the clients involved in the load test scenarios. One test may define more than one type of components in order to distinguish users of different categories or scenarios.

```
type component UserType {
  port ConnectionType connection;
  timer respTime;
  var integer fail := 0;
}
```

The specification of all test components, ports, connections and test system interface involved in a test case is called *test case configuration*. Every test case has one Main Test Component (MTC) which is the component on which the behaviour of the test case is executed. The MTC is created automatically by the test system at the start of the test case execution. The other test components defined for the test case are called parallel test components (PTC) and are created dynamically during the execution of the test case. The tested entity is called System under Test (SUT) and the interface to communicate with it is the Abstract Test System Interface (system).

The behaviour of a test component is defined by a function. A function is used in load tests to specify client activities within a test scenario. An SUT client may behave in different ways when interacting with the SUT, thus the test system may have different functions emulating different client behaviours.

```
function clientBehavior(in integer cid)
runs on UserType {
  // user behavior
  // communication with SUT
}
```

TTCN-3 supports message-based and procedure-based communication. The communication operations can be grouped into two parts: stimuli which send information to the SUT (send, call, reply, raise) and responses used to describe the expected reaction from the SUT (receive, getcall, getreply, catch). To apply a sending operation (stimuli) there shall be specified a port used to send the data, the value to be transmitted, and optionally an address to identify a particular connection if the port is connected to many ports. Additionally, for procedure based communication the response and exceptions are needed; they are specified by using the getreply and catch operations.

```
connection.send(aRequest(uid));
respTime.start();
alt {
  [] p.receive(correctResponse(uid)) { }
  [] p.receive { }
  [] respTime.timeout { }
}
```

Timers are a further essential feature in the development of load tests with TTCN-3 in order to evaluate the performance of the SUT. The operations with timers are start, stop, read (to read the elapsed time), running (to check if the timer is running) and timeout (to check if timeout event occurred). The start command may be used with parameter (the duration for which the timer will be running) or without parameter (when the default value specified at declaration is used). For load testing purpose, we define timers on test components and use them in the test behaviour to measure the time between sending a stimuli and the SUT response. If the SUT answer does not come in a predefined period of time, the fail rate statistics should be correspondingly updated.

Another important mechanism provided by TTCN-3 is the inter-component communication which allows connecting components to each other and transmitting messages between them. This mechanism is used in load testing for synchronization of actions (i.e. all components behaving as clients start together after receiving a synchronization token) or for collecting statistical information at a central point.

The handling of verdicts in load tests is different from the traditional verdict handling procedure in functional testing. In functional testing, we use the build-in concept of *verdict* which is always set when an action influences significantly the execution of the test (for example, if the SUT gives the correct answer we set the verdict pass; if the response timer expires we set the verdict inconc or fail). Load tests have also to maintain a verdict which should be presented to the tester at the end of a test case execution. However, the verdict in this case has rather a statistical meaning than only a functional one: Still, the verdict should be a sum of all verdicts reported by client test components. In our approach, the verdict is set by counting the rate of fails during one execution; i.e. if during the test more than a threshold percentage of clients behave correctly we consider the test passed. The percentage of correct behaviours in a tests must be configured by the tester himself and must be adapted to each SUT and test separately.

The collection of statistical information like fails, timeouts, successful transactions can be implemented by using counter variables on each component. These numbers can be communicated at the end of the test to a central entity (i.e. MTC) which computes the final results of the test. If the test needs to control the load based on the values of these variables, that central entity must be periodically updated.

3.2 Load Test Specification Patterns

Test patterns are generic, extensible and adaptable test definitions. Reusable test patterns are (as an analogy of software patterns [13][14][15]) derived from test methods, test solutions and target system technologies. They are available in form of software libraries and/or code generators which offer the tester ready to use code.

3.2.1 Workload Unit Specification
Even though the TTCN-3 language is very flexible and allows for various ways to write a test, we believe that load test designers follow at least one of the specification patterns presented in the following. The major role of a load test is to emulate the parallel behaviour of multiple clients (or users) interacting with the SUT. In literature, the SUT's clients are also called WLUs (workload units) and they are implemented as parallel processes or threads. Nevertheless, a parallel process may emulate the behaviour of more than one user at a time. In TTCN-3, the test component is the building block to be used to emulate one or more WLUs at a time. The parallelism is realized by running a number of test components concurrently on a number of test nodes. The methods to specify user behaviours as test components can be classified into the following patterns:

(a) The most obvious pattern to define a client is to define a component emulating only one client, i.e. the *one client per component pattern*. On this component we start a function which describes the actions the client interchanges with the SUT. Despite the easiness to write load tests using this technique, two main drawbacks exist. Firstly, load control is difficult to realize when the tester wants to keep a constant number of parallel users. The controller needs to control continuously the number of component acting in parallel and whenever a component terminates a new one has to be created. Secondly, the creation, start and termination of components are very expensive operations with respect to CPU on a test node. Because of this, it is preferable to *reuse* the existing test components to emulate more than one client on one test component. But, as we will see in the next section, from the distribution point of view the one-client-per-component specification style turns out to be an advantage for the application of a large number of distribution strategies since the distribution unit is small and the balancing of the load can often be reconfigured.

(b) Another pattern for the specification of load tests is the reuse of components to emulate a new client once the current client terminates. This pattern implies that one component repeats sequentially in a loop the behaviour of a client, but for each client a new set of data (id, request data, client reaction times etc) is used. This pattern, named *sequential repetition of clients per component pattern,* has the advantage that only a fixed number of test components are created and thus no additional time is spent on handling the test components. The disadvantage of this approach is that a test component can be distributed only once at the beginning and no further (re-)balancing is possible.

(c) An extension of the previous pattern is the interleaving of more than one client on a test component. In this way, a test component is able to simulate in parallel a number of clients. This pattern has the name *interleaved client behaviours per component pattern*. Unfortunately, the mixture of parallel behaviours on one component is complicate to specify and most of the time the TTCN-3 code loses its readability and becomes difficult to maintain. The approach has the same disadvantage as the previous pattern that the component can be distributed only once, at the beginning and no further (re)balancing is possible.

3.2.2 Differentiate Component Types

The different client types are usually defined in TTCN-3 as distinct test component types. This is in fact a recommended pattern in test specifications which helps to recognize easier the different types of components at distribution time. Most of the variables used during a test are usually defined directly as part of the test component type so that they can be directly accessed from any function being started on a test component of that type. This approach helps also at distribution time since at the instantiation of a test component most of the memory required by the test component is known right at creation time.

3.2.3 Test Architectures

With respect to test architectures, we identify at least two specification patterns of how to group different kinds of clients interacting with the SUT:

(a) Most of the load tests create independent client components which depend only on the interaction with the SUT. These components do not depend on other components and therefore their distribution on different test nodes is by no means constrained.

(b) Another category of test architectures requires pairs of clients or caller-callee clients to interact with the SUT. Typically, these types of tests need extra communication for coordination between the caller and callee component. Therefore, for these types of tests, the distribution strategy should consider that it is more efficient to install both components on the same test node since the local communication is faster than the communication between two test nodes.

3.2.4 Load Control

In order to control the volume of the created load, the test case has to control the number of users running in parallel. Such a mechanism is called *load control* and it is usually implemented as a separate test component (most of the times it is the MTC) which interacts with all other test components in order to increase or decrease the number of interactions with the SUT. The load is controlled by increasing/decreasing either the number of test components or the number of users emulated by one component. In both cases, the increase of load will bring additionally more need for hardware resources and therefore, increasing the number of components or users emulated after a certain level of the load, the test system will not be able of increasing its load but rather decrease it. Therefore, the tester should take care about this aspect when tuning the test and observe continuously the load of the test system. If it happens that the test system reaches the maximum producible load, then the tester should either try a more efficient distribution strategy for the test components or upgrade/extend the hardware resources.

3.2.5 Synchronization

The synchronization of the parallel test components is realized by passing coordination messages. In general, load tests require synchronization only at the start or stop of the parallel components and/or at increasing or decreasing the level of load. Moreover, the synchronization does not have constraints with respect to the time needed to realize the notification of all test components. TTCN-3 allows the tester to connect all parallel test components, participating as workload units, to a central component (i.e. MTC) which coordinates the synchronized activities.

3.3 Factors Influencing Test Distribution

Resource sharing (CPU, memory, disk or bandwidth) in parallel and distributed computing has been intensively researched over the last decades being the activity of efficient utilization of computing resources by partitioning and balancing the computational load among computing nodes [1]. Load distribution is the strategy to allocate parts of a bigger task to parallel workers (computers or processors) and, thus, to decrease the execution time of a program. Many algorithms have been researched and applied to particular problems. Depending on the problem, the algorithms work better or worse.

Very often, the parallel processes communicate with each other. Granularity is a parallelism measure which characterizes the inter-process communication. We say that the parallelism has big granularity in case of rare communication or has a fine granularity in case of high frequency of communication between processes.

The class of operations to be performed by parallel processes is a further factor which influences the performance of balancing algorithms. A non-exhaustive overview of classes of operations may classify them into: computational operations (i.e. floating point operations), memory access operations, operations with databases, files operations or communication with other computers.

Synchronization of activities of parallel processes is often needed. A process can be considered a sequence of atomic actions where each action transforms the state of the process. Some of these actions have to be synchronized with actions of other processes. Depending on the used synchronization mechanism, balancing algorithms may perform better or worse. For example, the clock synchronization in difference to message-passing based synchronization avoids the overhead added by the inter-process communication. If the message-passing synchronization is applied, the balancing algorithms should be aware also about the bandwidth consumed for synchronization.

Load testing of hardware components or applications is a resource consuming process which coupes also with the resource sharing discipline. Most of the times, in order to run high performance tests against a hardware component or an application (or just parts of it) many computers have to be involved in the test process so as to create enough traffic to evaluate the behaviour of the SUT under load conditions. In this respect, the tester has to be aware about the possible distribution algorithms and be able to decide which algorithm to apply.

For the distribution of TTCN-3 test components, a number of factors have to be considered when selecting the distribution algorithm. The distribution unit used for test distribution is the parallel test component which can simulate the behaviour of one or more clients. If the component emulates only one user, the component is relatively small and terminates after execution of the test scenario. This design pattern presents the advantage that the balance of resources can be performed at each component instantiation. The creation of components happen at small intervals of times since when a component terminates a new one is created in order to maintain the same number of users. If the component emulates sequentially or interleaved behaviours of more than one user, the component will live for a very long period of time (sometimes until the end of the load test). In this case, the algorithm does not have too much flexibility to balance the load except the creation of the component. In such situations, the recommended strategy should be based on resource consuming predictions.

The existing load on the underlying hardware is a further factor to be taken into consideration when applying a distribution strategy. The solution we foresee for distribution of component targets deployment of load tests on heterogeneous hardware which besides the test application may also run other tasks. Therefore, a continuous observation of the hardware usage is recommended while the distribution strategy considers the resource availability at any component instantiation.

4 TTCN-3 Test Distribution Realization

4.1 Test Component Distribution Language

The distribution strategy defines how the components are distributed among test nodes and thus it plays a major role in the efficiency of a test system. Test distribution defines which components are to be distributed and where they should be deployed. Distribution of components is a mathematical function of different parameters which is applied at deployment time separately for each test component in order to assign it to a home location where it will be executed. In the following function definition, D is the distribution function, p1, p2, ..., pn are the parameters which influence the distribution and h is the home where the test component should be distributed.

h = D (p1, p2, ..., pn)

There are two types of parameters which are taken into consideration when distributing test components: *external parameters* like bandwidth, CPU, memory and *internal parameters* like the number of components, type of components, type of behaviours, connections. The external parameters are application independent parameters whose values depend on the execution environment and are constant for all applications running on that environment. The internal parameters are related to the test component based application itself and are different for each test case.

Unfortunately, in TTCN-3 it is not possible to recognize a component by its id. This problem appears when creating test components like in the following example[1]:

```
for (i := 0; i < 100; i := i + 1) {
        var PTCType c := PTCType.create;
        map(c:port1, system:port1);
        c.start(someBehavior1());
}
```

In this example, the component variable c refers to the currently created test component, but is overwritten at each create operation, so that the execution environment has no differentiation of the test components. But there are some other characteristics of test components in TTCN-3 which can be used during execution to identify them. These characteristics are of two categories: *behaviour independent* and *behaviour dependent*. The behaviour independent ones concern parameters which can be accessed at the creation phase of the test component: the component type, the instance number and the port types which belong to that component. The behaviour dependent characteristics imply the use of characteristics of the test component we can gather after the component is started or executed (i.e. which Id will receive the test component from the SUT). The distribution mechanisms used in this case are based on analyzing the TTCN-3 code before starting the execution and decide upon execution monitoring where the test components should be deployed. This approach requires running a calibration behaviour in which an instance of a test component is created and its execution is monitored. The observed information is then used during the "real" test in order to decide where to distribute that test component.

---

[1] Please note that the new version of TTCN-3 being approved summer 2005 offers the assignment of explicit names to test components during their creation, however, this was not available for the presented work.

A minimal language for defining the distributions of test components has been defined. To help understanding the concepts related to test component distribution, some examples written in this language are presented here. The distribution specification is the process of assembling test components to hosts. The assembling process groups all components to be deployed, in a big set while the assembling rules shall define sub-sets of components with a common property (i.e. all components of the same type). A (sub-)set defines a selector of components and the homes where the selected components are placed. The filtering criteria of the selector handle component types or component instance numbers. The homes are the possible locations where the test components may be distributed; the homes reflect the user constraints for distribution.

The next XML code is an example of a component assembly file. The `special` tag indicates the host where the MTC component is deployed. The `selector` defines a filter to select all components of type `ptcType`. The selected components can be deployed either on `container1` or on `container2`. One can define deployment constraints for each container (for example, do not allow deployment of more than 100 components on `container2`). The user can also constrain the memory usage, the CPU load, the number of components etc.

```
<component_assembly>
    <description>Example to use TCDL language</description>
    <special container="container1"/>
    <set>
        <component_selectors>
            <componenttype>ptcType</componenttype>
        </component_selectors>
        <homes distribution="round-robin">
            <container id="container1">
                <max_components>10</max_components>
            </container>
            <container id="container2"/>
                <max_components>100</max_components>
            </container>
        </homes>
    </set>
</component_assembly>
```

Usually, the definition of constraints is a difficult task; for complex setups it may be very difficult to describe an efficient distribution. Therefore, the task of identifying hardware options and constraints should be realized by the test execution environment itself. It should provide services, which implement distribution algorithms that are designed to be efficient for a certain type of problems. The task of the user remains to select the algorithm which solves the problem best.

The code below shows a set which deploys the components of types `ptcType2`, `ptcType3` and the instances 1, 2 and 5 of type `ptcType4` on the `container2` and `container3`, according to a round-robin algorithm.

```
<set>
        <component_selectors>
            <componenttype>ptcType2</componenttype>
            <componenttype>ptcType3</componenttype>
            <instance type="single">
                <componenttype>ptcType4</componenttype>
                <number>1</number>
                <number>2</number>
                <number>5</number>
            </instance>
        </component_selectors>
        <homes distribution="round-robin">
            <container id="container2"/>
            <container id="container3"/>
        </homes>
</set>
```

The components which are not accepted by any set selector are deployed in a default home. This home is defined by collector tag.

```
<collector>
        <container id="container1"/>
</collector>
```

4.2 TTCN-3 Architecture Design for Distributed Execution

For deploying and executing distributed tests, we have designed and implemented the architecture depicted in Figure 1. This architecture follows the ETSI standard architecture [10],[11] for realizing distributed tests . The platform consists of a set of interacting entities which execute the code generated from a TTCN-3 specification, realize the distributed communication between test nodes, realize the communication with the SUT, implement external functions and handle timer operations.

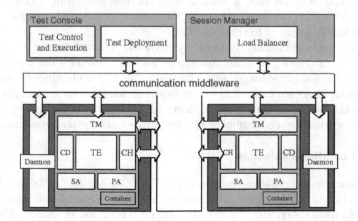

Fig. 1. Distributed test architecture

The Test Console handles the management operations to create test sessions, deploy test components into containers and control the test execution. The tests are deployed, configured and executed in the context of a test session. One of the most important functionality of the session manager is the load balancing one, which coordinates the distribution algorithms (compute the hosts of the components according to assembly rules, performance requirements, distribution algorithms etc). To supply the dynamic algorithms with the necessary information for distribution computation, the Session Manager provides an interface to the daemons in order to gather the resource consuming level.

Test Daemons are standalone processes installed on any hosts which manage the test containers. Containers intercede between Test Console and test components, providing services transparently to both of them, including transaction support and resource pooling. The containers are the hosts of Test Executable; they manage installation, configuration and removal of the parallel test components. Moreover, containers are the target operational environment and comply with the TCI standard for TTCN-3 test execution environment. Within the container, we find the specific test system entities: TM (Test Management), CD (Coder-Decoder), TE (Test Executable), CH (Component Handler), SA (System Adapter) and PA (Platform Adapter). For more information on the API and interactions between these entities, we refer [6]. The container subsystems are functionally bound together by the TCI interfaces and communicate with each other via the CORBA platform.

The distributed handling of the test components is realized within CH. CH distributes TTCN-3 configuration operations like create, start and stop of test components, the connection between test components (connect and map), and inter-component communication like send, call and reply among two TTCN-3 executables participating in the test session. The CH is not implementing the core TTCN-3 functionality – this is done by the TE, for example a test component is created, etc. Next, CH asks the Session Manager for a location for the new component. Based on the decision of the Session Manager, the request for the creation of a component will be either transmitted to the local TE or to a remote participating one if the component has to be created on the remote TE. The remote TE will create the TTCN-3 component and will provide a handle back to the requesting (local) TE. The requesting (local) TE can then operate on the remote created test component via the component handle given by the remote TE.

Hardware load monitoring is used by dynamic algorithms for the balancing decisions. The monitoring tools are running on each host used for the tests and is able to provide to the SessionManager an evaluation of the current hardware consumption. The monitoring tasks are controlled by the SessionManager over a specially designed interface which allow activating/deactivating of different sensors, setting the update refresh rate or counting a performance key parameter out of several parameters.

4.3 Test Execution Evaluation

The intensive use of hardware resources (i.e., 100% CPU) during the test execution leads very often to malfunctions of the test system which ends up running slower than expected. Consequently, the test results can be wrong as an effect of erroneous evaluation of SUT's responses. We encounter such a situation when, for example, the test system creates too many parallel processes which share the same CPU. The processes wait in a queue until (according to the used scheduling algorithm) they acquire the CPU. Hence, the bigger the number of processes is, the more time a process has to wait in the queue until it acquires the CPU. Since the execution of critical operations (like timer evaluation, data encoding or decoding, template matching) is automatically also delayed, the test system may consider an operation timed out while, in reality, it did not. The same phenomenon has a considerable impact also on load producing by decreasing the number of interaction per second.

The evaluation of load test results turns into a problem of determining whether the SUT is that slow as the results reveal or rather the test system is overloaded by its testing activities and cannot produce the necessary load and/or reacting in time. The answer to this question can only be given after analyzing the quality of the execution. To detect such problems we observe several parameters which help the tester to validate the test execution.

One of these parameters is the duration of the execution of critical tasks. We assign temporal dimensions to all operations to be executed sequentially in a test which might influence the evaluation of SUT's performance. For example when receiving a message from SUT and this message is used to validate the reaction of SUT to a previous request, the test system has to decode and match the received message only in a small amount of time, otherwise the additional computation time will be counted as the SUT reaction time. A further interesting parameter is the quantity of the demanded resources. If the test system requires constantly the maximum of the resources the underlying hardware can allocate to them, this is a first sign that the test might not be valid. Another parameter is the deviation average from load shape. If the load does fluctuate very often moving from lower to higher values, it proves that the test system might be overloaded.

We consider that a performance test is valid only if the platform satisfies the performance parameters of the workload. The quality of the load test execution is guaranteed if the test tool fulfils the requirements with respect to execution of the critical operations like decoding, matching, timer processing.

5   Balancing Algorithms Applied to Test Distribution

The literature differentiates load balancing algorithms by several criteria. Load balancing algorithms can be static or dynamic; the difference is made by the distribution decision which is known before actually running the test in case of static algorithms, while the decision depends upon the state of the system when dynamic algorithms are considered. The static algorithms work very well when the test nodes have more or less the same resources (same memory, CPU etc) and the usage during the tests is not influenced by other applications running on that hardware. According to our experience, round-robin algorithm works very well in such situations. The distribution function D, mentioned in section 4.1 is an incremental function over the number of hosts, which selects sequentially the next host for deployment. In the case of test nodes with different capacities, static algorithms do not work well anymore because of hardware limitations. One may try to use round-robin algorithm with empirically chosen constraints for the number of deployed components on each test node, but this method is difficult to use since the constraints have to be counted any time the number of components is increased. However, the obvious way to handle such hardware configuration is using of dynamic algorithms.

The criterion to distinguish dynamic algorithms is the adaptation to system load. These algorithms monitor and use information about the load of the system before making the distribution decision. The distribution function D is in this case a maximum function over the memory and CPU availability, which selects the next home the one with maximal resource availability. Some of them, the heuristic algorithms even change their policies according to the load of the system; in this case D takes in account a resource consuming threshold. The dynamic algorithms base on thresholds imposed on resource consuming. The threshold can by either the consuming level of a single resource (i.e. memory) or a key performance parameter counted according to a formula which considers several parameters. Dynamic algorithms require, unfortunately, extra activities (i.e., hardware monitoring) on the test nodes, hence the overhead is also bigger than for static algorithms. Also the updates on hardware consuming add some communication overhead when a new component is instantiated. The update on hardware usage may be realized only before a component creation or periodically according to an update rule. The periodical update might be combined with heuristic methods to count the refresh rate, for example the more loaded nodes should have longer delay between updates than nodes which have fewer loads. These algorithms work very well for tests using one client per component since the distribution may take into account the hardware consuming level before any component creation. If the sequential of interleaved behaviour pattern are used, it is recommended to wait a short period of time between component creations until the component reach the average resource consuming level. This approach is based on the assumption that the maximal (or average) resource consuming level for a component remains constant at emulation of sequential clients.

Another category of algorithms are the prediction based algorithms where the decision of deploying test components, formally defined as distribution function D, is based on some preliminary predicted information. For prediction purpose, we have to decide before the start of the test, how many test components to deploy on each node. The preliminary information should also reflect the test component resource consuming. In order to provide this information to the scheduler at the beginning of the test, we should run a small preliminary test to learn something about the behaviour of the test components. From this preliminary test we can measure parameters like: the amount of memory that the test process allocate on each host, the time needed to execute the test behaviour, the maximum amount of memory that a component allocates (the hot-spot). Considering these parameters we may distinguish between two categories of algorithms that can be implemented: memory based prediction algorithms and time based prediction algorithms. Memory management is very important in distributed testing because test components deal with important memory consuming. In this case, running a test on a host which does not provide the necessary amount of memory for test process could lead to a slower execution or a run out of memory exception. For time prediction based algorithms the decision criteria is based on a time factor proportional with the time duration of the component behaviour. To obtain this duration we should measure the time duration of a test component on each node. It is very important to measure the time duration of the same behaviour on each node. The preliminary test can be executed with one or more components on each node and after every execution an estimated time value (average value) should be profiled from each node. Based on these values for each node, the distribution should be made proportional with the time factor which indicates the number of components deployed on each node.

The control of a load balancing algorithm can be centralized, distributed or semi-distributed. A centralized approach is quite efficient as long the load balancer does not get overwhelmed itself by the request handling task. The distributed approach involves multiple load balancers in decision making. The semi-distributed approaches combine the centralized and distributed approaches; there are several load-balancers which group together multiple server instances and manage them in a centralized way. In our environment we experienced only with the centralized approach since for load testing purpose, the decision making does not add too much computational overhead. We used for our test the sequential or interleaved behaviour specification patterns, which imply that the components are created sequentially. This approach does not necessary require the component to be created very fast, since actually the most important issue is to use efficiently the resources and reach the load level after an undefined period of time.

Depending on the algorithm, the communication overhead is added by the algorithm for distribution and, consequently, the test system requires more resources. In the implementation architecture of the execution platform we presented, the SessionManager is the central entity responsible for the balancing of the test components. We implemented the different distribution strategies within the SessionManager which provides a distribution interface to all test daemons. This interface permits daemons to ask the SessionManager before each test component creation where to deploy that component. Therefore, the distribution operations using the static and prediction based algorithms add only a small communication overhead represented by the request for home location. The dynamic algorithms add a considerable communication overhead since the SessionManager has to be updated by each test node with the level of resource consumption.

6 An Example

In order to experiment with different categories of distribution algorithms we considered a Web server application and designed a load test suite.

The SUT application is a small Web application running on an Apache server. The application presents to its clients two different search forms: for cars and for houses. The information requested by the client is searched in a MySql [12] database. In a typical scenario, as depicted in Figure 2, the user accesses the main page of the SUT. The SUT time, on SUT side, is the time the Web server needs to deliver the main page to the user. The client thinking time on the user side is the time the client needs to read the main page and decide to search a car or a house. When the selection is made, a new request is sent to SUT which delivers back the search form. After another thinking time (to fulfil the form) the user sends the fulfilled search form to SUT. The SUT performs a search in the database and organizes the list of found items in a HTML page. Finally, the result is returned to the client. The search operation can obviously repeat for several times for any client.

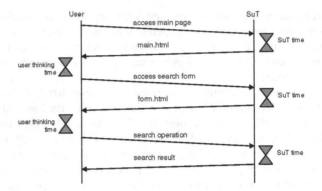

Fig. 2. The sequence of interactions between user and SUT

The design of the load test has the goal to emulate the parallel behaviour of a number of clients which is given as parameter to the test. Any client follows the interaction scenario presented before, but any client has arbitrary thinking times or number of searches in the database. The load (number of requests per second) is controlled by the MTC component which increases or decreases the number of components.

We implemented four distribution algorithms in order to experience with different categories of algorithms.

- *Round-robin (RR).* is a static algorithm which selects the hosts in a sequential order. It proved to be a good algorithm when used on homogeneous environments.
- *Memory threshold combined with round-robin based algorithm (MT).* This is a memory based algorithm that evaluates the percentage of free memory from the Java virtual machine. The decision criterion is based on the available memory for the JVM process on each host and it always deploys a new component on the host with the most available memory. The memory threshold based algorithms used alone could lead to the decision to deploy all components on the same node if the number of components is relatively small and the memory of one host is fairly bigger than on any other host. To avoid situations where all components would be deployed on the same node, the memory threshold algorithm should be combined with round-robin distribution in order to ensure that components will be distributed.
- *Memory factor based (MF).* This algorithm considers the number of test components to be deployed on each host to be proportional with the amount of memory on that host. The rule of deploying components is based on a memory factor that indicates how many components to deploy on each host. For obtaining the memory factor it is necessary to execute a preliminary calibration test for profiling the memory hot-spot.
- *Execution time factor based (TF).* The decision criterion of this algorithm is based on a time factor associated to the behaviour of a client running on a component. To

obtain this duration we should measure the execution time duration of a test component on each node. It is very important to measure the time duration of the same behaviour on each node. The preliminary test can be executed with one or more components on each node and after every execution an estimated time value (average value) should be profiled from each node. Based on these values for each node the distribution should be made proportional with the timeFactor which indicates the number of components deployed on each node in a sequence.

To compare the distribution algorithms we run the load tests and measure the computation time needed by the Test System between receiving a response from the SUT and processing it. This time usually increases with the number of components deployed on the same host. Depending on the algorithm and hardware resources, this time increases differently on the test nodes. The evaluation criterion considers that the best algorithm is the one which makes the computation time stay as small as possible on each node and the computation times grow up uniformly on the test nodes. All the graphs presented next have represented on the vertical axis the time needed for computation and on the horizontal axis the number of components deployed on that host. The test nodes evolution curves are associated with test nodes through dashed lines. The tests are executed on three computers with different hardware resources: TestNode1 (mem=512Mb, cpu=1.9 Ghz), TestNode2 (mem=2G, cpu= 2 x 3.5 Ghz), TestNode3 (mem=1G, cpu=3.5Ghz).

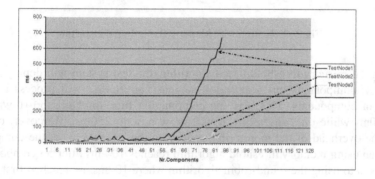

Fig. 3. Execution with Round-Robin algorithm

The Round-Robin algorithm distributes equally the number of components on the three test nodes. We observe that the computation times on the TestNode2 and TestNode3 grow very slowly to a negligible value while on TestNode1 they start growing up after deploying 60 components reaching at the end a computation time of 700ms. This delay may considerable influence the behaviour of the test making possible that some timers will timeout due to a long delay at processing the information received from SUT.

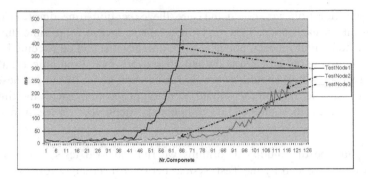

Fig. 4. Execution with Memory Threshold based algorithm

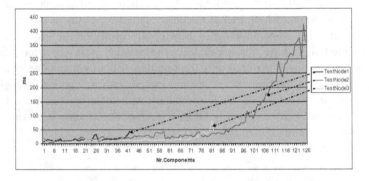

Fig. 5. Execution with Memory Factor based algorithm

Using the memory threshold based algorithm, the test system will deploy a bigger number of components on `TestNode2` so that `TestNode1` will process a smaller number of components. This way the computation time on `TestNode1` will reach only 450ms while on `TestNode2` it will grow now to 250ms. Moreover, one may notice the overhead of the monitoring system on the `TestNode1`; the computation time when using memory threshold algorithm reaches 100 ms after 53 components in comparison to using the round-robin algorithm where 100ms are reached first after 64 components.

A better result is obtained with Memory Factor based algorithm which will deploy a very small number of components on `TestNode1` and `TestNode2`. `TestNode3` will process 126 components but because of its good hardware resources the computation time will reach only 450 ms.

Even better results, are obtained by using the Time Factor based algorithm, which puts more components (almost the same number) on `TestNode2` and `TestNode3` so that `TestNode1` remains with a small number of components. This strategy affects again the performance on `TestNode1` but the computation time reaches only 400 ms which is less than the maximum obtained by the other algorithms.

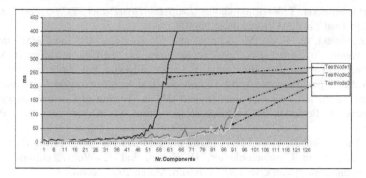

Fig. 6. Execution with execution time factor algorithm

7 Summary

This paper presents a study on applying the TTCN-3 technology for load testing. It introduces the language elements of TTCN-3 which can be used in test specification and discusses several patterns to specify load tests. As far as the execution of TTCN-3 load tests is concerned, the distribution of parallel test components on different test nodes is considered. The distribution is an interesting research topic since many strategies to balance the load can be applied and the balancing algorithms may influence the overall execution of a test. In this respect, we presented various factors which influence the efficiency of test component distribution and discuss different categories of load balancing algorithms. An emerging research subject is to establish theoretically which algorithms are better for special cases of test patterns.

An implementation architecture of the execution environment is also described. In order to experiment with several load balancing algorithms, a load test for a Web server application was performed. The results of experiments show how the distribution strategy influences the overall performance of the test system.

References

[1] B. A. Shirazi and K. M. Kavi and A. R. Hurson, *Scheduling and Load Balancing in Parallel and Distributed Systems*, 1995, ISBN = 0818665874, IEEE Computer Society Press

[2] ETSI ES 201 873-1 V3.1.1, Methods for Testing and Specification (MTS); The Testing and Test Control Notation version 3; Part 1: TTCN-3 Core Language, Sophia Antipolis, France, July 2005.

[3] ETSI : TTCN-3 Homepage, http://www.TTCN-3.org, June 2005.

[4] H. W.Neukirchen, *Languages, Tools and Patterns for the Specification of Distributed Real-Time Tests*, Dissertation, Universität Göttingen, November 2004 (electronically published on http://Webdoc.sub.gwdg.de/diss/2004/neukirchen/index.html and archived on http://deposit.ddb.de/cgi-bin/dokserv?idn=974026611)

[5] I. Schieferdecker, B. Stepien, A. Rennoch: *PerfTTCN, a TTCN language extension for performance testing*, in IWTCS'97 Proceedings, Cheju Island, Korea

[6]   I. Schieferdecker, T. Vassiliou-Gioles: *Realizing Distributed TTCN-3 Test Systems with TCI*. In TestCom 2003 Proceedings: 95-109

[7]   J. Grabowski, B. Koch, M. Schmitt and D. Hogrefe, *SDL and MSC Based Test Generation for Distributed Test Architectures*, In: 'SDL'99 - The next Millenium' (Editors: R. Dssouli, G. v. Bochmann, Y. Lahav), Elsevier, June 1999.

[8]   T. Walter, I. Schieferdecker and J. Grabowski, *Test Architectures for Distributed Systems - State of the Art   and Beyond*, Invited talk in: Testing of Communicating Systems (Editors: A. Petrenko, N. Yevtuschenko), volume 11, Kluwer Academic Publishers, 1998.

[9]   Z.R. Dai, J. Grabowski, and H. Neukirchen. TIMEDTTCN-3 -- A Real-Time Extension for TTCN-3. In I. Schieferdecker, H. Konig, and A. Wolisz, editors, Testing of Communicating Systems, volume 14, Berlin, March 2002. Kluwer.

[10] ETSI ES 201 873-5 V1.1.1: "The Testing and Test Control Notation version 3; Part 5: TTCN-3 Runtime Interface (TRI)", February 2003.

[11] Draft ETSI ES 201 873-6 V1.0.0: "The Testing and Test Control Notation version 3; Part 6: TTCN-3 Control Interfaces (TCI)", March 2003

[12] MySQL Homepage, http://www.mysql.com

[13] R. Binder: Testing Object-Oriented Systems: Models, Patterns and Tools. Addison-Wesley, 2000.

[14] C. Alexander, S. Ishikawa, M. Silverstein, M. Jacobson, I. Fiksdahl-King, and S. Angel. A Pattern Language: Towns, Buildings, Construction. Oxford University Press, 1977

[15] E. Gamma, R. Helm, R. Johnson, and J. Vlissides. Design Patterns. Elements of Reusable Object-Oriented Software. Addison Wesley, 1995

Analyzing the Impact of Protocol Changes on Tests

Mahadevan Subramaniam[1] and Zoltán Pap[2]

[1] Computer Science Department,
University of Nebraska at Omaha,
Omaha, NE 68182, USA
msubramaniam@mail.unomaha.edu

[2] Department of Telecommunications and Media Informatics – ETIK,
Budapest University of Technology and Economics,
Magyar tudósok körútja 2, H-1117, Budapest, Hungary
pap@tmit.bme.hu

Abstract. Protocols governing communication among system components evolve during design and maintenance and need to be re-tested. For faster testing turnaround time, it is important that the consistency of the testing infrastructure with the protocol be preserved across changes. In this paper, we propose a state exploration based approach to identify the impacts of protocol changes on a given set of protocol tests. Protocols are modeled as a network of communicating finite state machines exchanging messages over bounded queues. Each machine denotes the behavior of an individual protocol component (*controller*). A protocol test is modeled as a sequence of inputs from the environment to the protocol controllers in an execution starting from a stable protocol state. A notion of *consistency* of a test relative to a protocol is introduced. Conditions under which a protocol change requires changing a test to preserve the consistency of the test are identified. Changes consisting of multiple atomic updates are analyzed to remove redundancies and their impact on tests is studied. A by-product of the proposed approach is a classification of tests based on how they are impacted by protocol changes, which can help users in regression test selection.

Keywords: Changes, evolution, protocol, communicating finite state machines, test consistency.

1 Introduction

Testing of protocols has been extensively studied [1,2,3,4] due to the central role played by protocols in both software and hardware. Typically, a protocol may undergo several changes during the system design and maintenance phases and may need to be repeatedly re-tested. Generating and running tests in response to every protocol change is time consuming and may also not lead to good cumulative coverage of protocol functionality across the changes. To re-test a changed protocol it is important to analyze effects of changes on existing tests.

M.Ü. Uyar, A.Y. Duale, and M.A. Fecko (Eds.): TestCom 2006, LNCS 3964, pp. 197–212, 2006.

Changes to a protocol may affect existing tests in several ways. Several existing tests may be re-usable with changed protocol. However, only some of these tests may exercise changed behaviors (in addition to original ones) whereas the rest may exercise only original behaviors. Determining whether a re-usable test exercises a changed behavior may be useful in generating new tests. Some existing tests may also become unusable for certain changes. It may be possible to use some of these tests to test the changed protocol after patching while others may have to be simply discarded. Given that a large number of tests are typically needed to validate even simple protocols, manually determining impacts of changes on the existing tests is tedious and error prone. Determining such impact may not be easy even for a single test since this may require considering all executions that can happen in the test, which may be numerous.

In this paper, we propose a white-box[1] approach to automatically determine the impact of protocol changes on protocol tests. We model protocols as a network of interacting communicating finite state machines (*CFSM*s) with bounded queues [5, 6, 7]. The transition relation of each *CFSM* describes the behavior of an individual protocol controller. The *CFSM*s interact with each other by message exchanges over the queues. They also interact with an external environment by exchanging messages over the queues to/from the environment to the controllers. We consider protocol changes at the transition level. Protocol changes add/delete/replace one or more transitions from one or more protocol controllers and are explicitly represented as change specifications [8].

We formalize a protocol test as *a protocol stable state along with a set of environment inputs*. The environment inputs are used in the executions starting from the associated protocol stable state. A simple notion of *consistency* of a test relative to a protocol is proposed. Informally, a test is consistent with a protocol if it exactly specifies the environment inputs needed in each execution that can happen in the test starting from the protocol stable state. At least one execution must happen in each consistent test. Further, since a consistent test exactly specifies the inputs used in their executions each execution can happen in at most one consistent test.

Changes can potentially produce protocols whose executions cannot happen in any consistent test. We consider such changes uninteresting. A notion of *consistently testable* protocol is introduced. In a consistent testable protocol, it is possible to design a consistent test for each execution such that the execution happens in the consistent test. Consistent testability ensures that each protocol execution can happen in exactly one consistent test.

We constrain changes to preserve consistent testability of protocols. We characterize the impact of a change on an existing test in terms of whether the change preserves the consistency of the test. To check whether a change preserves the consistency of an existing test, it is enough to just find one execution

---

[1] A testing environment consists of a set of protocol tests, along with a set of runtime monitors derived from protocol specifications and/or correctness properties. The runtime monitors observe sequences of global states during execution and check if each protocol execution in each protocol test conforms to the specification.

of the changed protocol that can happen in the test such that the execution uses all the inputs specified in the test. Otherwise, the test is *inconsistent*. This is because for a consistent testable protocol, if a test is consistent for a single execution that can happen in the test then the test must be consistent for all executions that can happen in the test since otherwise, that execution cannot happen in any consistent test. Consistent testability makes it easier to check whether consistency of a test is preserved by a change since it is enough to check this for just one execution of the changed protocol. For certain changes it may not be necessary to analyze even a single execution. For instance, see the discussion for changes that add transitions below.

Tests whose consistency is preserved by a change may either be *independent* of the change or they may be *re-usable*. Informally, a test is *independent* of a change if none of the executions that can happen in the test are affected by the change. Independent tests may be discarded to minimize the testing effort across changes. All other tests whose consistency is preserved by the change are re-usable tests. Re-usable tests exercise the changed behavior and can be used without any modifications to test the changed protocol.

State exploration is used to determine whether tests whose consistency is preserved by a change are independent. This is done by computing an *interaction context* for a given protocol transition. Interaction context of a given protocol transition is the set of all consistent tests such that the transition appears in some execution of each test. Informally, if a test whose consistency is preserved by a change does not belong to the interaction context of transitions being changed, then it is not affected by the change and is independent. If the test belongs to the interaction contexts of a new transition introduced by a change then it is re-usable.

If a test is made inconsistent by a change and the protocol stable state of the test is a stable state of the changed protocol then the test is patched by changing its inputs to be the external inputs used in the executions starting from that stable state of the changed protocol. A patched test is generated for each such execution with distinct environment inputs. Inconsistent tests whose protocol stable state is not a stable state of the changed protocol are not patchable and hence discarded. Interaction contexts of protocol transitions are used to patch an inconsistent test whenever possible to do so. We also describe a state exploration based procedure based on interaction contexts for checking consistent testability.

We first consider single transition changes and show that a change that adds a new transition t'_i to a controller, preserves the consistency of all existing tests. This is because such a change preserves all the executions of the original protocol in addition to preserving consistent testability. An existing consistent test γ is re-usable with respect to such a change if it belongs to the interaction context of the newly added transition t'_i, which is computed over the changed protocol. Otherwise, the test γ is independent of the change.

Consider a change replacing a transition t_i by another transition t'_i in the same controller. To determine the impact of such a change on an existing consistent test γ, the interaction context of transition t'_i and the interaction contexts of the transitions t_js in the original protocol that are distinct from t_i are computed. All the contexts are computed by state over the changed protocol. The test γ is re-usable if it belongs to the interaction context of t'_i and γ is independent if it does not belong to the context of t'_i but belongs to the context of some t_j. In this case, an execution of the original protocol without the replaced transition t_i can happen in the test γ. Otherwise, the test γ is inconsistent. The test is patched, as described above, if the protocol stable state of γ belongs to any of the interaction contexts of t_j's. The test γ is discarded otherwise.

It is in practice inevitable to deal with complex changes consisting of multiple atomic updates. Our method of analyzing the impact of multiple changes on tests is based on examining the effect of the comprised individual add / delete / replacement changes. However, the approach developed for the single transition change cases can not be used as it is, some modifications are needed to handle certain types of independence of the different atomic changes. If all the atomic updates are only additions then the effect on a test can be determined by considering the individual interaction contexts independently. However, for replacements and deletions effects on a test can not be determined independently since the inconsistency of a test whose executions contain a group of replaced protocol transitions may not be detected. To avoid this problem, the combined interaction contexts of the protocol transitions being replaced is used to determine the impact on a test.

As multiple changes are usually defined during the development process in an ad hoc manner, they are often confusing and redundant. The proposed approach also includes a novel approach to reduce complex changes and create the shortest – or in an other sense optimal – sequence of update rules inducing the required modifications.

The paper is organized as follows. After a brief discussion on the related literature in Section 2, we introduce the background of the current research including our model of protocols in Section 3. Protocol tests and the notion of test consistency are introduced in Section 4. Section 5 presents a framework for analyzing the impact of transition changes on tests and studies the effect of single changes. Section 6 describes handling and impact of multiple transition changes. Section 7 concludes the paper.

2   Related Work

Both test generation and software change impact analysis have been active areas of research in the past. Testing is an indispensable phase of a system development lifecycle, yet it has turned out to be a difficult task for the increasingly complex protocols. Because of their practical importance, significant effort has been devoted to the development of automatic test generation methods for protocols to overcome the inefficiency of manual testing [1,2,3,4]. Furthermore, with the constant increase in complexity, protocol design and implementation has become an increasingly evolutionary and iterative process. Motivated by this trend there have been some studies investigating changes and their impacts [9,10,11].

Several researchers have focused earlier on evaluating the effect of system changes on tests [10,11]. However, surprisingly, not much attention has been paid to evolution of protocols and the impact of protocol changes on tests. To the best of our knowledge, this is perhaps the first attempt towards analyzing the impact of CFSM-based protocol changes on protocol tests. Further, most of the earlier work on change impact analysis has been based on conservative static analysis and hence does not provide precise answers. In contrast, the proposed approach is based on formal approach based on state exploration that allows for accurate analysis of change impacts and also enables us to provide more useful feedback by synthesizing new tests that are guaranteed to be consistent with the changes.

The approach proposed in this paper builds on our earlier work in [8, 12]. In [8] we have developed a systematic approach to specify and consistently incorporate changes to CFSM-based protocols. This approach has been applied to several protocols including industrial-strength cache coherence protocols. In [12] we showed how the impact of consistent protocol changes on runtime monitors can be automatically evaluated and new monitors can be synthesized. Typically, in white-box testing environments, monitors are used in conjunction with tests to observe and flag errors in protocol executions. This paper shows how change impacts can be analyzed for tests and in this sense fills an important gap for evaluating impact of protocol changes on the testing infrastructure.

3   Preliminaries

A protocol $P = (P_1, \cdots, P_n, \epsilon)$ is a network of *CFSM*s [5,6], where each P_i is a protocol controller and ϵ is the environment. Communication among P_i's and ϵ is achieved by exchanging messages over a network of bounded queues.

Each protocol controller P_i is a *CFSM*, a 4-tuple, (S_i, I_i, M_i, T_i) where S_i is a finite set of states, $I_i \in S_i$ is an initial state, M_i are the messages sent and received by P_i, and $T_i \subseteq M_i \times S_i \mapsto S_i \times 2^{M_i}$ is a deterministic transition relation. Each set $M_i = M_{i,j} \cup M_{j,i}$, for $\{i,j\} \in \{1, \cdots, n, \epsilon\}$, $i \neq j$, where $M_{i,j}$ is the set of messages sent by the controller P_i to the controller P_j and $M_{j,i}$ is the set of messages received by controller P_i from P_j. Each message in $M_{i,j}$ is written as $m_k(i,j)$ where m_k is the message label and (i,j) denotes the message queue from controller P_i to controller P_j. The set $M_{i,i}$ is empty for all i; we assume that no P_i exchanges messages with itself.

Each protocol transition t_i in T_i is of the form $m_0(j,i)$, $s_i \mapsto s_i'$, $\{m_1(i,k_1)$, $\cdots, m_l(i,k_l)\}$ where m_0 is the input message, s_i and s_i' are the input and output states respectively, and $m_1, \cdots, m_l$ are the output messages to the controllers $k_1 \cdots k_l$, $1 \leq l \leq n$, $l \neq i$. Let $S = \bigcup_{1 \leq i \leq n} S_i$ be the set of protocol states.

A *global protocol state*, $g = \langle u, v \rangle$ is a pair where u is an n-tuple of the individual controller states and v represents the messages, if any, in the protocol message queues. Let $u[j]$ stand for the state of the controller P_j in u and $v[i,j]$ stand for the messages in the queue (i, j) in v. For ease of exposition, we will only show non-empty queues in v in a global protocol state. We write $v = \langle \rangle$ when we all queues are empty in the state g.

The *initial global protocol state* $g_0 = \langle u_0, v_0 \rangle$ is a pair where $u_0[i] \in I_i$ for all i, and $v_0[i, j] = \langle \rangle$ if $i \neq \epsilon$ and $j \neq \epsilon$. So, the only non-empty queues in an initial global protocol state are those containing messages from/to the environment ϵ.

A *stable global protocol state* is a global protocol state $g = \langle u, \langle \rangle \rangle$ where all the queues are empty. We call the elements of S in u in a stable global protocol state as *stable controller states*. Note that the initial global protocol state is a stable global protocol state but not vice versa.

A transition t_i: $m_0(j, i)$, $s_i \mapsto s_i'$, $\{m_1(i, k_1), \cdots, m_l(i, k_l)\}$ is *enabled* in global state $g = \langle u, v \rangle$ if $u[i]$ is the input state s_i and the top of queue $v[j, i]$ is the input message $m_0(j, i)$, and all the output messages $m_1(i, k_1), \cdots, m_l(i, k_l)$ can be enqueued (there is space to do so) in the corresponding queues in v.

An *execution step*, $g \to^{t_i} g'$ using the transition t_i enabled in the global protocol state g produces a global protocol state g' with the controller P_i in the state s_i'; other controller states are unchanged. The input message $m_0(j, i)$ is dequeued from the queue $v[j, i]$ and the output messages $m_1(i, k_1), \cdots, m_l(i, k_l)$ are enqueued to the appropriate queues in v to produce g'. If t_i is enabled in g then we say that the global protocol state g' is enabled by t_i. The state $g' = g$ if t_i is not enabled in g.

An *executable* path of P, $r_1 = g_0 \to^{t_0} g_1 \cdots \to^{t_k} g_{k+1}$. We will simply represent the path r_1 by a sequence of transitions $[t_0, \cdots, t_{n-1}]$.

A protocol *run* is an executable path starting and ending in stable global protocol states. Each protocol run starts from a stable global protocol state by processing the environment input messages leading to further message exchanges until a stable global protocol state is reached.

Protocol P is *consistent* if and only if i) no two protocol transitions t_i and t_j have the same input state and same input message, ii) every t_i appears in a protocol run, and iii) every executable path from a stable global state reaching a state enabling a transition t_i is a prefix of a protocol run.

Henceforth, we will assume that protocols are consistent. Note that in our model, protocol controllers are not input-enabled since they communicate with each other by exchanging messages over bounded queues. Our model thus more closely relates to the process algebraic than the I/O model of concurrency.

Changes to protocols are specified as a finite set of rules, *update rules* (ur), that may add, delete, or replace one or more transitions in one or more controllers. ur_j: $t_i \Rightarrow t_i'$ replaces a transition t_i by a new transition t_i', ur_j: t_i deletes a transition t_i, and ur_j: t_i' adds a transition t_i' in controller P_j. We consider changes with both single as well multiple rules.

It is assumed that all changes preserve the consistency of a protocol.

4   Tests and Consistently Testable Protocols

We primarily focus on white-box testing in this paper [13]. In our testing model, a protocol implementation under test is stimulated by the environment ϵ by inputting messages. The implementation processes the messages one at a time, starting from a stable global protocol state resulting in a protocol test run.

To ensure conformance of the implementation, a runtime monitor [14, 15] is added to the implementation. The monitor observes certain global protocol states in each run in a test. Starting from a conforming monitor state, the monitor transitions with each observed global protocol state ending in a conforming monitor state if the test run is behaviorally conforming to the protocol specification; the monitor ends in an error state for non-conforming test runs. The global protocol states observed by the monitor in a test run depend on the conformance properties being checked.

A protocol *test* $\gamma = \langle u, v \rangle$ is a pair where u is an n-tuple of controller stable states and v is an n-tuple of queues containing input messages from environment ϵ to the n individual protocol controllers. For ease of exposition, we will only show non-empty queues in v while describing a test γ.

A *test run* of the test γ is a protocol run starting from the stable global protocol state $\langle u, \langle\rangle \rangle$ that processes all the environment input messages appearing in v. In general, test γ may have several test runs based on the different interleavings of the protocol transitions.

A test γ *exercises* a transition t_i if t_i appears in a test run of γ.

A test γ is *consistent* relative to a protocol if there is at least one test run for γ and every executable path starting from γ is extensible to a test run of γ.

Note that if the test γ is consistent then $v \neq \langle\rangle$ since protocol runs always start with an environment input message. Further, the consistency of a test ensures that every execution path starting from the stable state corresponding to that test can be extended to a run that uses all the environment input messages specified in test γ. More importantly, consistency of a test ensures that consistent tests that share a test run must be identical.

Proposition 1. *Consistent test $\gamma_1 = \gamma_2$ if test run of γ_1 is a test run of γ_2.*

Proof Sketch. Let r_1 be the common test run that starts from the stable global stable state u. Let v be the external environment input messages used in run r_1. Let $\gamma_1 = \langle u_1, v_1 \rangle$ and $\gamma_2 = \langle u_2, v_2 \rangle$. Since r_1 is a test run of both γ_1 and γ_2, $u_1 = u_2 = u$ and since both γ_1 and γ_2 are both consistent test, $v_1 = v_2 = v$. $\square$

From Proposition 1, it follows that a protocol run is a test run of at most one consistent test. An example of a consistent and an inconsistent test can be found in Section 5.2 and Section 4.1, respectively.

In our testing model, the possible outcomes of a test are all the observable sequences of global protocol states that can happen in all the test runs. Since the global protocol states are directly accessible to the runtime monitor they are not a part of a test. Our notion of consistency of a test is closely related to the notion of *valid tests* that has been extensively studied earlier in the testing of protocols and FSMs. An overview of FSM and protocol testing may be found in [13]. A valid test is a consistent test whose outcome conforms to the protocol specification. However, a consistent test need not be a valid test since consistency does not require a conforming test outcome. A consistent test however, precludes invalid or inconclusive tests [13] with inputs that cannot be processed in any run.

Further, in our model, the verdict associated with a test is determined by the runtime monitor, based on the global states observed by the monitor. Hence on changing an implementation, it may be necessary to change both the monitor as well as the tests. In this paper, we assume that any change to the protocol implementation is preceded by a change to a specification and leads to an appropriate change to the monitor[2]. We only focus on impact of implementation changes on the test itself. Henceforth, in this paper, we do not explicitly distinguish a protocol implementation under test from its specification and simply refer to the former as the protocol.

Our choice of the above white-box testing model is largely motivated by the experience of first author in designing real cache coherence, network, and I/O protocol products. In practice, many protocols work over bounded resources and hence are not input-enabled. For protocols with input-enabled components, every test is trivially consistent. In such cases, the approach in [12] may be used to determine how the change affects the test purpose.

4.1   Consistently Testable Protocols

As mentioned above, consistency of protocol tests ensures that any protocol run is a test run of at most one consistent test. To test the conformance of a protocol, we must also make sure that every protocol run is a test run of some consistent test, *i.e.*, it must be possible to devise a consistent test to check each protocol run. A protocol is *consistently testable* if each protocol run is a test run of a consistent test. Consequently, each protocol run of a consistently testable protocol is a test run of exactly one consistent test.

A protocol may be consistent but it may not be consistently testable.

Example 1: Consider the following protocol with the controllers P_1 and P_2 with respective controller stable states $\{s_0\}$ and $\{t_0\}$ with transitions,

$$P_1 : 1. \; m_0(\epsilon, 1), \; s_0 \mapsto s_1, \; m_1(1, 2), \quad 2. \; m_2(2, 1), \; s_1 \mapsto s_0, \; m_3(1, 2),$$
$$ 3. \; m_9(2, 1), \; s_1 \mapsto s_0, \; m_4(1, \epsilon).$$
$$P_2 : 4. \; m_1(\epsilon, 2), \; t_0 \mapsto t_2, \; m_9(2, 1), \quad 5. \; m_1(1, 2), \; t_0, \mapsto t_1, \; m_2(2, 1),$$
$$ 6. \; m_3(1, 2), \; t_1 \mapsto t_0, \; m_4(2, \epsilon), \quad 7. \; m_1(1, 2), \; t_2 \mapsto t_0, \; m_5(2, \epsilon),$$
$$ 8. \; m_5(\epsilon, 2), \; t_1 \mapsto t_1, \; m_6(2, \epsilon).$$

It can be verified that the above protocol is consistent. However, the protocol is not consistently testable since the protocol run $r_1 = [1, 4, 3, 7]$ is not a test run of any consistent test. This is because for r_1 to be a test run, the test must contain the environment input messages $m_0(\epsilon, 1)$ and $m_1(\epsilon, 2)$. Such a test is not consistent since it can lead to the run $r_2 = [1, 5, 2, 6]$, which does not use the input message $m_1(\epsilon, 2)$ and cannot be further extended to a run ending in a stable global protocol state while doing so.

---

[2] In an earlier paper [12], we have shown how impact of protocol changes on runtime monitors can be automatically evaluated. The procedure described there uses selective state exploration guided by protocol states observed by the monitor to identify and automatically synthesize monitors for protocol changes. For more details the reader may please refer to that paper.

In general, consistent testability imposes additional constraints on the runs of a consistent protocol, which can be used to check whether a given protocol is consistently testable.

In a consistent testable protocol, for any two protocol runs r_1 and r_2 that start by processing the same environment input message in the same stable state, the set of environment inputs processed by r_1 (r_2) should be a subset of those processed by r_2 (r_1). If the environment messages used in r_1 (r_2) is a proper subset of r_2 (r_1) then the extra messages in r_2 (r_1) should only transition the protocol from one stable controller state to another.

In principle, to verify consistent testability of a given protocol, it suffices to find two runs r_1 and r_2 that violate the above condition. For instance in the above protocol, the runs r_1 and r_2 start from the same global stable state $\langle\langle s_0, t_0\rangle, \langle\rangle\rangle$, and process the same environment input message $m_0(\epsilon, 1)$ in that stable state. However, the environment messages processed by r_2 is a proper subset of those processed by r_1; run r_1 additionally processes the environment input message $m_1(\epsilon, 2)$, which does not transition the protocol from one stable state to another. Hence the protocol is not consistently testable as shown above.

A more feasible approach for verifying consistent testability of a given protocol is to compute the interaction contexts of the transitions of the protocol by state exploration and analyze these contexts for the above described condition by considering messages of context elements with the same stable state. This is described in the next section.

5   Impact of Single Transition Changes on a Test

In this section, we describe how to determine the impact of addition, replacement, and deletion of single protocol transitions on an existing protocol test. Conditions under which the consistency of a test is preserved by a change are identified. If such a test exercises the change then it may be re-used without any additional modifications. Otherwise, the test is independent of the change and may not be included to test the changed protocol[3].

In this paper, we assume the changes are incorporated into protocols only if they preserve the consistent testability. In principle, this can be checked by considering the runs of the changed protocol starting from the same controller stable state and checking whether their environment inputs are a subset of each other as described above.

5.1   Interaction Context of Transitions

To determine the effect of changes on tests an *interaction context* is associated with each protocol transition.

The interaction context of a transition t_i, $IC(t_i) = \{\langle u_j, v_j\rangle\}$ is the set of all tests $\langle u_j, v_j\rangle$ exercising the transition t_i.

---

[3] Of course, such tests may be included in regression testing of the changed protocol based on several coverage criteria and these are not considered here.

The interaction context $IC(t_i)$ is computed by doing repeated backward and forward image computations over the global protocol state space starting respectively from the global state enabling and the state enabled by the transition t_i. The computations stop once all reachable global states with controller stable states and only environment inputs are obtained. The set of global stable protocol states produced by the forward and the backward computations are then matched to produce global stable states and the environment inputs and form one pair of the interaction context. The context $IC(t_i)$ is the set contains all such pairs produced by the matching stable global protocol states.

For instance, consider computing the interaction context $IC(5)$ of the transition 5 in the protocol example described in the previous section. The backward image computation starts from the global state enabling transition 5, $g_p = \langle\langle xs_1, t_0\rangle, \langle m_1(1, 2)\rangle\rangle$ where xs_1 is a symbolic state variable denoting the controller $P_1's$ state. One step image computation of state g_p with transition 1 gives the global state $g_1 = \langle\langle s_0, t_0\rangle, \langle m_0(\epsilon, 1)\rangle\rangle$ and this step also instantiates the variable xs_1 to value s_1 to give the instantiated state $g_p = \langle\langle s_1, t_0\rangle, \langle m_1(1, 2)\rangle\rangle$. Since g_1 is a stable global protocol state with only environment inputs and there are no other predecessors of g_p, the backward image computation stops.

Similarly, the forward image computation starts with the state $g_s = \langle\langle xs_2, t_1\rangle, \langle m_2(2, 1)\rangle\rangle$, and after two image computation steps using the transitions 2 followed by transition 6 stops with the global state $g_2 = \langle\langle s_0, t_0\rangle, \langle\rangle\rangle$. The variable xs_2 is instantiated with value s_1 to produce the instantiated $g_s = \langle\langle s_1, t_1\rangle, \langle m_2(2, 1)\rangle\rangle$. The states g_1 and g_2 match since $g_p \rightarrow^5 g_s$ is an execution step for the instantiated states g_p and g_s. Hence the initial state g_1 is included in the context to produce $IC(5) = \{g_1 = \langle\langle s_0, t_0\rangle, \langle m_1(\epsilon, 1)\rangle\rangle\}$.

Interaction contexts may be used to check whether a protocol is consistently testable. To do so, we union the interaction contexts of the protocol transitions. Then, for each pair $\langle u_1, v_1\rangle$ and $\langle u_2, v_2\rangle$ such that $u_1 = u_2$ if v_1 and v_2 have the same prefix of environment inputs then we check that v_1 (v_2) is a subset of $v_2(v_1)$. If one is proper subset of the other then it is ensured that each extra message m_i is processed by a transition whose input and output states are controller stable states. This guarantees that the extra messages only transition the protocol among global stable protocol states.

For each change, we consider the transitions appearing in the change and compute their interaction contexts. The interaction context may be computed either based on the original or computed based on the changed protocol depending on whether we are adding, deleting or replacing a protocol transition. As explained below, for replacement changes the interaction context of the transition being replaced is computed based on the original protocol whereas that of the transition being added is computed using the changed protocol.

5.2   Adding a Transition

Consider an update rule $ur: t_i'$, that adds transition $t_i': m(j, i), s_i \mapsto s_i', m'(i, k)$ to controller P_i of a protocol P. The update ur allows the controller P_i to process the input message $m(j, i)$ in state s_i, which is not possible in the original protocol, and produces a consistent and consistent testable changed protocol. It should be clear that every protocol run of the original protocol is a run of the changed protocol since all transitions of P are also present in P'. Further, since the changed protocol must be consistent it also follows that the transition t'_i appears in at least one changed protocol run.

Let $\gamma = \langle u, v \rangle$ be any consistent protocol test of the original protocol. Since γ is consistent, a run of the original protocol and therefore, a run of the changed protocol is a test run of γ. Now, since the changed protocol is consistently testable, there is exactly one consistent test for each changed protocol run. Hence it follows that γ is a consistent test of the changed protocol.

The effect of such a change on the test γ is determined by computing the interaction context $IC(t'_i)$ of the newly added transition t'_i, by performing state exploration over the changed protocol as described above. If the test γ belongs to $IC(t'_i)$ then the test γ exercises the newly added transition t'_i. In this case γ is re-usable. If the test does not appear in $IC(t_i)$ then none of the executions of γ contain the transition t'_i and hence γ is independent of the change.

Example 2: Consider the following protocol with controllers P_1 and P_2 with stable states $\{s_0, s'_0\}$ and $\{t_0\}$ respectively, with the transitions,

$$P_1 : 1.\ m_1(2,1),\ s_0 \mapsto s_1,\ m_2(1,2),\quad 2.\ m_3(2,1),\ s_1 \mapsto s_2,\ m_4(1,2),$$
$$3.\ m_5(2,1),\ s_2 \mapsto s'_0,\ m_6(1,\epsilon),\quad 4.\ m_0(\epsilon,1),\ s'_0 \mapsto s_0,\ m_0(1,e).$$
$$P_2 : 5.\ m_0(\epsilon,2),\ t_0 \mapsto t_0,\ m_1(2,1),\quad 6.\ m_2(1,2),\ t_0, \mapsto t_0,\ m_3(2,1),$$
$$7.\ m_4(1,2),\ t_0 \mapsto t_0,\ m_5(2,1).$$

A consistent protocol test is $\gamma = \langle \langle s'_0, t_0 \rangle, \{ m_0(\epsilon, 2),\ m_0(\epsilon, 1) \} \rangle$; a test run for γ is $r_1 = [4, 5, 1, 6, 2, 7, 3]$. Suppose we add transition 8: $m_1(2,1),\ s'_0 \mapsto s_1,\ m_2(1,2)$ to controller P_1. It can be verified that this change preserves both the consistency and the consistent testability of the protocol. The interaction context of this new transition, $IC(8) = \{\langle \langle s'_0, t_0 \rangle, \{ m_0(\epsilon, 2),\ m_0(\epsilon, 1) \}, \langle \langle s'_0, t_0 \rangle, \{ m_0(\epsilon, 1) \} \}$, includes the test γ and hence the test γ is re-usable for the changed protocol. A test run of γ with the new transition is $[5, 8, 6, 2, 7, 3, 4]$.

Alternatively, we can add transition, $9 : m_7(\epsilon,2),\ t_0 \mapsto t_0,\ m_8(2,\epsilon)$ to the controller P_2 of the above protocol while preserving its consistency and consistent testability. Then, the interaction context $IC(9) = \{\langle \langle s_0, t_0 \rangle, \langle m_7(\epsilon, 2) \rangle, \langle s'_0, t_0 \rangle, \langle m_7(\epsilon, 2) \rangle \}$, does not include the test γ and it can be verified that test γ does not exercise transition 9 and hence is independent of this change.

Note that each pair in the interaction context $IC(t'_i)$ corresponds to a consistent test that exercises the newly added transition t'_i. For changes that add a single transition, we simply determine, which of these tests already exist for the original protocol. The remaining pairs may be used as new tests.

5.3   Replacement and Deletion of Transition

Consider an update rule $ur: t_i \Rightarrow t'_i$ that replaces a transition t_i in controller P_i with a new transition t'_i in the same controller. The update ur produces a consistent changed protocol in which there are runs containing the newly added transition t'_i. It also ensures that every transition t_j that appears in an original protocol run with the replaced transition t_i appears in some other runs not containing t_i. The changed protocol is also consistently testable.

To determine the effect of the update ur on a consistent test γ of the original protocol, we first determine whether γ exercises the new transition t'_i in the changed protocol. To do so, the interaction context $IC(t'_i)$ is computed over the changed protocol. If γ belongs to $IC(t'_i)$ then a test run of γ containing t'_i is a protocol run of the changed protocol. In this case, the update ur must preserve the consistency of γ since there is a run with t'_i in the changed protocol that can be tested only by using the test γ. Hence γ is re-usable.

However, if γ does not belong to $IC(t'_i)$ then it may no longer be a consistent test of the changed protocol. As an example, consider changing the protocol described in the previous subsection, using the rule ur: 4. $m_0(e,1)$, $s'_0 \mapsto s_0$, $m_0(1,e)$, $\Rightarrow 8$: $m_1(2,1)$, $s'_0 \mapsto s_1$, $m_2(1,2)$ that replaces transition 4 by transition 8. The changed protocol is consistent and consistently testable. However, the existing test $\gamma = \langle\langle s'_0, t_0\rangle, \langle m_0(\epsilon, 2), m_0(\epsilon, 1)\rangle\rangle$ is no longer consistent since no transition in the changed protocol can process the input message $m_0(\epsilon, 1)$. Note that a consistent test that exercises the new transition 8 is $\langle\langle s'_0, t_0\rangle, \langle m_0(\epsilon, 2)\rangle\rangle$.

In general, the replacement change ur makes a test γ inconsistent only if every test run of γ in the original protocol contains the replaced transition t_i. We can determine this by extending the interaction context computation to include the executable paths. Then, it can be checked that every executable path in the context $IC(t_i)$ computed over the original protocol contains the transition t_i.

Alternatively, we can consider each transition t_j distinct from t_i in the original protocol and compute the contexts $IC(t_j)$ and check that the test γ belongs to one of these contexts. This ensures that γ exercises a transition t_j different than t_i. To ensure that γ does not exercise t_i we simply compute these contexts over the changed protocol, where transition t_i has been replaced.

If the test γ does not belong to any $IC(t_j)$ then it is inconsistent for the changed protocol. In this case, we consider each pair $\langle u_j, v_j\rangle$ in each context $IC(t_j)$. If $u = u_j$ for some pair then we patch γ to generate a new test $\gamma_j = \langle u, v_j\rangle$. If no such pair is found then γ cannot be patched and is discarded.

Obviously, if γ belongs to neither $IC(t_i)$ nor $IC(t'_i)$ then no runs in the original protocol containing transition t_i are test runs of γ and no runs in the changed protocol containing t'_i are test runs of γ. As these are the only runs affected by the change, the test γ is consistent with respect to the changed protocol. In this case, the test γ is independent of the replacement change and may be discarded from the tests used for the changed protocol.

As an example, consider the protocol obtained from the protocol previous subsection after addition of the transition 8. This protocol can be changed using the replacement rule ur: $8 \Rightarrow 9$, that replaces transition 8 by the transition 9 described there. The change produces a consistent and consistently testable protocol. It can be verified that this replacement change preserves the consistency of the existing test $\gamma = \langle\langle s'_0, t_0\rangle, \langle m_0(\epsilon, 2), m_0(\epsilon, 1)\rangle\rangle$, since [4, 5, 1, 6, 2, 7, 3] is a run of the original protocol from γ not containing the replaced transition 8.

The impact of changes that perform deletion of a transition t_i is also determined by computing the interaction contexts of transitions t_j distinct from t_i as described above. The test γ is re-usable across such a change if it belongs to some context $IC(t_j)$; otherwise, γ is inconsistent. An inconsistent test γ is patched in the same way as described above.

6   Impact of Multiple Transition Changes on a Test

It is in practice inevitable to deal with multiple updates simultaneously. Some of the more general changes – for example the introduction of new states – are too complex to be specified by a single update rule, thus multiple rules – sequences of atomic rules – have to be used to define these changes. Furthermore, in any development process it is not practical to modify and/or analyze the test suite for the given protocol at each atomic update. Instead, test suites are revised at certain stages of the development, typically after some substantial changes have been introduced to the system.

Let $ur: \{t'_1, t'_2, \cdots, t'_i, t_{i+1} \Rightarrow t'_{i+1}, \cdots t_m \Rightarrow t'_m\}$ be any protocol update, denoting the set of transition changes to a given protocol P, where (primed) transitions t'_j's are added and (unprimed) transitions t_k's are deleted.

The effect of change ur on a protocol test γ is determined by considering the interaction contexts of the transitions in ur. The contexts of the (primed) transitions t'_j's are computed over the changed protocol and that of (unprimed) transitions t_k's are computed over the original protocol.

The protocol transitions t'_j's being added may be considered individually and the effect on the test γ may be determined as described in the previous subsection on addition of single transitions. However, determining the effect of deletion of transitions by considering transitions t_k's individually does not work since inconsistencies arising due to test runs containing multiple deleted transitions may be missed. For instance, let t_1 and t_2 be any two transitions in ur that are being deleted. Assume that every test run of γ contains either t_1 or t_2 but not both. Hence in this case, removal of both t_1 and t_2 must make the test γ inconsistent.

However, this will not be the case if we individually consider deletions since while considering deletion of t_1, the test γ will be consistent since there is a run with t_2. Similarly, individually considering t_2 will also lead to γ being consistent since there is a test run with t_1.

To handle this problem the procedure for handling individual replacements is modified to consider all the interaction contexts of the protocol transitions being replaced. Let $IC_r = \bigcup_{t_l \in ur} IC(t_l)$ be the union of all the interaction contexts of the transitions t_l's in the original protocol that are not being replaced. The multi-controller ur preserves the consistency of a test γ only if it belongs to IC_r.

6.1   Handling Redundancies in Multiple Updates

According to the discussion above, we sometimes have to consider significant changes to a controller involving large sequences of individual update rules. As thc changes are defined in an ad hoc development process, they are often unnecessarily complex and contain redundancies. In such cases the impact of the given update on tests can not be analyzed efficiently based on the original change specification. Instead, an equivalent multiple update is constructed, which is optimal in the sense that it is the best suitable for the analysis.

The essence of our approach is as follows: Let us consider that we are given a redundant update specification with multiple update rules, and we have to analyze its impact on a given test. We apply the specified update to compute a changed protocol and ensure that this protocol is consistent and consistently testable. But then we do not immediately move on to analyze the impact of the change based on the original update specification. Instead we first apply a method to reduce the update, i.e., to determine equivalent set of atomic changes that are producing the same changed protocol and that are more appropriate for evaluating the impacts. The reduction in the most straightforward case brings on the removal of redundancies, but in a more sophisticated approach it creates an update that is optimal with respect to the cost of evaluating the impacts. Finally, we apply the method described in the first part of this section to determine the impact of the change on the test considering each atomic change of the optimized multiple update.

We consider the previously discussed three types of atomic update rules: Addition, deletion and replacement of transitions. The problem of determining the best equivalent update can be stated as follows: Let us consider an update δ with multiple rules turning CFSM P_i to CFSM P_i'. Identify the shortest equivalent sequence of update rules changing CFSM P_i to P_i'.

In a more sophisticated approach – if some update rules are preferred over others – a cost function may be assigned to update rules. Let ρ be a cost function that assigns a nonnegative real number $\rho(ur)$ to each update rule. We constrain ρ to be a distance metric. That is, it satisfies the following three properties: $\rho(ur) \geq 0$ and $\rho(ur^0) = 0$ (nonnegative definiteness); $\rho(ur) = \rho(ur^{-1})$ (symmetry); $\rho(ur_{13}) \leq \rho(ur_{12}) + \rho(ur_{23})$ for any three operations with the following property: $P_i \rightarrow P_i'$ via ur_{12}, $P_i' \rightarrow P_i''$ via ur_{23} and $P_i \rightarrow P_i''$ via ur_{13} (triangle inequality). Furthermore, let the cost of an update with multiple rules $\delta = \{ur_1, ur_2, ..., ur_k\}$ be $\rho(\delta) = \sum_{i=1}^{k} \rho(ur_i)$.

The cost of an update rule – in general – may represent any practical property of the given atomic change. In our case costs reflect the impact of the given atomic update rule on the test set; updates that are likely to induce inconsistent tests are assigned a higher cost than others. For instance, consider a multiple controller change that includes – among others – the addition of two interdependent transitions t_1' and t_2', such that t_2' occurs in a run in the changed protocol iff t_1' also occurs. Obviously, the two update rules have the same interaction contexts, thus we only have to consider one of them to analyze the impacts. This interdependency can be taken into account for example by setting the cost of one of the update rules to 0.

As our costs are defined as distance metrics, the problem of optimizing multiple updates can be restated as finding the (edit) distance between two CFSMs

P_i and P_i', where the distance between P_i and P_i' is defined to be the minimum cost of all sequences of edit operations that change P_i to P_i':[4]

Definition 1. $dist(P_i, P_i') = min\{\rho(\delta) \mid \delta \text{ is an update changing } P_i \text{ to } P_i'\}.$

With this approach we have turned the problem of reducing multiple updates to an approximate graph matching problem [16]. Thus the tools and algorithms of the graph matching theory can be used to generate an update with the following properties:[5] It is equivalent to the original update specification, i.e., it induces the required modifications; it is the lowest-cost update, i.e., the impact of the given change on tests can be most effectively calculated based on it.

7   Conclusion

An automatic approach for determining impact of protocol changes on existing protocol tests in a white-box testing model is proposed. Protocols are modeled as a network of CFSMs that interact by message passing over bounded queues. Protocol changes add/replace/delete one or more protocol transitions in one or more controllers. Protocol tests are formalized as a protocol stable state along with a set of external environment inputs. Notions of consistent tests and consistently testable protocols are introduced. Changes must preserve consistently testability of protocols so that it is still possible to test all the runs of the changed protocol by using consistent tests. It is shown how symbolic state exploration over the changed protocol can be used to ensure that the protocol is consistently testable. The impact of a change on a test is characterized in terms of whether the consistency of the test is preserved by the change. For tests, whose consistency is preserved, we further show how state exploration can be used to determine whether the test exercises the changed behavior in which case it is re-usable; otherwise, the test is independent of change. We showed that single transition additions always preserve consistency of existing tests. Single transition replacements may make tests inconsistent if every test run exercises the deleted transition. We have shown how the approach can be extended to deal with more complex changes where protocol transitions in multiple controllers are simultaneously changed. We also describe a novel approach to reduce complex changes and create the shortest – or in an other sense optimal – sequence of changes inducing the required modifications. To the best of our knowledge, this is perhaps the first paper to formally address the effect of changes to CFSM-based protocols on protocol tests. We plan to extend this approach to use static analysis to analyze the protocol transition dependencies [19] to make it more useful in practice. We also plan to investigate augmenting existing tests with addition information such as states where inputs are issued to further facilitate the change impact analysis in practice.

---

[4] The original problem identifying the shortest sequence of update rules is a special case of the latter with all update rules having equal costs.

[5] For the algorithms and their application considering CFSMs see [17] and our earlier paper [18].

References

1. Bochmann, G.V., Petrenko, A.: Protocol testing: review of methods and relevance for software testing. In: ISSTA '94: Proceedings of the 1994 ACM SIGSOFT international symposium on Software testing and analysis, New York, NY, USA, ACM Press (1994) 109–124
2. Linn, R.J., Uyar, M.., eds.: Conformance testing methodologies and architectures for OSI protocols. IEEE Computer Society Press, Los Alamitos, CA, USA (1995)
3. Lee, D., Yiannakakis, M.: Principles and methods of testing finite state machines – a survey. Proceedings of the IEEE **84**(8) (1996) 1090–1123
4. Duale, A.Y., Uyar, M..: A method enabling feasible conformance test sequence generation for efsm models. IEEE Trans. Comput. **53**(5) (2004) 614–627
5. D. Brand, A.M., Zafiropulo, P.: On communicating finite state machines. In: Journal of Associating Computing Machinery, JACM. Volume 30(2). (1983)
6. Peng, W., Purushothaman, S.: Data flow analyses of communicating finite state machines. In: Transactions on Programming Languagaes and Systems TOPLAS. Volume 13. (1991)
7. Holzmann, G.J.: Design and validation of computer protocols. Prentice-Hall, Inc., Upper Saddle River, NJ, USA (1991)
8. Subramaniam, M., Chundi, P.: Preserving consistency and executability of protocols across updates. In: Proceedings of the 6th International Conference on Formal Engineering Methods, ICFEM. Volume LNCS. (2004)
9. Arnold, R.S.: Software Change Impact Analysis. IEEE Computer Society Press, Los Alamitos, CA, USA (1996)
10. Ryder, B.G., Tip, F.: Change impact analysis for object-oriented programs. In: Proceedings of PASTE-01. (2001)
11. Rothermal, G., Harrold, M.J.: A safe, efficient regression test selection technique. In: ACM Transactions on Software Engineering and Methodology. Volume 6(2). (6(2), 1997)
12. Subramaniam, M.: Preserving consistency of runtime monitors across protocol changes. In: Proc. of Tenth IEEE International Conference on Engineering of Complex Computer Systems ICECCS. (2005)
13. Schmitt, M.: Automatic Test Generation Based on Formal Specifications. Ph.d., Georg-August-University of Goettingen (2003)
14. M. Kaufmann, A.M., Pixely, C.: Design constraints in symbolic model checking. In: Proc. of Intl. Conference on Computer-Aided Verification CAV. Volume LNCS. (1998)
15. K. Shimizu, D. L. Dill, A.J.H.: Monitor-based formal specification of pci. In: Proc. of Intl. Conference on Formal Methods in Computer-aided design, FMCAD. Volume LNCS 1954. (LNCS 1954, 2000)
16. Bunke, H.: Graph matching: Theoretical foundations, algorithms, and applications. In: Proceedings of Vision Interface 2000, Montreal. (2000) 82–88
17. Wang, J.T.L., Zhang, K., Chirn, G.W.: Algorithms for approximate graph matching. Information Sciences **82**(1-2) (1995) 45–74
18. Pap, Z., Csopaki, G., Dibuz, S.: On the theory of patching. In: Proceedings of the 3rd IEEE International Conference on Software Engineering and Formal Methods, SEFM. (2005) 263–271
19. Subramaniam, M., Shi, J.: Using dominators to extract protocol contexts. In: Proceedings of the 3rd IEEE International Conference on Software Engineering and Formal Methods, SEFM. (2005)

Detecting Observability Problems in Distributed Testing

Jessica Chen[1] and Hasan Ural[2]

[1] School of Computer Science, University of Windsor,
Windsor, Ontario, Canada N9B 3P4
xjchen@uwindsor.ca
[2] School of Information Technology and Engineering,
University of Ottawa,
Ottawa, Ontario, Canada K1N 6N5
ural@site.ottawa.ca

Abstract. Application of a test or checking sequence in a distributed test architecture often requires the use of external coordination message exchanges among multiple remote testers for eluding potential controllability and observability problems. Recent literature reports on conditions on a given finite state machine (FSM) under which controllability and observability problems can be overcome without using external coordination messages. However, these conditions do not guarantee that any test/checking sequence constructed from such FSMs are free from controllability and observability problems. For a given test or checking sequence, this paper investigates whether it is possible to eliminate the need for external coordination messages and proposes algorithms to identify or construct subsequences either within the given sequence or as an extension to the given sequence, respectively.

Keywords: Finite state machine, testing, distributed test architecture, observability, controllability.

1 Introduction

In a *distributed test architecture*, there is one tester at each interface/*port* of the system under test (SUT) N. These testers participate in applying a given test sequence [1, 15, 16] or checking sequence [7, 9, 11, 19] which is a sequence of input/output pairs, constructed from the specification M of the SUT N. The use of multiple remote testers in a distributed architecture brings out the possibility of controllability and observability problems during the application of a test or checking sequence. A controllability problem arises when a tester is required to send the current input and because it did not send the previous input and did not receive the previous output it cannot determine when to send the input. An observability problem arises when a tester is expecting an output in response to either a previous input or the current input and because it is not the

M.Ü. Uyar, A.Y. Duale, and M.A. Fecko (Eds.): TestCom 2006, LNCS 3964, pp. 213–226, 2006.

sender of the current input, it cannot determine when to start and stop waiting for the output.

These problems and their solutions have been studied in the context where M is a Finite State Machine (FSM) and N is a state-based system whose externally observable behavior can also be represented by an FSM. Much of the previous work has been focused on automatically generating test or checking sequences from FSMs that causes no controllability or observability problems during its application in a distributed test architecture (see, for example, [2, 6, 8, 10, 13, 17, 18, 20]). For some FSMs, there have been test/checking sequences in which the coordination among testers can be achieved indirectly via their interactions with N [14, 16]. For some others, it may be necessary for testers to communicate directly by exchanging external coordination messages among themselves over a dedicated channel for overcoming the controllability and observability problems encountered during the application of the test/checking sequence [2, 3, 17]. Using external coordination messages introduces delays and the necessity to set up a dedicated communications channel among testers. Thus, the emphasis of the recent work is to minimize the use of external coordination message exchanges among testers [3, 10] or to identify conditions on a given FSM M under which controllability and observability problems can be overcome without using external coordination messages [4, 5].

Such conditions lead to the algorithms for identifying paths within a given FSM M that provide evidence for the possibility of eliminating the controllability and observability problems [4, 5]. [4] gives conditions on M so that each transition involved in an observability problem can be independently verified at port p. By *verified at port p*, it is meant that one can conclude that the output of this transition at port p is correct if one observes the correct output sequence on a certain path within M. By *independently*, it is meant that the above conclusion regarding the output at port p for a transition does not rely on the correctness of any other transitions. Since the notion of independence may not be required in some cases, the above condition on M can be weakened in these cases. [5] gives an algorithm that determines whether M satisfies this weaker condition and when it does so, identifies paths within M that check the output of the transitions.

In this paper, we assume that the given FSM M satisfies the condition in [5]. Then, we pose the following problem and solve it in a restricted setting: Given an FSM M and a synchronizable test or checking sequence τ_0 starting at the initial state of M, extend τ_0 with minimal number of subsequences to form a synchronizable test or checking sequence τ^* such that the detectability of the observability problems in τ_0 is guaranteed without using external coordination messages exchanged among remote testers.

The rest of the paper is organized as follows. Section 2 introduces the preliminary terminology. Section 3 gives a formal definition of the general problem and defines a restricted version of this problem. Section 4 presents our solution. Section 5 concludes the paper with our final remarks.

2   An n-Port FSM and Directed Graphs

An n-port *Finite State Machine* M (called henceforth an FSM M) is defined as $M = (S, I, O, \delta, \lambda, s_0)$ where S is a finite set of states; $s_0 \in S$ is the initial state; $I = \bigcup_{i=1}^{n} I_i$, where I_i is the set of input symbols of port i, and $I_i \cap I_j = \emptyset$ for $i, j \in [1, n]$, $i \neq j$; $O = \prod_{i=1}^{n}(O_i \cup \{-\})$, where O_i is the set of output symbols of port i, and $-$ means null output; δ is the transition function that maps $S \times I$ to S; and λ is the output function that maps $S \times I$ to O. Each $y \in O$ is a *vector of outputs*, i.e., $y = \langle o_1, o_2, ..., o_n \rangle$ where $o_i \in O_i \cup \{-\}$ for $i \in [1, n]$. A *transition* of an FSM M is a triple $t = (s_1, s_2, x/y)$, where $s_1, s_2 \in S$, $x \in I$, and $y \in O$ such that $\delta(s_1, x) = s_2$, $\lambda(s_1, x) = y$. s_1 and s_2 are called the *starting state* and the *ending state* of t respectively. The *input/output pair* x/y is called the *label* of t. $p \in [1, n]$ will denote a port and we use $y \mid_p$ or $t \mid_p$ to denote the output at p in output vector y or in transition t respectively. We use $\mathcal{T}$ to denote the set of all transitions in M.

A *path* $\rho = t_1 t_2 \ldots t_k$ $(k \geq 0)$ is a finite sequence of transitions such that for $k \geq 2$, the ending state of t_i is the starting state of t_{i+1} for all $i \in [1, k-1]$. We say t is *contained in* (or simply *in*) ρ if t is a transition along path ρ. When the ending state of the last transition of path ρ_1 is the starting state of the first transition of path ρ_2, we use $\rho_1 \rho_2$ to denote the *concatenation* of ρ_1 and ρ_2. The *label* of a path $(s_1, s_2, x_1/y_1)$ $(s_2, s_3, x_2/y_2) \ldots (s_k, s_{k+1}, x_k/y_k)$ $(k \geq 1)$ is the sequence of input/output pairs x_1/y_1 $x_2/y_2 \ldots x_k/y_k$ which is an *input/output sequence*.

When ρ is non-empty, we use *first*(ρ) and *last*(ρ) to denote the first and last transitions of path ρ respectively and *pre*(ρ) to denote the path obtained from ρ by removing its last transition.

Given an FSM M and a path $t_1 t_2 \ldots t_k$ $(k > 1)$ of M with label x_1/y_1 $x_2/y_2 \ldots x_k/y_k$, a *controllability* (also called *synchronization*) *problem* occurs when, in the labels x_i/y_i and x_{i+1}/y_{i+1} of two consecutive transitions, there exists $p \in [1, n]$ such that $x_{i+1} \in I_p$, $x_i \notin I_p$, $y_i \mid_p = - (i \in [1, k-1])$. If this controllability problem occurs then the tester at p does not know when to send x_{i+1} and the test/checking sequence cannot be applied. Consecutive transitions t_i and t_{i+1} form a *synchronizable pair* of transitions if t_{i+1} can follow t_i without causing a synchronization problem. Any path in which every pair of consecutive transitions is synchronizable is called a *synchronizable path*. An input/output sequence is synchronizable if it is the label of a synchronizable path.

We assume that for every pair of transitions (t, t') there is a synchronizable path that starts with t and ends with t'. If this condition holds, then the FSM is called *intrinsically synchronizable*.

Suppose that we are given an FSM M and a synchronizable path $\rho = t_1 t_2 \ldots t_k$ of M with label $x_1/y_1 x_2/y_2 \ldots x_k/y_k$. An *output shift fault* in an implementation N of M exists if one of the following holds for some $1 \leq i < j \leq k$:

a) For some $p \in [1, n]$ and $o \in O_p$, $y_i \mid_p = o$ in M and for all $i < l \leq j$, $y_l \mid_p = -$ in M whereas for all $i \leq l < j$, N produces output $-$ at p in response to x_l after $x_1 \ldots x_{l-1}$, and N produces output o at p in response to x_j after $x_1 \ldots x_{j-1}$.

b) For some $p \in [1, n]$ and $o \in O_p$, $y_j \mid_p = o$ in M and for all $i \leq l < j$, $y_l \mid_p = -$ in M whereas for all $i < l \leq j$, N produces output $-$ at p in response to x_l after $x_1 \ldots x_{l-1}$, and N produces output o at p in response to x_i after $x_1 \ldots x_{i-1}$.

In a) the output o shifts from being produced in response to x_i to being produced in response to x_j and the shift is from t_i to t_j (i.e., a *forward* shift). In b) the output o shifts from being produced in response to x_j to being produced in response to x_i and the shift is from t_j to t_i (i.e., a *backward* shift).

An instance of the observability problem manifests itself as a *potentially undetectable output shift fault* if there is an output shift fault related to $o \in O_p$ in two transitions t_i and t_j in ρ with labels x_i/y_i and x_j/y_j, such that $x_{i+1} \ldots x_j \notin I_p$. The tester at p will not be able to detect the faults since it will observe the expected sequence of interactions in response to $x_i \ldots x_j$. Both t_i and t_j are said to be *involved* in the potentially undetectable output shift fault. When $j = i + 1$, we also call it potentially undetectable *1-shift* output fault.

In the following, τ_0 is a given test/checking sequence, which is the label of path $\rho_0 = t_1 t_2 \ldots t_m$. We will use $\mathcal{T}_{\rho_0, p}$ to denote the set of transitions of M that can be involved in potentially undetectable output shift faults in ρ_0. Thus $t \in \mathcal{T}_{\rho_0, p}$ if there exists a transition t' and a synchronizable path $t \rho t'$ or $t' \rho t$ such that both t and t' are involved in a potentially undetectable output shift fault when we apply τ_0 to N.

Let t be a transition, and $\mathcal{U}$ a set of transitions in M. ρ is an *absolute verifying path upon $\mathcal{U}$ for (t, p)* if

- ρ is a synchronizable path;
- t is contained in pre(ρ);
- first(ρ) and last(ρ) and only these two transitions in ρ have input at p;
- $t \notin \mathcal{U}$ and for all t' contained in pre(ρ), either $t' \in \mathcal{U}$ or $t' \mid_p = - \Leftrightarrow t \mid_p = -$ [5].

Note that given t and ρ we will typically consider a *minimal* set $\mathcal{U}$ that satisfies the above conditions: if $t' \mid_p = - \Leftrightarrow t \mid_p = -$ then $t' \notin \mathcal{U}$.

Suppose that $\mathcal{U}$ is a set of transitions of M, $\mathcal{R} \subseteq \mathcal{U} \times \mathcal{U}$ is a relation, and $\mathcal{P}$ is a function from $\mathcal{U}$ to synchronizable paths of M. Let p be any port in M. The set $\mathcal{U}$ of transitions is *verifiable at p under $\mathcal{R}$ and $\mathcal{P}$* if the following hold [5].

(a) For all $t \in \mathcal{U}$, $\mathcal{P}(t)$ is an absolute verifying path upon $\{t' \mid (t, t') \in \mathcal{R}\}$ for (t, p);
(b) $\mathcal{R} \cup \{(t, t) \mid t \in \mathcal{U}\}$ is a partial order.

Where such $\mathcal{R}$ and $\mathcal{P}$ exist we also say that $\mathcal{U}$ is verifiable at p.

Let $\mathcal{T}_p$ be the set of all transitions involved in some potentially undetectable output shift faults in M at port p. In this paper, we assume that $\mathcal{T}_p$ is verifiable at p for all $p \in [1, n]$.

A *directed graph (digraph)* G is defined by a tuple (V, E) in which V is a set of vertices and E is a set of directed edges between the vertices. An edge e from vertex v_i to vertex v_j is represented by (v_i, v_j). A *walk* is a sequence of pairwise adjacent edges in G. A digraph is *strongly connected* if for any ordered pair of vertices (v_i, v_j) there is a walk from v_i to v_j.

3   The Problem Definition

Given a deterministic, minimal, and completely specified FSM M which is intrinsically synchronizable, and a synchronizable test/checking sequence τ_0 starting at the initial state of M, we consider the problem of constructing a synchronizable test/checking sequence τ^* that can be applied to resolve observability problems in τ_0 without using external coordination message exchanges by identifying the subsequences within τ_0 or to be appended to τ_0.

Clearly, for each $t \in T_{\rho_0,p}$, we should verify its output at port p. As we discussed in [5], to verify the output of transition t at port p, we can construct an *absolute verifying path* upon a *set* $\mathcal{U}$ of transitions whose outputs at p are verified. Such a path ρ has the following properties:

- it is synchronizable;
- we are able to determine the output sequence of ρ at p by applying the label of ρ from the starting state of ρ;
- from the correct output sequence of ρ at p we can determine that the output of t at p is correct.

This is because (i) no matter how ρ is concatenated with other subsequences, we can always determine the output sequence produced at p in response to the first $|pre(\rho)|$ inputs in the label of ρ since this output sequence is immediately preceded and followed by input at p; (ii) the condition *for all t' contained in pre(ρ), either $t' \in \mathcal{U}$ or $t'|_p = - \Leftrightarrow t|_p = -$* allows us to determine the correct output of (t,p) from the correct output sequence of ρ at p (Proposition 2 in [5]).

Thus, to verify the outputs of the transitions in $T_{\rho_0,p}$ at port p, we search for an acyclic digraph of transitions such that all transitions in $T_{\rho_0,p}$ are present, and each transition has an absolute verifying path upon a set of transitions that appear as its successors in the digraph. In other words, we search for $\mathcal{R}$ and $\mathcal{P}$ such that set $T_{\rho_0,p}$ of transitions is verifiable at p under $\mathcal{R}$ and $\mathcal{P}$.

It is possible that ρ_0 contains some absolute verifying paths for transitions in $T_{\rho_0,p}$. Let Q_p be the set of all those paths in $codomain(\mathcal{P})$ but not as subsequences in ρ_0. τ^* will be the label of a path ρ^* which contains both ρ_0 and all paths in Q_p.

Clearly, for efficiency reasons,

- We should maximize the images of $\mathcal{P}$ in ρ_0. That is, whenever possible, we should define $\mathcal{P}(t)$ as a subsequence in ρ_0 for any $t \in T$.
- No path in Q_p should appear as a subsequence of another path in Q_p. This is always true as the absolute verifying paths have input at port p only in its first and last transitions.
- There is no redundant path in Q_p. An absolute verifying path ρ is redundant in Q_p if we can modify $\mathcal{P}$ (and $\mathcal{R}$ correspondingly) by changing the mapping of all transitions whose image is ρ under $\mathcal{P}$ to some other paths in Q_p while keeping the property that $T_{\rho_0,p}$ is verifiable at p under the modified definitions of $\mathcal{P}$ and $\mathcal{R}$. Figure 1(a) shows a case where $\{t_1, t_2, t_3\}$ is verifiable at p under $\mathcal{P}$ and $\mathcal{R}$ where $\mathcal{P}(t_i) = \rho_i$ for $i = 1, 2, 3$. Suppose that ρ_2 is also an absolute verifying path upon $\{t_3\}$ for (t_1, p), then Figure 1(b) shows an

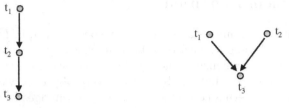

ρ_1: an absolute verifying path upon $\{t_2\}$ for (t_1, p)
ρ_2: an absolute verifying path upon $\{t_3\}$ for (t_2, p)
ρ_3: an absolute verifying path upon ϕ for (t_3, p)

ρ_2: an absolute verifying path upon $\{t_3\}$ for (t_1, p) and (t_2,p)
ρ_3: an absolute verifying path upon ϕ for (t_3, p)

(a)   (b)

Fig. 1. An example of reducing paths in Q_p

alternative way to verify $\{t_1, t_2, t_3\}$ which requires less paths in Q_p to be considered in constructing τ^*: $\mathcal{P}(t_1) = \mathcal{P}(t_2) = \rho_2$, $\mathcal{P}(t_3) = \rho_3$.

4   Our Proposed Solution

Now we present our solution to construct Q_p and τ^*.

4.1   Identifying Transitions Involved in Observability Problems

Recall that $\tau_0 = x_1/y_1 \ x_2/y_2 \ \ldots \ x_m/y_m$ is a test/checking sequence of M which is the label of a path $\rho_0 = t_1 t_2 \ldots t_m$. First we need to calculate $\mathcal{T}_{\rho_0,p}$, the set of transitions involved in potentially undetectable output shift faults at port p in ρ_0, for all $p \in [1, n]$. Figure 2 shows an algorithm for this purpose. It scans τ_0 and uses *emptyPointer* and *nonEmptyPointer* as auxiliary variables. We do not consider the case when $|\tau_0| = 0$ which is meaningless. Suppose we are currently considering $x_i/y_i \in \tau_0$.

emptyPointer is the minimal index of the transitions in τ_0 such that

- $\forall k \in [emptyPointer + 1, i - 1]$. $x_k \notin I_p$ and
- $\forall k \in [emptyPointer, i - 1]$. $y_k \mid_p = -$

nonEmptyPointer is the index of the transitions in τ_0 such that

- $y_k \mid_p \neq -$ for $k = nonEmptyPointer$ and
- $\forall k \in [nonEmptyPointer + 1, i - 1]$. $x_k \notin I_p \wedge y_k \mid_p = -$

If neither *emptyPointer* nor *nonEmptyPointer* is *null*, then for all $k \in [non\text{-}EmptyPointer, i - 1]$, t_k is involved in a potentially undetectable forward output shift fault. Furthermore, in the case $x_i \notin I_p$ and $y_i \mid_p = -$, t_i is also involved in a potentially undetectable forward output shift fault.

If *emptyPointer* is not *null*, no matter whether *nonEmptyPointer* is *null* or not, t_k is involved in a potentially undetectable backward output shift fault for all $k \in [emptyPointer, i]$ when $x_i \notin I_p$ and $y_i \mid_p \neq -$.

1: **input**: an FSM M, a port p, a test/checking sequence $\tau_0 = x_1/y_1\ x_2/y_2\ \dots x_m/y_m$
   of M
2: **output**: $\mathcal{T}_{\rho_0,p}$
3: $nonEmptyPointer := null$
4: $emptyPointer := null$
5: $i := 1$
6: **while** $i < m$ **do**
7:     **if** $x_i \notin I_p$ **then**
8:         **if** $y_i \mid_p \neq\ -$ **then**
9:             **if** $emptyPointer \neq null \wedge nonEmptyPointer \neq null$ **then**
10:                 add $t_{nonEmptyPointer}, \dots, t_i$ to $\mathcal{T}_{\rho_0,p}$
11:             **end if**
12:             **if** $emptyPointer \neq null \wedge nonEmptyPointer = null$ **then**
13:                 add $t_{emptyPointer}, \dots, t_i$ to $\mathcal{T}_{\rho_0,p}$
14:             **end if**
15:             $nonEmptyPointer := i$
16:             $emptyPointer := null$
17:         **else**
18:             **if** $nonEmptyPointer = i - 1$ **then**
19:                 $emptyPointer = i$
20:             **end if**
21:         **end if**
22:     **else**
23:         **if** $emptyPointer \neq null \wedge nonEmptyPointer \neq null$ **then**
24:             add $t_{nonEmptyPointer}, \dots, t_{i-1}$ to $\mathcal{T}_{\rho_0,p}$
25:         **end if**
26:         **if** $y_i \mid_p \neq\ -$ **then**
27:             $nonEmptyPointer := i$
28:             $emptyPointer := null$
29:         **else**
30:             $nonEmptyPointer := null$
31:             $emptyPointer := i$
32:         **end if**
33:     **end if**
34:     $i := i + 1$
35: **end while**
36: **if** $emptyPointer \neq null \wedge nonEmptyPointer \neq null$ **then**
37:     add $t_{nonEmptyPointer}, \dots, t_m$ to $\mathcal{T}_{\rho_0,p}$
38: **end if**
39: **if** $emptyPointer \neq null \wedge nonEmptyPointer = null$ **then**
40:     **if** $x_m \notin I_p$ **then**
41:         add $t_{emptyPointer}, \dots, t_m$ to $\mathcal{T}_{\rho_0,p}$
42:     **else**
43:         add t_m to $\mathcal{T}_{\rho_0,p}$
44:     **end if**
45: **end if**
46: **if** $emptyPointer = null \wedge nonEmptyPointer \neq null$ **then**
47:     **if** $x_m \notin I_p$ and $y_m \mid_p = -$ **then**
48:         add t_{m-1}, t_m to $\mathcal{T}_{\rho_0,p}$
49:     **else**
50:         add t_m to $\mathcal{T}_{\rho_0,p}$
51:     **end if**
52: **end if**

Fig. 2. Algorithm 1: Construction of $\mathcal{T}_{\rho_0,p}$

Note that some transitions at the end of ρ_0 that are not involved in any potentially undetectable output shift fault in ρ_0 may be involved in such faults in the constructed ρ^*. All these transitions are also added into $T_{\rho_0,p}$ in lines 36-52 which specifically handle the case when $i = m$.

The execution of Algorithm 1 can be done in $\mathcal{O}(|\tau_0|)$ time.

4.2   Identifying Verifiable Transitions

By definition, the transitions in $T - T_{\rho_0,p}$ all have correct output at p. On the other hand, not all transitions in $T_{\rho_0,p}$ need to be verified for its output at p with additional subsequences. This is based on the following two observations:

- A transition in $T_{\rho_0,p}$ may appear in a different place in ρ_0 where it is not involved in any potentially undetectable output shift faults at p in ρ_0, and thus its output at p is verified in ρ_0.
- Given a transition $t \in T_{\rho_0,p}$, there may exist an absolute verifying path upon $T - T_{\rho_0,p}$ for (t,p) in ρ_0.

In general, before constructing additional subsequences to be appended to τ_0, we would like to find $\mathcal{R}_0$, $\mathcal{P}_0$ and $\mathcal{U}_0 \subset T_{\rho_0,p}$ such that

- $\mathcal{U}_0$ is verifiable at p under $\mathcal{R}_0$ and $\mathcal{P}_0$ *in* ρ_0, in the sense that $\mathcal{U}_0$ is verifiable at p under $\mathcal{R}_0$ and $\mathcal{P}_0$, and the paths in $codomain(\mathcal{P}_0)$ are all in ρ_0;
- $\mathcal{U}_0$ is maximized, in the sense that for any $\mathcal{R}_0'$, $\mathcal{P}_0'$ and $\mathcal{U}_0'$ such that $\mathcal{U}_0'$ is verifiable at p under $\mathcal{R}_0'$ and $\mathcal{P}_0'$ in ρ_0, $\mathcal{U}_0' \subseteq \mathcal{U}_0$.

The following proposition follows directly from the definition.

Proposition 1. *Let ρ be a synchronizable path with input at p only in* first(ρ) *and* last(ρ)*, and $t \in$* pre(ρ)*. Let $D_{t,\rho}$ be the set of transitions in* pre(ρ) *such that for any $t' \in D_{t,\rho}$, $t' \mid_p = - \Leftrightarrow t \mid_p \neq -$. Then ρ is an absolute verifying path upon $D_{t,\rho}$ for (t,p).*

Let ρ be a subsequence in ρ_0 with input at p both at the beginning and at the end. Based on the above proposition, if the set of all those transitions in ρ with empty output at p is verifiable, then the set of all transitions in ρ is verifiable using ρ as an absolute verifying path. Analogously, if the set of all those transitions in ρ with non-empty output at p is verifiable, then the set of all transitions in ρ is verifiable.

Thus, we can derive from ρ_0 a set of so-called *counter-pairs* (L_1, L_2) of sets of transitions. Each counter-pair (L_1, L_2) corresponds to a *potential candidate* of absolute verifying path in ρ_0 that can be used in defining $\mathcal{P}$. It is obtained in this way: for any subsequence ρ of ρ_0 with input at p both at the beginning and at the end (and no other input at p in it), there is a counter-pair (L_1, L_2) where L_1 contains all transitions in $pre(\rho)$ with empty output at p, and L_2 contains all transitions in $pre(\rho)$ with non-empty output at p. Such counter-pairs hold the following property: for any set A of transitions in T, the outputs of all transitions in L_1 are verifiable upon A implies the outputs of all transitions

1: **input**: an FSM M, a port p, a test/checking sequence $\tau_0 = x_1/y_1 x_2/y_2 \ldots x_m/y_m$ of M, and $\mathcal{T}_{\rho_0,p}$
2: **output**: a set $\mathcal{U}_0$ of transitions that is verifiable at p in ρ_0, and a set Θ of counter-pairs of p
3: $\quad \Theta := \emptyset$
4: Let $r \le m$, s.t. $x_r \in I_p$ and $\forall k, 1 \le k < r, x_k \notin I_p$
5: **while** $\exists j. \ r < j \le m$ s.t. $x_j \in I_p$ and $\forall k, \ r < k < j, \ x_k \notin I_p$ **do**
6: $\quad$ let j be such that $r < j \le m$, $x_j \in I_p$ and $\forall k, \ r < k < j, \ x_k \notin I_p$
7: $\quad$ **if** $\exists r \le k < j$ s.t. $t_k \in \mathcal{T}_{\rho_0,p}$ **then**
8: $\quad\quad$ $L_1 := \emptyset$
9: $\quad\quad$ $L_2 := \emptyset$
10: $\quad\quad$ **for** $k, \ r \le k < j$ **do**
11: $\quad\quad\quad$ **if** $y_k \mid_p = -$ **then**
12: $\quad\quad\quad\quad$ add t_k to L_1
13: $\quad\quad\quad$ **else**
14: $\quad\quad\quad\quad$ add t_k to L_2
15: $\quad\quad\quad$ **end if**
16: $\quad\quad$ **end for**
17: $\quad\quad$ add (L_1, L_2) to Θ
18: $\quad$ **end if**
19: $\quad$ $r = j$
20: **end while**
21: $(\mathcal{U}', \Theta') := counterPairsUpdate(\mathcal{T} - \mathcal{T}_{\rho_0,p}, \Theta)$
22: **return** $\mathcal{U}'$ and Θ'

Fig. 3. Algorithm 2: Construction of U_0 and Θ

in L_2 are verifiable upon $A \cup L_1$; and the outputs of all transitions in L_2 are verifiable upon A implies the outputs of all transitions in L_1 are verifiable upon $A \cup L_2$. Consequently, for any $t \in L_1$, the path corresponding to (L_1, L_2) can be used as an absolute verifying path upon $\mathcal{U}$ for (t, p) if $L_2 \subseteq \mathcal{U}$. Conversely, for any $t \in L_2$, the path corresponding to (L_1, L_2) can be used as an absolute verifying path upon $\mathcal{U}$ for (t, p) if $L_1 \subseteq \mathcal{U}$.

Figure 3 gives an algorithm to calculate set $\mathcal{U}_0$ of transitions whose outputs at p are verifiable in ρ_0. Set Θ contains those counter-pairs that correspond to potential candidates of absolute verifying paths. Given a set $\mathcal{U}_0$ of transitions that is verifiable at p under $\mathcal{R}_0$ and $\mathcal{P}_0$ in ρ_0, we can check if any potential candidate of absolute verifying path can be used to extend $\mathcal{U}_0$. This operation is performed in Figure 4. Counter-pairs whose corresponding paths will no more be used during the construction of $\mathcal{R}_0$ and $\mathcal{P}_0$ are removed from Θ.

Note that if there is no input in τ_0 that will be given at port p, then we are not able to construct an absolute verifying path for any output at p. Since we assume that $\mathcal{T}_p$ is verifiable, this implies that $\mathcal{T}_{\rho_0,p} = \emptyset$, and thus there is no need for the subsequences to be appended to ρ_0 for port p. Hence we consider there is at least one input at p in τ_0.

At the end of Algorithm 2, we have that (i) $\mathcal{U}_0$ is verifiable at p under $\mathcal{R}_0$ and $\mathcal{P}_0$ in ρ_0, and it is maximized; (ii) all potential absolute verifying paths in ρ_0 for further use have their correspondence in Θ.

```
 1: input: U and Θ
 2: output: updated U and Θ
 3:   change := true
 4: while change = true do
 5:     for each (L₁, L₂) ∈ Θ do
 6:         L₁ := L₁ − U
 7:         L₂ := L₂ − U
 8:     end for
 9:     change := false
10:     for  each (L₁, L₂) ∈ Θ do
11:         if L₁ = ∅ then
12:             add all transitions in L₂ to U
13:             remove (L₁, L₂) from Θ
14:             change := true
15:         end if
16:         if L₂ = ∅ then
17:             add all transitions in L₁ to U
18:             remove (L₁, L₂) from Θ
19:             change := true
20:         end if
21:     end for
22: end while
```

Fig. 4. Procedure of *counterPairsUpdate*

We know that $\Sigma_{(L_1,L_2)\in\Theta}(|L_1| + |L_2|) \leq |\tau_0|$, and $|\mathcal{U}| \leq |\mathcal{T}|$. So in Figure 4, the first for-loop will be executed maximally $|\tau_0| \times |\mathcal{T}|$ times, and the second for-loop will be executed maximally $|\tau_0|$ times. The while-loop each time removes at least one counter-pair from Θ. So in total it takes $\mathcal{O}(|\tau_0| \times |\mathcal{T}| \times |\Theta|)$ time to perform *counterPairsUpdate*. Consequently, it takes $\mathcal{O}(|\tau_0| \times |\mathcal{T}| \times |\Theta|)$ time to run Algorithm 2.

4.3   Identifying Subsequences to Be Added to τ_0

Given an initial set $\mathcal{U}_0$ of transitions that is verifiable at p in ρ_0, and a set Θ of counter-pairs corresponding to some potential absolute verifying paths, we define $\mathcal{P}$ and $\mathcal{R}$ such that $\mathcal{T}_{\rho_0,p}$ is verifiable at p under $\mathcal{R}$ and $\mathcal{P}$; the images of $\mathcal{P}$ in ρ_0 is maximized; there is no redundant path in $\mathcal{U}$. This leads to the construction of Q_p that we want.

Figure 5 gives an algorithm to construct Q_p. Here *checkset* is used to keep the transitions that we may need to construct additional subsequences to verify their output at p. Since we assume that $\mathcal{T}_p$ is verifiable at port p, $\mathcal{T}_{\rho_0,p} - \mathcal{U}$ is also verifiable. So for each iteration of the outer while-loop, we can surely find an absolute verifying path upon $\mathcal{U}$ for some $t \in$ *checkset* before *checkset* becomes empty.

Whenever we find an absolute verifying path upon $\mathcal{U}$ for some $t \in$ *checkset*, we add to $\mathcal{U}$ all transitions in $pre(\rho)$ such that they have empty output at p if and only if t has empty output at p. This is because if ρ is an absolute verifying path upon $\mathcal{U}$ for (t, p), then ρ is an absolute verifying path upon $\mathcal{U}$ for (t', p) for

```
1: input: p, 𝒯ρ₀,p, Θ and 𝒰₀
2: output: Q_p
3: 𝒰 = 𝒰₀
4: while 𝒯ρ₀,p − 𝒰 ≠ ∅ do
5:    checkset := 𝒯ρ₀,p − 𝒰
6:    found := false
7:    while found = false do
8:       let t ∈ checkset
9:       if there exists an absolute verifying path upon 𝒰 for (t, p) then
10:          let ρ be a minimal-length absolute verifying path upon 𝒰 for (t, p)
11:          add ρ to Q_p
12:          for each transition t' ∈ pre(ρ) s.t. t' |_p= − ⇔ t |_p= −, add t' to 𝒰
13:          (𝒰, Θ) := counterPairsUpdate(𝒰, Θ)
14:          found := true
15:       else
16:          checkset := checkset − {t}
17:       end if
18:    end while
19: end while
```

Fig. 5. Algorithm 3: Construction of Q_p

all $t' \in pre(\rho)$ such that $t' |_p= - \Leftrightarrow t |_p= -$ (Proposition 1 in [5]). This also guarantees that when we search for an absolute verifying path upon $\mathcal{U}$ for (t, p), we do not need to check whether previously constructed subsequences in Q_p can be re-used. Consequently, there is no redundant path in $\mathcal{U}$.

Whenever an additional sequence is constructed and added to Q_p, $\mathcal{U}$ is updated. Correspondingly, we call procedure *counterPairsUpdate* to check if based on the updated $\mathcal{U}$ any potential absolute verifying path in ρ_0 can be used. As the initial value of $\mathcal{U}$ is from Algorithm 2, this guarantees that for any $\rho \in Q_p$, ρ is not a subsequence of ρ_0. Thus, the images of $\mathcal{P}$ in ρ_0 is maximized.

¿From [5], we know that if ρ is an absolute verifying path upon $\mathcal{U}$ for (t, p), then when we apply the label of ρ from a state in N similar to the starting state of ρ, then we can verify that the output of t at p is correct. So, when we have $\mathcal{T}_{\rho_0,p} - \mathcal{U} = \emptyset$ at the end of the algorithm, we know that if we apply τ_0 from the initial state of N and apply the label of ρ from a state similar to the starting state of ρ for all $\rho \in Q_p$, then we can verify that there is no undetectable output shift faults occurred in applying τ_0 to N.

To find a minimal-length absolute verifying path upon $\mathcal{U}$ for (t, p), similar as in [5], we can construct $G[t, \mathcal{U}]$ which is obtained from G by removing all edges except those corresponding to a transition t' in one of the following cases:

− t' has input at p;
− $t' |_p= -$ if and only if $t |_p= -$;
− $t' \in \mathcal{U}$

We then use breadth-first search to construct minimal-length synchronizable path in $G[t, \mathcal{U}]$ that starts with input at p and ends with input at p. Note that there may exist more than one such path with minimal-length.

Note also that while more transitions are added to $\mathcal{U}$, there may exist shorter path for a transition whose image under $\mathcal{P}$ was previously added to Q_p.

Now we turn to the complexity of the algorithm. For each outer while-loop, $\mathcal{U}$ is augmented by at least one transition. So the outer while-loop will be executed at most v times where v is the number of transitions to be verified. For the inner while-loop, we need to check if we can find an absolute verifying path upon $\mathcal{U}$ for some $t \in$ *checkset* where $|checkset| \le v$. This can be realized by trying to construct an absolute verifying path upon $\mathcal{U}$ for each t in *checkset* until such a path is found. This takes at most $|checkset|$ times of effort for each attempt. For each attempt to construct an absolute verifying path upon $\mathcal{U}$ for a given transition t, it takes $\mathcal{O}(w \times |T|)$ times where w is the number of states in M. In summary, the time complexity of Algorithm 3 is $\mathcal{O}(v^2 \times w \times |T|)$.

4.4   Adding Subsequences to τ_0

Finally, given ρ_0 and Q_p for each p, we need to construct a minimal-length test/checking sequence τ^* so that (i) it is synchronizable; (ii) it starts with τ_0 and it contains all the input/output sequences of the paths in Q_p for each $p \in [1, n]$. Figure 6 gives such an algorithm. It generates a synchronizable path ρ^* and its label τ^*.

1: **input**: M, τ_0, and Q_p for each $p \in [1, n]$
2: **output**: test/checking sequence τ^*
3: Let $Q = \cup_{p \in [1,n]} Q_p \cup \{\rho_0\}$
4: Let graph G contain one vertex v_ρ for each path ρ in Q
5: **for** each ordered pair $(\rho_1, \rho_2) \in Q$ such that $\rho_1 \ne \rho_2$ **do**
6:    find a shortest path ρ' in M such that $last(\rho_1)$ ρ' $first(\rho_2)$ is a synchronizable path.
7:    In G, add an edge $e = (v_{\rho_1}, v_{\rho_2})$, with $|\rho'|$ as its weight
8:    let $f_1(e) = \rho_1$, $f_2(e) = \rho_1\rho'$, $f_3(e) = \rho_1\rho'\rho_2$
9: **end for**
10: Find a walk $r = e_1 e_2 \ldots e_k$ in G that visits all vertices at least once with minimal cost, and that $f_1(e_1) = \rho_0$
11: Let $\rho^* = f_2(e_1)f_2(e_2)\ldots f_2(e_{k-1})f_3(e_k)$
12: Let τ^* be the label of ρ^*

Fig. 6. Algorithm 4: Addition of elements of Q_p to ρ_0 to form ρ^*

As we assume that M is intrinsically synchronizable, G is a strongly-connected digraph. This guarantees the existence of r. In general, the time complexity of Algorithm 4 is equivalent to that of finding a travelling saleman tour in a digraph. Efficient heuristics exist for the solution of Travelling Saleman Problem, cf. [12].

Note that ρ^* may introduce new observability problems. However, since each path in Q_p has input at p in its first and last transitions, a new observability problem cannot happen between a transition in a *connecting path*, i.e. a path used to connect paths in Q_p, and a transition in an absolute verifying path in Q_p. It can only happen (i) within a connecting path; (ii) within an absolute verifying path; or (iii) between a transition in ρ_0 and a transition in a connecting path. The new observability problems occurred in cases (i) and (ii) do not affect the ability of τ^* to verify that there is no undetectable output shift faults when τ_0 is applied to N. The new observability problems in case (iii) are resolved because we have included into $\mathcal{T}_{\rho_0,p}$ all transitions that may possibly get involved in some potentially undetectable output shift fault between a transition in ρ_0 and a transition in a path concatenated to the end of ρ_0 (cf. Algorithm 1).

5   Conclusions and Final Remarks

We have presented a method for eliminating the use of external coordination message exchanges for resolving observability problems in a given test/checking sequence constructed from an FSM satisfying conditions given in [5]. There are various optimization problems remaining to be solved. First, the existence of multiple minimal-length absolute verifying paths can be used to optimize the total length of ρ^*. Second, in our solution, the order of generating the subsequences will have an effect on the final set of additional subsequences. It will be interesting to find approaches for eliminating this effect. Third, our solution only considers the subproblem of constructing the subsequences for each port p individually. It remains as an interesting problem to consider the global optimization problem among all ports. Fourth, it will be quite interesting to incorporate some of the algorithms proposed here into a checking sequence construction method to construct a checking sequence in which there are no external coordination message exchanges. It is anticipated that the complexity of the last two optimization problems will be very high.

Acknowledgements

This work is supported by Natural Sciences and Engineering Research Council (NSERC) of Canada under grant RGPIN 976 and 209774.

References

1. A. V. Aho, A. T. Dahbura, D. Lee, and M. U. Uyar. An optimization technique for protocol conformance test generation based on UIO sequences and Rural Chinese Postman Tours. In *Protocol Specification, Testing, and Verification VIII*, pages 75–86, Atlantic City, 1988. Elsevier (North-Holland).
2. S. Boyd and H. Ural. The synchronization problem in protocol testing and its complexity. *Information Processing Letters*, 40:131–136, 1991.
3. L. Cacciari and O. Rafiq. Controllability and observability in distributed testing. *Information and Software Technology*, 41:767–780, 1999.
4. J. Chen, R. M. Hierons, and H. Ural. Conditions for resolving observability problems in distributed testing. In *24rd IFIP International Conference on Formal Techniques for Networked and Distributed Systems (FORTE 2004)*, volume 3235 of *LNCS*, pages 229–242. Springer-Verlag, 2004.

5. J. Chen, R. M. Hierons, and H. Ural. Resolving observability problems in distrib-uted test architecture. In *25rd IFIP International Conference on Formal Tech-niques for Networked and Distributed Systems (FORTE 2005)*, volume 3731 of *LNCS*, pages 219–232. Springer-Verlag, 2005.
6. W. Chen and H. Ural. Synchronizable checking sequences based on multiple UIO sequences. *IEEE/ACM Transactions on Networking*, 3:152–157, 1995.
7. A. Gill. *Introduction to the Theory of Finite-State Machines*. New York: McGraw-Hill, 1962.
8. S. Guyot and H. Ural. Synchronizable checking sequences based on UIO sequences. In *Proc. of IFIP IWPTS'95*, pages 395–407, Evry, France, September 1995.
9. F.C. Hennie. Fault detecting experiments for sequential circuits. In *Proc. of Fifth Ann. Symp. Switching Circuit Theory and Logical Design*, pages 95–110, Princeton, N.J., 1964.
10. R. M. Hierons. Testing a distributed system: generating minimal synchronised test sequences that detect output-shifting faults. *Information and Software Technology*, 43(9):551–560, 2001.
11. D. Lee and M. Yannakakis. Principles and methods of testing finite–state machines – a survey. *Proceedings of the IEEE*, 84(8):1089–1123, 1996.
12. S. Lin and B. W. Kernighan. An effective heuristic algorithm for the traveling-salesman problem. *Operations Research*, 21(2):498–516, March-April 1973.
13. G. Luo, R. Dssouli, and G. v. Bochmann. Generating synchronizable test sequences based on finite state machine with distributed ports. In *The 6th IFIP Workshop on Protocol Test Systems*, pages 139–153. Elsevier (North-Holland), 1993.
14. G. Luo, R. Dssouli, G. v. Bochmann, P. Venkataram, and A. Ghedamsi. Test gen-eration with respect to distributed interfaces. *Computer Standards and Interfaces*, 16:119–132, 1994.
15. K.K. Sabnani and A.T. Dahbura. A protocol test generation procedure. *Computer Networks*, 15:285–297, 1988.
16. B. Sarikaya and G. v. Bochmann. Synchronization and specification issues in protocol testing. *IEEE Transactions on Communications*, 32:389–395, April 1984.
17. K.C. Tai and Y.C. Young. Synchronizable test sequences of finite state machines. *Computer Networks*, 13:1111–1134, 1998.
18. H. Ural and Z. Wang. Synchronizable test sequence generation using UIO se-quences. *Computer Communications*, 16:653–661, 1993.
19. H. Ural, X. Wu, and F. Zhang. On minimizing the lengths of checking sequences. *IEEE Transactions on Computers*, 46:93–99, 1997.
20. Y.C. Young and K.C. Tai. Observation inaccuracy in conformance testing with multiple testers. In *Proc. of IEEE WASET*, pages 80–85, 1998.

Compositional Testing of Communication Systems

Reinhard Gotzhein[1] and Ferhat Khendek[2]

[1] Computer Science Department, University of Kaiserslautern,
Postfach 3049, D-67653 Kaiserslautern, Germany
gotzhein@informatik.uni-kl.de
[2] ECE Department, Concordia University, 1455 de Maisonneuve Blvd. W.,
Montréal, Québec, Canada H3G 1M8
khendek@ece.concordia.ca

Abstract. In this paper, we propose the *compositional test method (C-method)*, which exploits the structure of component-based communication systems. The C-method first tests each component separately for output and/or transfer faults, using one of the traditional test methods, then checks for composability, and finally tests the composite system for composition faults. To check for composability and to derive the test suite for the detection of composition faults, it is not required to construct the global state machine. Instead, all information is derived from the component state machines, which avoids a potential state explosion and lengthy test cases. Furthermore, the test suite checks for composition faults only. This substantially reduces the size of the test suite and thus the overall test effort.

1 Introduction

Systematic methods for testing protocol implementations have a long and successful record. The relevance and the potential of protocol testing are first recognized in [16], which has initiated a research stream that has produced a diversity of test methods with different foci. These methods usually assume that the design of the protocol implementation to be tested is given in the form of a finite state machine (FSM), and that this state machine is minimal, completely specified, and fully connected. Some methods further assume the FSM to be deterministic [3,7,20], while others relax this constraint [12]. Recently, the focus has shifted to real-time systems testing [6,17,18], interoperability testing [1,2,4,5] and testing in context [14,15].

On the other hand, component-based software engineering is becoming an important trend among practitioners. This approach aims at shortening the development process and therefore reducing the cost. Once developed and tested, components are reused and glued together in different contexts. The testing of such systems formed by reused components remains an open and challenging issue [21], mainly because components are developed and reused by different people without or with very little information sharing.

The purpose of this paper is to propose a formal approach for testing component-based communicating systems, which we call *compositional testing (C-method)*. Here, communication systems are perceived as being built from components that can be modeled as FSMs. Each of these components is tested using well-proven

M.Ü. Uyar, A.Y. Duale, and M.A. Fecko (Eds.): TestCom 2006, LNCS 3964, pp. 227–244, 2006.

techniques, such as the UIOv-method [20] or the Wp-method [7]. However, when these components are composed, no monolithic FSM is constructed in order to derive test cases for the composite system, which would lead to lengthy test cases, large test suites, and a repetition of tests already performed on component level. Instead, the composite system is only tested for *composition faults*, i.e., faulty composition code (also called *glue code*) - a new type of fault that extends and complements the classical fault model. We position our compositional testing approach among the existing and related techniques that also view systems as a set of interacting components, such as interoperability testing, testing in context and other compositional testing techniques.

In this paper, we will develop these ideas up to a certain point, and illustrate them through examples. We focus on a specific type of composition, called *concurrent com-position*. However, other types of composition may be considered as well. Section 2 defines the concurrent composition of asynchronously communicating FSMs, and states necessary conditions for composability. The *compositional test method* (*C-method*) is defined in Section 3. An application of the C-method is shown in Section 4. In Section 5, related work is reviewed and the contributions of this paper are positioned. We draw conclusions and indicate future research topics in Section 6.

2   Concurrent Composition

In this section, we define the concurrent composition of two FSMs. Further types of composition such as sequential composition are perceivable, for instance, in the context of micro protocols [8] or general component-based software systems. At specification level, composition can be expressed by defining a *composition operator*. At implementation level, this operator is usually realized by a piece of code that we call *glue code*.

Concurrent composition may be applied to put local and/or remote components together. From the conceptual viewpoint, this should not make any difference. For instance, we may compose protocol entities $PE_{1,1}$ and $PE_{1,2}$ as well as $PE_{1,2}$ and $PE_{2,2}$

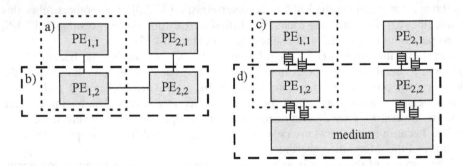

Fig. 1. Concurrent composition of protocol entities

concurrently, as shown in Figure 1a and b, respectively. For the local composition, the glue code may consist of internal data structures and operations to add signals to the input queue of the other protocol entity (Figure 1c). For the remote composition, the glue code may comprise an entire logical communication medium, which may in turn be a composite system (Figure 1d). From the practical viewpoint, the usual constraints concerning observability and controllability apply, which may be handled by external coordination procedures.

In this paper, we use the standard definition of FSM, and a derived notion:

<u>Definition 1</u>: A *finite state machine (FSM)* M is a tuple (S,I,O,s_0,λ_e) with:

- S is a finite set of states.
- I is a finite input alphabet.
- O is a finite output alphabet.
- $s_0 \in S$ is the initial state.
- $\lambda_e \subseteq S \times I \times O \times S$ defines the *transitions* of M.

A finite state machine is completely specified, if for each state and each input, a transition is defined. There exist several ways to extend a given FSM to a completely specified machine, e.g., by assuming implicit transitions (cf. SDL [10]). The standard definition of FSMs (see Definition 1) does not distinguish between explicit and implicit transitions. We consider explicit transitions as regular behavior. Implicit transitions are undesired behavior, but included to enhance testability of the implementation. In this paper, we adopt this interpretation, but the proposed test method does work for any interpretation of implicit transitions.

<u>Definition 2</u>: A *completely specified finite state machine (csFSM)* $N = (S,I,O_e,s_0,\lambda)$ is derived from an FSM $M = (S,I,O,s_0,\lambda_e)$ as follows:

- S, I, s_0 as in M.
- $O_e = O \cup \{e\}$, where $e \notin O$ is called *error output*.
- $\lambda = \lambda_e \cup \lambda_i$ is the transition relation of N. Tuples of λ are called *transitions* of N.
- λ_e defines the *explicit transitions* of N.
- $\lambda_i = \{ (s,i,e,s) \mid s \in S \land i \in I \land \neg \exists o \in O, s' \in S: (s,i,o,s') \in \lambda_e \}$ defines the *implicit transitions* of N.

In the rest of the paper, we omit the error output e and the relation λ_i for brevity.

To define the concurrent composition of csFSMs, we assume that they communicate by asynchronous reliable signal exchange, where sending and receiving of signals is modeled as output and input of the communicating csFSMs, respectively. Therefore, an input queue collecting signals that are delivered, but not yet consumed, is associated with each csFSM. Furthermore, each signal carries identifications of the sending and receiving machine, which may be evaluated as needed. The identifications are determined dynamically from the sending machine, the connection structure of the communicating csFSMs consisting of typed channels, and explicit addressing, if necessary.

<u>Definition 3</u>: Let $N_1 = (S_1,I_1,O_1,s_{0,1},\lambda_1)$ and $N_2 = (S_2,I_2,O_2,s_{0,2},\lambda_2)$ be csFSMs. Let $OI_{1,2} = O_1 \cap I_2$ $(OI_{2,1} = O_2 \cap I_1)$ be the set of signals exchanged between N_1 and N_2 (N_2 and N_1), called *internal signals*. The *concurrent composition* of N_1 and N_2, denoted $N_1 \parallel N_2$, is defined by the derived state machine $Q = (S,I,O,s_0,\lambda)$ with:

- $S = S_1 \times I_1^* \times S_2 \times I_2^*$ is the set of states.
- $I = (I_1 - OI_{2,1}) \cup (I_2 - OI_{1,2})$ is the (finite) input alphabet.
- $O = (O_1 - OI_{1,2}) \cup (O_2 - OI_{2,1})$ is the (finite) output alphabet.
- $s_0 = (s_{0,1},<>,s_{0,2},<>)$ is the initial state, consisting of the initial states of N_1 and N_2 and the initial states of *input queues* associated with N_1 and N_2, respectively.
- $\lambda \subseteq S \times I \times O \times S$ is the transition relation of Q. Tuples of λ are called *transitions* of Q. λ is derived from λ_1 and λ_2 as follows:

$$(s,i,o,s') \in \lambda \text{ with } s = (s_1,q_1,s_2,q_2) \text{ and } s' = (s_1',q_1',s_2',q_2') \text{ iff}$$
$$(\exists(s_1,i,o,s_1') \in \lambda_1: (q_1 = <i>^\frown q_1' \wedge q_2' = \text{if } o \in OI_{1,2} \text{ then } q_2^\frown<o>$$
$$\text{else } q_2 \wedge s_2 = s_2')) \vee$$
$$(\exists(s_2,i,o,s_2') \in \lambda_2: (q_2 = <i>^\frown q_2' \wedge q_1' = \text{if } o \in OI_{2,1} \text{ then } q_1^\frown<o>$$
$$\text{else } q_1 \wedge s_1 = s_1'))$$

This definition includes the concurrent composition of two independent csFSMs, i.e., two csFSMs that do not exchange signals. In this case, $OI_{1,2} = OI_{2,1} = \{\}$.

A csFSM can be represented as a labeled directed graph, where states correspond to nodes, and transitions correspond to edges labeled with input and output.

<u>Definition 4</u>: A *labeled directed graph* G is a tuple (V,L,E), consisting of a set of nodes V, a set of labels L, and a relation $E \subseteq V \times V \times L$, defining the directed edges of the graph. A *path* is a non-empty sequence of consecutive edges. A *tour* is a path that starts and ends at the same node. It is called *minimal*, if no edge is contained more than once in the tour. An *initial tour* is a tour that starts and ends at the initial node. A directed graph G is *strongly connected*, if for each pair of nodes (v,v'), where $v \neq v'$, there is a path from v to v'.

<u>Example 1</u>: Figure 2 shows the concurrent composition of deterministic, strongly connected csFSMs N_1 and N_2. Note that the error output as well as the implicit transitions are not shown in the figure. The machines interact via channel ch, which is typed by

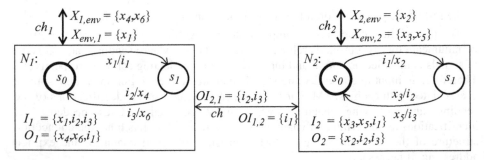

Fig. 2. Concurrent composition: component machines N_1 and N_2 (Example 1)

$OI_{1,2}$ and $OI_{2,1}$, and are connected to the environment by typed channels ch_1 and ch_2. The resulting behavior after composition (see Figure 3) can be represented by the state machine $Q = N_1 \| N_2$, where states are represented as tuples (s_1,q_1,s_2,q_2) denoting the states of N_1 and N_2, and of their input queues.

While it is syntactically possible to compose all kinds of csFSMs, this is not always meaningful. Which csFSMs to compose first of all depends on the intended global behavior, which is problem specific. However, some *general composition criteria* can be stated:

CC$_1$. Internal signals of either machine are eventually consumed by the other machine in an *explicit* transition, i.e., the composed system is *free of internal un-specified receptions*. This excludes transitions that have been added to obtain a completely specified state machine, i.e., implicit transitions yielding an error output (see Definition 2).

CC$_2$. The composed system is free of *internal deadlocks*. Since it is assumed that external signals can be produced in any order, this again restricts the internal interaction only.

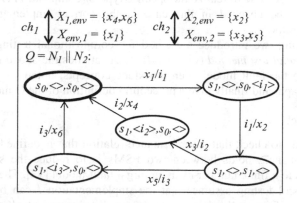

Fig. 3. Concurrent composition: derived machine Q (Example 1)

3 Compositional Testing of Concurrently Composed csFSMs

In this section, we will show how to derive test suites for testing the implementation of concurrently composed csFSMs. We make certain assumptions about the component csFSMs (e.g., strongly connected, deterministic) and their implementations (e.g., concerning the number of states), and we assume that the implementation of each csFSM can be tested using a test method that detects all output and transfer faults.

A direct approach to test the composite system would be to determine its global state machine, and then apply one of the existing test methods to derive test cases for this machine. This, however, has the following drawbacks:

- The state set of the global state machine may be very large. Firstly, this can consume considerable computational resources to determine the machine. Secondly, it can lead to a large test suite containing lengthy test cases, implying a testing effort that could quickly become unmanageable.
- The global state machine may be non-deterministic, due to the concurrency of the composite system, which reduces the applicability of existing test methods.
- All tests already executed at the component level are repeated. This is a severe drawback in general, and especially if components are to be reused in different protocol configurations.

To avoid these disadvantages, a test method satisfying the following properties is sought:

- It is not necessary to compute the global state machine.
- Only tests checking the correctness of the glue code of the csFSMs are derived.
- Tests already performed at the component level are not repeated.

These properties can only be satisfied if the implementations of the design components, which have been tested at the component level, remain unchanged. This means that only glue code to realize the specific type of composition is added, and all what remains to be checked in this case is the correct implementation of the composition operator.

In the following, we introduce a method for compositional testing - henceforth called *compositional test method* (*C-method*) - that satisfies the above properties. We start by defining the fault model, then introduce concepts, notations, and an initial tour coverage graph, and finally give a procedural definition of the C-method.

3.1 Fault Model

The common way to check that a conformance relation that is defined on an infinite set of input sequences holds between two FSMs is to reduce the set of possible implementations to a finite number by assuming a fault model [13]. The classical fault model for protocol testing assumes that the implementation I can be treated as a *mutant* of the specification S, where a mutant may be obtained by altering outputs of transitions (*out-put faults*), by altering tail states of transitions (*transfer faults*), by adding states up to a given number as well as extra transitions to and from these states. This general fault model is sometimes reduced to output and transfer faults by assuming that the number of implementation states is less than a given maximum number, and to deterministic implementations.

Implementations are tested by applying input sequences and observing the output sequences. An implementation fault is detected, if an observed output sequence differs from the expected output sequence. Whether this fault is an output fault or a transfer fault, or due to an extra state or an extra transition, depends on the fault model, on the diagnosis capability of the test method, and on the knowledge about the implementation at the time of test execution.

The classical fault model is usually applied to single components that are specified by an FSM, e.g., a single protocol entity. It may also be applied to a composite system, e.g., protocol entities and an underlying medium, if an FSM of that system can be constructed. This, however, causes the aforementioned problems (large state spaces, non-determinism, repetition of tests). In order to avoid these problems, we propose to take the structural aspect of the composition into account, and to distinguish the following fault categories:

- *component fault*: the implementation of a *component* does not satisfy its specification
- *composition fault*: the *glue code* does not satisfy its specification in the given contex

The problem of compositional testing can then be stated as follows:

Let N_1 and N_2 be the specifications of two components, and I_1 and I_2 be their implementations, where I_1 and I_2 satisfy their specifications N_1 and N_2, respectively. Then, derive a minimal test suite that is sufficient to check whether the system I consisting of I_1, I_2, and glue code satisfies the specification $N_1 \parallel N_2$.

As usual, implementations are tested by applying input sequences, and comparing the observed and the expected output sequences. Again, it depends on the fault model, the diagnosis capability of the test method, and the knowledge about the implementation at the time of test execution how a detected fault may be classified. For instance, if the components have already been tested successfully, and their implementations are reused in the composite system, then detected faults can be classified as composition faults.

To derive a minimal test suite that is sufficient to check the composed system, a model of the glue code is needed. In general, the glue code could be a component or a composite system itself, for instance, a logical communication medium, which may have further attached components. As testing would be unfeasible in this general setting, we make the following assumption:

i) Whenever I_1 and I_2 are both in their initial states, the glue code is in a determined state w.r.t. I_1 and I_2.

ii) The behavior of the glue code is deterministic w.r.t. I_1 and I_2.

iii) If the glue code interacts with other components, this has no effect on its behavior towards I_1 and I_2.

iv) The glue code is not creating messages for I_1 or I_2.

The first assumption limits the maximum length of test suites to the set of all initial tours, i.e., paths that start and end in the initial state. All assumptions together ensure that a finite number of test cases are sufficient.

Notice that if a model of the glue is given as an FSM, then the composition fault could be refined further into the same basic faults of an FSM based implementation.

3.2 Concepts and Notations

The following definitions recall and introduce some concepts and notations for testing:

Definition 5: A *test case tc* is a non-empty sequence of inputs $i_1.i_2.....i_n$. A *test suite ts* is a non-empty set of test cases $\{tc_1,tc_2,...,tc_m\}$. An *augmented test case atc* is defined as a non-empty sequence of transitions (also called *test elements*) $i_1/o_1.i_2/o_2.....i_n/o_n$. An *augmented test suite ats* is a non-empty set of augmented test cases $\{atc_1,atc_2,...,atc_m\}$.

Definition 6: Let atc_1 and atc_2 be augmented test cases (sequences of transitions) of deterministic csFSMs N_1 and N_2 that communicate via a common channel ch with sets $OI_{1,2}$ and $OI_{2,1}$ of internal signals. The *concurrent composition of atc_1 and atc_2*, denoted $atc_1 \| atc_2$, is one path $atc_{1,2}$ of the tree obtained by sequencing the test elements in atc_1 and atc_2 according to the following ordering constraints:

- the order of test elements of atc_1 and atc_2 is preserved;
- a test element of atc_1 (atc_2) triggered by an internal signal is constrained by the corresponding test element in atc_2 (atc_1) that produces this internal signal;
- the order of outputs is preserved.

Example 2: For the csFSMs N_1 and N_2 of Example 1, the following augmented test cases can be derived and composed:

- $atc_1 = x_1/i_1.i_2/x_4$
- $atc_2 = i_1/x_2.x_3/i_2$
- $atc_1 \| atc_2 = x_1/i_1.i_1/x_2.x_3/i_2.i_2/x_4$

In this case, the composition produces only one path because the test elements are totally ordered.

Definition 7: The concurrent composition of two augmented test cases is called *complete*, iff all their test elements are included, and the input queues of the corresponding csFSMs will be empty after their execution. Otherwise, it is called *incomplete*.

Example 3: The concurrent composition of atc_1 and atc_2 in Example 2 is complete. However, the concurrent composition of atc_1 and $atc_2' = i_1/x_2$ results in $x_1/i_1.i_1/x_2$, which is incomplete.

3.3 Initial Tour Coverage Tree

Selected augmented test cases of components form the basis for deriving a test suite for validating the correct implementation of their composition. These test cases are derived from a so-called *initial tour coverage tree*, reduced to the set of relevant test cases, and composed with matching test cases of the other component.

Definition 8: Let $N = (S,I,O_e,s_0,\lambda)$ be a csFSM with the underlying graph G, where G is strongly connected. An *initial tour coverage tree* T is a tree containing all minimal initial tours such that every edge is covered at least once and no tour is contained as a prefix or a suffix of another tour in the set.

The rationale behind this choice is that (i) transition coverage can be achieved this way[1], and that (ii) both automata should be synchronized at least in their initial states, a criterion for composability. The concept of initial tour coverage is different from minimal transition tour, which visits every transition once and only once, but which also relies on stronger conditions to exist. To construct an initial tour coverage tree, we use a tree that, for a given state, captures all cycle free paths to the initial state, called *hom-ing tree*:

<u>Definition 9</u>: Given a csFSM $N = (S, I, O_e, s_0, \lambda)$ and a state $s \in S$, where the underlying graph is strongly connected, a *homing tree* $H(s)$ is a minimal tree that covers all cycle-free paths of N leading from s to the initial state s_0.

We give algorithms for the construction of homing trees and initial tour coverage trees in Tables 1 and 2, respectively. Both algorithms are illustrated.

Table 1. Construction of a homing tree $H(s)$

Step 1: Start the construction of $H(s)$ with its root node n_r, labeled with s.

Step 2: Assume that $H(s)$ has been constructed up to level k, $k \geq 1$. Then level $k+1$ is built by examining the nodes of level k:

> Step 2.1: A node n of level k is terminated, if its label is identical to the label of a node on level j, where $1 \leq j < k$, or if it is identical to s_0.

> Step 2.2: Otherwise, let s denote the label of node n. Then, for all transitions (s, x, y, s'), attach a branch and successor node to the current node, labeled x/y and s', respectively.

Step 3: Prune the resulting tree by successively removing all leaf nodes that have a label $s \neq s_0$, and the corresponding edges.

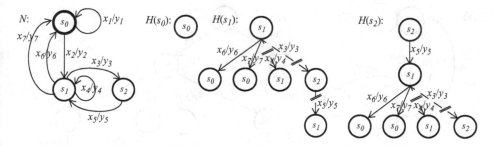

Fig. 4. Homing trees (example)

3.4 The C-Method

In Section 2, we have stated general composition criteria CC_1 and CC_2 that should be satisfied for a meaningful composition at the design level. First, the composed system

---

[1] Initial tour coverage is a reduced form of path coverage.

should be free of internal unspecified receptions, which means that receptions occurring during „normal operation" have to be consumed by explicit transitions. This excludes transitions that have been added for mere technical reasons to obtain fully specified state machines (see Definition 2). Also, the composed system should be free of internal deadlocks.

To check whether two csFSMs N_1 and N_2 meet these criteria, we assume that they are always capable to resynchronize in their initial states. In other words, if N_1 is in its initial state and stays there, N_2 should be able to reach its initial state without further

Table 2. Construction of an initial tour coverage tree T

Step 1: For each state s of N, construct a homing tree $H(s)$.

Step 2: Start the construction of T with the root node n_r, labeled with the initial state s_0 of N. This is level 1 of T.

Step 3: Assume that T has been constructed up to level k, $k \geq 1$. Then level $k+1$ is built by examining the nodes of level k:

Step 3.1: A node n of level k is terminated, if its label is identical to the label of a node on level j, where $1 \leq j < k$.

Step 3.2: Otherwise, let s denote the label of n. Then, for each transition (s,x,y,s'), attach a branch and successor node to the current node, labeled x/y and s', respectively.

Step 4: To each leaf node n, attach the homing tree $H(s)$ by merging the root node of $H(s)$ with n, where s denotes the label of n.

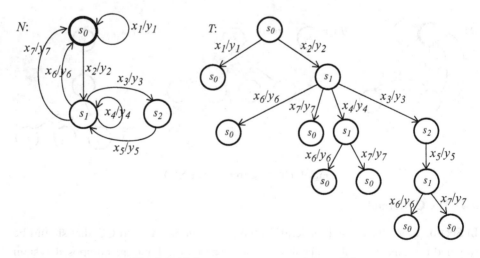

Fig. 5. Initial tour coverage tree (example)

Table 3. The C-method

<u>C-method</u>

Step 1: Test the implementations I_1 and I_2 of components N_1 and N_2.

    Step 1.1: Select a test method (e.g., DS [11], UIOv [20], Wp [12]).

    Step 1.2: Derive the test suites for N_1 and N_2.

    Step 1.3: Execute the tests. If all tests are successful, continue with Step 2. If not, correct the faults and repeat Step 1.

Step 2: Test the implementation of the concurrent composition of N_1 and N_2.

    Step 2.1: Remove all transitions of N_1 and N_2 that yield an error output. These transitions have already been tested during component testing, and need not be tested again.

    Step 2.2: Build the initial tour coverage trees for N_1 and N_2, and determine all maximal paths, i.e., all paths that start at the root node and end at a leaf node, constituting augmented test suites ats_1 and ats_2.

    Step 2.3: From the augmented test suites ats_1 (ats_2), remove all internally triggered test cases, i.e., those test cases that are triggered by N_2 (N_1).

    Step 2.4: From the augmented test cases, remove all local tours, i.e., (sub)sequences of test case elements that (1) start and end in the same state, and (2) contain only external inputs and outputs. They have already been checked during component testing, and need not be tested again.

    Step 2.5: Remove the maximum suffix that does not contain an interaction with the other component. These test elements have been checked already.

    Step 2.6: For each test case $atc_{1,j}$ of the augmented test suite ats_1 after Step 2.5, find an augmented test case $atc_{2,j}$ of N_2 from Step 2.2 such that $atc_{1,j} \parallel atc_{2,j}$ is complete, and determine $atc_{1,2,j} = atc_{1,j} \parallel atc_{2,j}$, yielding the concurrent augmented test suite $ats_{1,2}$. Analogously for each test case $atc_{2,j}$ of ats_2.

    Step 2.7: Based on ats_1, ats_2, and $ats_{1,2}$, check whether N_1 and N_2 meet the composition criteria CC_1 and CC_2, i.e., whether for each test case of ats_1 (ats_2), there is a matching test case of N_2 (N_1). Yes: continue with Step 2.8; no: stop.

    Step 2.8: For each test case in $ats_{1,2}$: merge adjacent test case elements in cases where (1) the internal output of the first matches the internal input of the second, and (2) the output is the only signal in the queue after being sent. Replace internal inputs and outputs by "-", and remove test case elements "-/-".

    Step 2.9: Execute the test.

interaction with N_1, and vice versa. If this assumption is satisfied, it suffices to consider the explicit initial tours of both automata, i.e., the explicit transition sequences starting and ending in the initial states, and to check whether for each explicit initial tour, there is a matching explicit initial tour of the other automaton such that their concurrent composition is complete. This design criterion can also be stated in terms of concurrent composition of augmented test suites, and thus be checked as a by-product of test case derivation.

In Table 3, the C-method is defined in a procedural style. We point out that in the course of applying the test procedure, it is checked whether N_1 and N_2 satisfy the composition criteria. This is a constraint imposed on design level, which should be checked before implementing the design and testing the implementation. Thus, all steps except Steps 1.2, 1.3, 2.8, and 2.9 should be executed in the design phase. Step 2.6 could be optimized further by reducing the number of considered compositions (see [5]).

As expected, the augmented test suites ats_1 and ats_2 are reduced to empty test suites in case N_1 and N_2 do not interact, i.e., in case of independent concurrent composition, which, among other things, satisfies the criterion for concurrent composability. The rea son is that all necessary testing has already been done on component level. Of course, one can argue that in the implementation, interaction of the two components may occur, and has to be excluded. This, however, is not covered by this type of tests. When protocol components are reused, it is sufficient to test them once, which means in a certain sense that testing is reused, too. In these cases, compositional testing starts with Step 2.

4   Application of the C-Method

To illustrate the C-method, we apply it to the Initiator Responder (InRes) protocol [9]. The InRes protocol is a connection-oriented communication protocol for the reliable exchange of message over an order-preserving, connection-less medium. It provides an asymmetrical service: the initiator requests connections and sends data, the responder accepts, refuses, and clears connections, and receives data. In this example, the InRes protocol entities I and R are the components that are composed concurrently, yielding a composite system $I \parallel R$. In the implementation of this system, the glue code is represented by the underlying medium. To be able to use this medium for the implementation of the $I \parallel R$, we assume that it does not lose messages.

Figure 6 shows the specifications I and R of the InRes protocol entities and their concurrent composition. Both automata contain further transitions that can be derived by applying Definition 2, and thus are fully-specified. To avoid cluttering, we have omitted these transitions in the figure. The underlying graphs are deterministic, and strongly connected. We assume that Step 1 of the C-method that tests the implementations of I and R separately has already been executed successfully. Below, we go through Step 2:

- Step 2.1: Removal of transitions yielding an error output
  These transitions have been omitted in the figure, therefore, starting point for Step 2.2 are the finite state automata shown in Figure 6.

- Step 2.2: Build initial tour coverage trees, and determine ats_I and ats_R.
  The initial tour coverage trees for I and R are shown in Figure 7. Test suites are:
  $ats_I = \{atc_{I,1}, atc_{I,2}, atc_{I,3}, atc_{I,4}\}$, with

  $\quad atc_{I,1}$ = ICONreq/CR . DR/IDISind

  $\quad atc_{I,2}$ = ICONreq/CR . CC/ICONcnf . DR/IDISind

  $\quad atc_{I,3}$ = ICONreq/CR . CC/ICONcnf . IDATreq/DT . DR/IDISind

  $\quad atc_{I,4}$ = ICONreq/CR . CC/ICONcnf . IDATreq/DT . AK/-. DR/IDISind

  $ats_R = \{atc_{R,1}, atc_{R,2}, atc_{R,3}, atc_{R,4}\}$, with

  $\quad atc_{R,a}$ = DT/$atc_{R,b}$ = CR/ICONind . IDISreq/DR

  $\quad atc_{R,c}$ = CR/ICONind . ICONrsp/CC . IDISreq/DR

  $\quad atc_{R,d}$ = CR/ICONind . ICONrsp/CC . DT/IDATind . -/AK . IDISreq/DR

- Step 2.3: Remove test cases triggered by internal inputs.
  All test cases of R are triggered by inputs of the Initiator and therefore removed:
  $ats_I{}' = \{atc_{I,1}, atc_{I,2}, atc_{I,3}, atc_{I,4}\}$
  $ats_R{}' = \{\}$

- Step 2.4: Remove external local tours.
  Not applicable in the InRes example.

- Step 2.5: Remove suffix containing external interaction only.
  Not applicable in the InRes example.

- Step 2.6: For each augmented test case in ats_I (ats_R) after Step 2.5, find an augmented test case of R (I) from Step 2.2 such that their concurrent composition is complete, and determine the concurrent augmented test suite $ats_{I,2}$.

  $ats_{I,2} = \{ atc_{I,1} \| atc_{R,a}, atc_{I,2} \| atc_{R,b}, atc_{I,3} \| atc_{R,c}, atc_{I,4} \| atc_{R,d} \}$, with:

  $atc_{I,1}$ = ICONreq/CR . DR/IDISind

  $\quad atc_{R,a}$ = CR/ICONind . IDISreq/DR

  $\quad atc_{I,1} \| atc_{R,a}$ = { ICONreq/CR . CR/ICONind . IDISreq/DR . DR/IDISind }

  $atc_{I,2}$ = ICONreq/CR . CC/ICONcnf . DR/IDISind

  $\quad atc_{R,b}$ = CR/ICONind . ICONrsp/CC . IDISreq/DR

  $\quad atc_{I,2} \| atc_{R,b}$ = { ICONreq/CR . CR/ICONind . ICONrsp/CC . CC/ICONcnf .
  $\qquad\qquad$ IDISreq/DR . DR/IDISind,
  $\qquad\qquad$ ICONreq/CR . CR/ICONind . ICONrsp/CC . IDISreq/DR .
  $\qquad\qquad$ CC/ICONcnf . DR/IDISind }

  $atc_{I,3}$ = ICONreq/CR . CC/ICONcnf . IDATreq/DT . DR/IDISind

  $\quad atc_{R,c}$ = CR/ICONind . ICONrsp/CC . IDISreq/DR . DT/-

  $\quad atc_{I,3} \| atc_{R,c}$ = { ICONreq/CR . CR/ICONind . ICONrsp/CC . CC/ICONcnf .
  $\qquad\qquad$ IDATreq/DT . IDISreq/DR . DR/IDISind . DT/-,
  $\qquad\qquad$ ICONreq/CR . CR/ICONind . ICONrsp/CC . CC/ICONcnf .
  $\qquad\qquad$ IDATreq/DT . IDISreq/DR . DT/- . DR/IDISind,
  $\qquad\qquad$ ICONreq/CR . CR/ICONind . ICONrsp/CC . CC/ICONcnf .
  $\qquad\qquad$ IDISreq/DR . IDATreq/DT . DT/- . DR/IDISind,
  $\qquad\qquad$ ICONreq/CR . CR/ICONind . ICONrsp/CC . IDISreq/DR .
  $\qquad\qquad$ CC/ICONcnf . IDATreq/DT . DT/- . DR/IDISind }

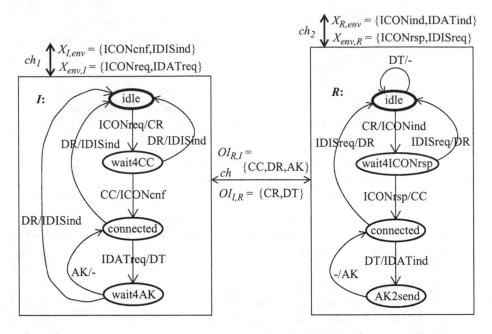

Fig. 6. InRes protocol entities *I* and *R*

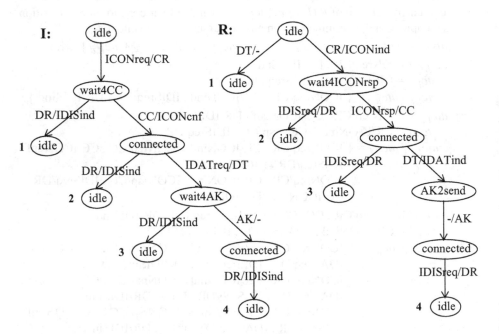

Fig. 7. Initial tour coverage trees of *I* and *R*

$atc_{I,4}$ = ICONreq/CR . CC/ICONcnf . IDATreq/DT . AK/- . DR/IDISind
$atc_{R,d}$ = CR/ICONind . ICONrsp/CC . DT/IDATind . -/AK . IDISreq/DR
$atc_{I,4}$ ‖ $atc_{R,d}$ = { ICONreq/CR . CR/ICONind . ICONrsp/CC . CC/ICONcnf .
                    IDATreq/DT . DT/IDATind . -/AK . AK/- . IDISreq/DR .
                    DR/IDISind, ICONreq/CR . CR/ICONind . ICONrsp/CC .
                    CC/ICONcnf . IDATreq/DT . DT/IDATind . -/AK .
                    IDISreq/DR . AK/- . DR/IDISind}

- Step 2.7: Check the composition criteria CC_1 and CC_2
  For each test case of ats_I, there is a test case of R such that their concurrent composition is complete. This trivially holds for ats_R, which is empty.
- Step 2.8: Merge test case elements, and replace internal inputs and outputs by „-"
  $ats_{I,2}$ = { $atc_{I,1}$ ‖ $atc_{R,a}$, $atc_{I,2}$ ‖ $atc_{R,b}$, $atc_{I,3}$ ‖ $atc_{R,c}$, $atc_{I,4}$ ‖ $atc_{R,d}$ }, with:
  $atc_{I,1}$ ‖ $atc_{R,a}$ = {ICONreq/ICONind . IDISreq/IDISind }
  $atc_{I,2}$ ‖ $atc_{R,b}$ = {ICONreq/ICONind . ICONrsp/ICONcnf . IDISreq/IDISind }
  $atc_{I,3}$ ‖ $atc_{R,c}$ = { ICONreq/ICONind . ICONrsp/ICONcnf .
                    [IDATreq/- ‖‖ IDISreq/-] . -/IDISind }
  $atc_{I,4}$ ‖ $atc_{R,d}$ = {ICONreq/ICONind . ICONrsp/ICONcnf . IDATreq/IDATind .
                    IDISreq/IDISind }

Note that test case $atc_{I,3}$ ‖ $atc_{R,c}$ requires that test input IDATreq and IDISreq are to be applied concurrently to stimulate this behavior. This is expressed by the notation [tce_1 ‖‖ tce_2]. The resulting test suite $ats_{I,2}$ consists of 4 test cases, with 14 test case elements. In addition, component tests are to be performed.

5 Related Work

The purpose of this section is not to review deeply all the rich literature on FSM-based testing, but to position the proposed C-method with respect to the related types of testing such as interoperability testing, testing in context and compositional testing.

5.1 Interoperability Testing

Interoperability testing [1,2,4,5] aims at checking if two implementations, which are conforming to a common specification, interact correctly and provide a required service when interconnected through a communication medium. The communication medium, i.e. the glue between the two protocol entities, is assumed to behave correctly. This is different from the C-Method, where we assume that any integration problem or fault is coming from the glue, once the components have been individually tested. In addition, the C-Method also checks whether a required service is provided (see Step 2.7 - composition criteria checking). However, these general composition criteria are checked at the specification level in case of the C-Method, while it is done at testing time in case of interoperability testing.

5.2 Testing in Context

Testing in context [14,15] consists of testing a component Cp in a given context Cx formed by other components, with the purpose of detecting faults in Cp. The component Cp is generally not directly observable from the environment or only partially. The specifications of context Cx and component Cp are both available.

Testing in context is about testing the component Cp, not the behavior of the whole system. However, since the component is not directly accessible, or only partially, but it is tested through its context. Therefore, we select the observable behavior of the global system that will stimulate the behavior of the component as much as possible, and interpret the system output. This system reaction is generally coming from the context following a reaction from the component under test. The global state space is generally not constructed.

The aim of the C-method is to test a composed system that consists of n components by testing individually each component against its specification, by checking the composability of these components, and by testing the glue, which is putting all these components together to obtain a particular system. It aims at validating the whole system instead of the glue in context only, but by testing only portions of the system behavior. Once the components are tested successfully, their behavior is not questioned anymore. If an error happens, only the behavior of the glue is in question.

5.3 Compositional Testing

An approach for compositional testing has been proposed in [19]. It is based on *ioco* and therefore on a synchronous communications setting. The aim of this approach is to find the conditions under which the conformance of the components to their respective specification leads automatically, without any extra testing, to the conformance of the system implementation to the system specification. The operator considered so far is parallel composition. There is no glue code in this approach.

6 Conclusions and Future Work

In this paper, the *compositional method* (*C-method*) for testing communicating systems has been introduced. The C-method first tests each protocol component separately for component faults (output and/or transfer faults), using one of the traditional test methods, and then checks their composition for composition faults.

To apply the C-method, it is not necessary to compute the global state machine. Instead, composition tests are derived from local initial tour coverage trees. Only tests checking the glue code are derived. We have introduced and justified a fault model for the glue code that leads to manageable composition test suites.

The work on compositional testing has been triggered by the component based software engineering trend and our results on micro protocols [8], a concept to structure communication systems and to foster reuse of protocol designs. Micro protocols are protocols with a single (distributed) functionality and the required collaboration among protocol entities. To develop customized communication systems, micro protocol designs are selected from a library, composed to yield a complete design, and implemented. We expect that the C-method will contribute to the testing of customized communication systems that are composed of micro protocols.

The results presented in this paper leave room for further work. The following improvements and enhancements are perceivable:

- So far, only composition of two FSMs has been considered. It would be useful to extend the C-method to compositions of more than two FSMs, and also to the composition of composites that have already been tested successfully.
- Other types of compositions, for instance, concurrent composition with shared variables, or composition through inheritance, are perceivable. Again, this requires extensions to the C-method.
- The justification of the C-method and its benefits should be treated more rigorously, developing a test theory rich enough to provide a formal proof that the derived test suite is both necessary and sufficient to detect composition faults.
- The complexity of the C-method in comparison to other testing approaches should be formally assessed. Since the C-method exploits the structure of the system under test to reduce the number and length of test cases, we expect significant improvements.

Finally, a generic testing approach, where interoperability testing, testing in context, and compositional testing are seen as specific instances with different goals and assumptions, will be an interesting research issue to pursue.

References

[1] R. Castanet, O. Kone: *Deriving Coordinated Testers for Interoperability*, Protocol Test Systems, Volume VI C-19, Pau, France, 1994

[2] R. Castanet and O. Kone: *Test Generation for Interworking Systems*, Computer Communications, Elsevier, Vol. 23, 2000, pp. 642 652.

[3] T. S. Chow: *Testing Software Design Modeled by Finite-State Machines*, IEEE Transactions on Software Engineering, Vol. 4, No. 3, 1978, pp. 178-187

[4] A. Desmoulin, C. Viho: *Quiescence Management Improves Interoperability Testing*, Proceedings of TestCom 2005, LNCS 3502, Springer, pp. 365-379

[5] A. Desmoulin, C. Viho: *Formalizing interoperability testing: Quiescence management and test generation*, Proceedings of FORTE'2005, LNCS 3731, Springer, pp. 533-537

[6] A. En-Nouaary, R. Dssouli, F. Khendek: *Timed Wp: Testing Real-Time Systems*, IEEE Transactions on Software Engineering, Vol. 28, No. 11, November 2002, pp. 1023-1038

[7] S. Fujiwara, G. v. Bochmann, F. Khendek, M. Amalou, A. Ghedamsi: *Test Selection Based on Finite State Models*, IEEE ToSE, Vol. 17, No. 6, June 1991, pp. 591-603

[8] R. Gotzhein, F. Khendek, P. Schaible: *Micro Protocol Design: The SNMP Case Study*, SDL and MSC Workshop (SAM'2002), LNCS 2599, Springer, 2002

[9] D. Hogrefe: *OSI Formal Specification Case Study: The InRes Protocol and Service, revised*, Report No. IAM-91-012, Update May 1992, University of Berne, May 1992

[10] ITU-T Recommendation Z.100 (11/99) - *Specification and Description Language (SDL)*, International Telecommunication Union (ITU), 1999

[11] Z. Kohavi: *Switching and Finite Automata Theory*, McGraw Hill, USA, 1978

[12] G. Luo, G. v. Bochmann, A. Petrenko: *Test Selection Based on Communicating Nondeter-ministic Finite-State Machines Using a Generalized Wp-Method*, IEEE Transactions on Software Engineering, Vol. 20, No. 2, 1994, pp. 149-162

[13] A. Petrenko, G. v. Bochmann, M. Yao: *On Fault Coverage of Tests for Finite State Specifications*, Computer Networks and ISDN Systems, Vol. 29, 1996, pp. 81-106

[14] A. Petrenko, N. Yevtushenko, G. v. Bochmann, R. Dssouli: *Testing in Context: Framework and Test Derivation*, Computer Communications, Elsevier, Vol. 19, 1996, pp. 12361249

[15] A. Petrenko, N. Yevtushenko, G. v. Bochmann: *Fault Models for Testing in Context*, Proceedings of FORTE'96, pp. 163-178

[16] B. Sarikaya, G. v. Bochmann: *Some Experience with Test Sequence Generation for Proto-cols*, Proceedings of the 2nd International Workshop on Protocol Specification, Testing, and Verification, North Holland, 1982, pp. 555-567

[17] J. Springintveld, F. W. Vaandrager, P. R. D'Argenio: *Testing timed automata*, Theoretical Computer Science, Vol. 254 (1-2), 2001, pp. 225-257

[18] M. Ü. Uyar, Y. Wang, S. S. Batth, A. Wise, M. A. Fecko: *Timing Fault Models for Systems with Multiple Timers*, Proceedings of TestCom 2005, LNCS 3502, Springer, 192-208

[19] M. van der Bijl, A. Rensink, J. Tretmans: *Compositional Testing with ioco*, Proceedings of FATES 2003, LNCS 2931, 2003

[20] S. T. Vuong, W. W. L. Chan, M. R. Ito: *The UIOv-Method for Protocol Test Sequence Generation*, 2nd International Workshop on Protocol Test Systems, Berlin, Germany, 1989

[21] E. J. Weyuker: *Testing Component Based Software: A Cautionary Tale*, IEEE Software, September/October 1998, pp.54-59

FSM Test Translation Through Context

Khaled El-Fakih[1], Alexandre Petrenko[2], and Nina Yevtushenko[3]

[1] American University of Sharjah, UAE
[2] Centre de recherche informatique de Montreal (CRIM), Montreal, Canada
[3] Tomsk State University, Russia
kelfakih@aus.edu, petrenko@crim.ca, yevtushenko@elefot.tsu.ru

Abstract. In this paper, we define a formal approach for translating internal tests derived for a component embedded within a modular system into external tests defined over the external observable alphabets of the system. The system is represented as two communicating complete deterministic finite state machines, an embedded component machine to be tested and a context machine that represents the remaining part of the system. The context is assumed to be fault free and the interactions between the component machines are observable. When an internal test can not be translated in the given context, we demonstrate how another test with the guaranteed fault detection power could be determined (if such a test exists) that can be translated in the given context.

1 Introduction

The problem of testing in context is about testing a component embedded within a modular system that is usually represented as two communicating machines, an embedded component machine and a context machine that models the remaining part of the system and is assumed to be correctly implemented.

A number of test derivation methods have been proposed for testing in context [5,6,7,9,10] when the system components are modeled as Finite State Machines (FSMs). Some of these methods derive test suites with the guaranteed fault coverage directly from the embedded component [7,9,10]. However, such tests are generated in the form of input/output sequences defined over the input/output alphabets of the embedded machine. These tests have then to be translated into external tests defined over the external observable alphabets of the overall system. The problem of translating internal tests into external ones is known as the *fault propagation* or *test translation problem*. Different approaches for solving the translation problem for the case when the internal interactions between the component machines are unobservable are given in [2,7].

In this paper, we formally define and solve the test translation problem for the case when the interactions between the component FSMs are observable. Given an internal test for the embedded component, we present necessary and sufficient conditions for this test to be translated in the given context and show how to translate internal tests into external tests with the same fault detection power (if it is possible). When internal interactions are observable an external test that is a translation of an internal test has the same fault detection power as an internal test, i.e., it detects every faulty implementation of the embedded component that is detectable by the internal test in

M.Ü. Uyar, A.Y. Duale, and M.A. Fecko (Eds.): TestCom 2006, LNCS 3964, pp. 245–258, 2006.

isolation. If an internal test cannot be translated within the given context, we derive (when possible) another internal test with the same fault detection power that can be translated within the given context. For this purpose, a so-called observable equivalent of the embedded component is derived. The notion of the observable equivalent is close to the notion of the embedded equivalent in [10]. However, in that work, the observable equivalent is derived under the assumption that the internal channels are not observable; in fact, in this paper, the observable equivalent refines a so-called conforming part of the embedded component [10] by restricting it to internal alphabets. Any internal test case derived from the observable equivalent can be translated in the given context.

The paper is organized as follows. Section 2 contains definitions of IOTS, FSM, and other preliminaries. Section 3 includes a formal definition and a method for test translation with simple application examples. Section 4 presents a method for deriving, when possible, internal test suites with the guaranteed fault coverage that can be translated in the given context. Section 5 concludes the paper.

2   Preliminaries

2.1   Input Output Transition Systems and Finite State Machines

We assume in this paper that components of a modular system are FSMs, however, we find it more convenient to compose state machines by encoding them into IOTSs.

An *Input/Output Transition System* is a quintuple $A = \langle S, I, O, \lambda_A, s_0 \rangle$, where S is a finite nonempty set of states with the initial state s_0, $I \cup O$ is an alphabet of input and output actions such that $I \cap O = \varnothing$, and $\lambda_A \subseteq S \times (I \cup O \cup \{\tau\}) \times S$ is a transition relation, where $\tau \notin I \cup O$ is the internal action. We say that there is a *transition* from a state s to a state s' labeled with an action $v \in (I \cup O \cup \{\tau\})$ if and only if the triple (s, v, s') is in the transition relation λ_A.

For IOTS $A = \langle S, I, O, \lambda_A, s_0 \rangle$, we use $init(s)$ to denote the set of actions enabled in state $s \in S$, i.e., $init(s) = \{a \in (I \cup O \cup \{\tau\}) \mid \exists s' \in S\ ((s, a, s') \in \lambda_A)\}$. We use $in(s) \subseteq init(s)$ $(out(s) \subseteq init(s))$ to denote input (output) actions in state s. State $s \in S$ is called *stable* or *quiescent* if no output or internal actions are enabled in s: $init(s) \cap (O \cup \{\tau\}) = \varnothing$. Otherwise, s is called *unstable*.

State $s \in S$ with no action enabled, i.e., $init(s) = \varnothing$, is called a *deadlock* state. IOTS A *deadlocks* if there is a deadlock state reachable from the initial state. An IOTS is *deterministic* if it contains no internal action and the transition relation is a function, i.e., $(s, a, s_1), (s, a, s_2) \in \lambda_A$ for $a \in I \cup O$ implies $s_1 = s_2$.

As usual, the transition relation λ_A of the IOTS A is extended to sequences over the alphabet V. These sequences are usually called *traces* of the IOTS A. Given a state s of the IOTS A, the set of traces $Tr(s) = \{\alpha \in V^* \mid \exists s_i \in S\ ((s, \alpha, s_i) \in \lambda_A)\}$ is called the *language generated at the state s*. The language generated by the IOTS A at the initial state is called the *behavior of* or *language generated by* the *IOTS A*, denoted by $Tr(A)$. As usual, given a language L over the alphabet V, the *prefix closure* $\langle L \rangle$ contains each prefix of each sequence of L. The language is *prefix closed* if the language and its prefix closure coincide. By definition, the language of an IOTS is prefix closed.

Given a trace α over alphabet V, the *U-restriction* of α, written $\alpha_{\downarrow U}$, is obtained by deleting from α all symbols that belong to the set $V\backslash U$. Correspondingly, the *U-restriction* of a set T of traces over alphabet V, written $T_{\downarrow U}$, is the set of all sequences $\alpha_{\downarrow U}$, $\alpha \in T$. Given an IOTS $A = \langle S, I, O, \lambda_A, s_0 \rangle$ and $U \subseteq I \cup O$, the *U-restriction* $A_{\downarrow U}$ of A is obtained by replacing actions in $(I \cup O)\backslash U$ with the internal action τ and by determinizing [3] the resulting IOTS.

Let $A = \langle S, I, O, \lambda_A, s_0 \rangle$ and $B = \langle Q, I, O, \lambda_B, q_0 \rangle$ be two IOTSs, state s of IOTS A and state q of IOTS B are (trace) *equivalent*, if $Tr(s) = Tr(q)$. IOTSs A and B are (trace) *equivalent* if $Tr(A) = Tr(B)$.

A *finite state machine* (FSM) is a 7-tuple $M = \langle S, I, O, D_M, \Delta_M, \Lambda_M, s_0 \rangle$, where S is a finite nonempty set of states with the initial state s_0, I and O are input and output alphabets, $D_M \subseteq S \times I$ is the specification domain and $\Delta_M: D_M \rightarrow S$ and $\Lambda_M: D_M \rightarrow O$ are the next state and the output functions. FSM M is called *complete* if $D_M = S \times I$; in this case, D_M can be omitted, i.e., a complete FSM is a 6-tuple $M = \langle S, I, O, \Delta_M, \Lambda_M, s_0 \rangle$. If M is not complete then it is *partial*. In usual way, the next state and output functions are extended to input sequences. Given state s and input sequence $i_1 \ldots i_k$, $\Delta_M(s, i_1 \ldots i_k) = s'$ while $\Lambda_M(s, i_1 \ldots i_k) = o_1 \ldots o_k$ if and only if there exist states $s_1', \ldots, s_{k+1}'$ such that $s_1' = s$, $s_{k+1}' = s'$, and $\Delta_M(s_j', i_j) = s_{j+1}'$ while $\Lambda_M(s_j', i_j) = o_j$ for each $j = 1, \ldots, k$. In this case, the sequence $i_1 o_1 \ldots i_k o_k$ is called an *I/O sequence* at state s. The set of all I/O sequences at the initial state of M is the *language* of FSM M. Each FSM can be represented as a deterministic IOTS with the same language by unfolding each transition [10].

We say that an IOTS has an *FSM behavior* if the IOTS is deterministic, has non-empty input and output sets, inputs are enabled only at stable states, and stable and unstable states alternate, i.e., for every stable state s and input $a \in in(s)$, $(s, a, s') \in \lambda_A$ implies that s' is an unstable state and for every unstable state s and output $a \in out(s)$, $(s, a, s') \in \lambda_A$ implies that s' is a stable state while the initial state is stable. If all input actions are enabled at every stable state, we say that the IOTS has a *behavior of a complete FSM*. If each input is followed by a single output, i.e., $|out(s)| = 1$ for each unstable state s, we say that the IOTS has a *behavior of a deterministic FSM*.

2.2 Parallel Composition of IOTSs

To compose complete FSMs we consider their IOTS counterparts. The joint behavior of k deterministic IOTSs $A_j = \langle S_j, I_j, O_j, \lambda_j, s_{j0} \rangle$, $j = 1, \ldots, k$, is described by the parallel composition of IOTSs. The *parallel composition* $\|A_j$ (written also as $A_1 \| A_2 \ldots \| A_k$) is the IOTS $\langle R, \bigcup I_j \backslash \bigcup O_j, \bigcup O_j, \lambda, s_{10} \ldots s_{k0} \rangle$, where the set of states $R \subseteq \times S_j$ and the transition relation λ are the smallest sets obtained by applying the following inference rules:

- $s_{10} \ldots s_{k0} \in R$;
- given $(s_1 \ldots s_k) \in R$, $(s_1' \ldots s_k') \in \times S_j$ and $a \in \bigcup I_j \cup \bigcup O_j$, $(s_1 \ldots s_k, a, s_1' \ldots s_k') \in \lambda$, if for each $j \in \{1, \ldots, k\}$ it holds that

if $a \in I_j \cup O_j$ then $(s_j, a, s_j') \in \lambda_j$ and if $a \notin I_j \cup O_j$ then $s_j' = s_j$.

Sometimes we need to hide some actions that are not observable in the resulting composition. This is achieved using the *U-restriction* defined above. In particular,

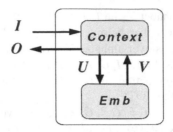

Fig. 1. Parallel Composition of the IOTSs *Context* and *Emb*

given a subset $I \subseteq \bigcup I_j \backslash \bigcup O_j$ and a subset O of the set $\bigcup O_j$, $I \cup O \neq \varnothing$, $(\|A_j)\downarrow_{I \cup O}$ is obtained by restricting the IOTS $\|A_j$ to the alphabet $I \cup O$.

In this paper, we consider a system of two complete deterministic FSMs, each of which is represented as an IOTS. The system consists of the context IOTS *Context* = $\langle S, I \cup V, O \cup U, \lambda_{Con}, s_0 \rangle$ and the embedded IOTS *Emb* = $\langle Q, U, V, \lambda_{Emb}, q_0 \rangle$, as shown in Figure 1. The alphabets I and O represent the *external* inputs and outputs of the system, while the alphabets V and U represent the *internal* interactions between the two IOTSs. As usual, for the sake of simplicity, we assume that the sets I, O, V, U are pair-wise disjoint. We also assume that the composition works in a slow environment, i.e., an external input can be applied to the composition after the latter has produced an external output to a previous external input. A behavior of such an environment can be represented by the IOTS *MAX* = $\langle \{p_0, p_1\}, I, O, \lambda_{Max}, p_0 \rangle$, $\forall\, i \in I$ $(p_0, i, p_1) \in \lambda_{Max}$ and $\forall\, o \in O\ (p_1, o, p_0) \in \lambda_{Max}$.

Therefore, the behavior of *Context* and *Emb* in the slow environment can be described by the parallel composition *MAX* ‖ *Context* ‖ *Emb*. We note that the IOTS *MAX* ‖ *Context* ‖ *Emb* does not have an FSM behavior. The reason is that the input set is empty. The following proposition states how the language of the IOTS *Context* ‖ *Emb* is constrained by a slow environment.

Proposition 1. The language of the IOTS *MAX* ‖ *Context* ‖ *Emb* is a subset of the prefix closure of the language $(I(UV)^*O)^*$.

Proposition 1 states that when an environment is slow, the component machines can execute a sequence of the set $(UV)^*$ before an external output is produced by the context in response to external input $i \in I$ received from the environment. Only after the context has produced an external output to a previous input, a next external input can be applied to the context.

3   Fault Propagation

3.1   Test Definitions

Definition 1. Given a specification IOTS $A = \langle S, I, O, \lambda_A, s_0 \rangle$, a *test case* (*test*) is a non-empty sequence over alphabet $I \cup O$. A test αb is said to be *reduced* (w.r.t. the given specification A) if α is the longest prefix of αb that is a trace of the specification.

Given an IOTS specification A, the set of all possible implementations of A that are IOTSs over the alphabet $I \cup O$, is called the *fault domain* of A, denoted by $\mathfrak{I}(A)$. When A is clear from the context, we use the notation $\mathfrak{I}$ instead of $\mathfrak{I}(A)$. The fault domain includes both, conforming and nonconforming implementations, where the trace equivalence of IOTSs is the conformance relation. Thus, a fault to be detected by a test occurs when an implementation IOTS has a trace that is not a valid trace of the specification IOTS. To be more specific, such invalid trace has always an output as its last symbol. This is true for any IOTS that encodes a complete FSM, as well as for an IOTS that describes the composition of such IOTSs. It is not difficult to demonstrate that for this class of IOTSs either only all input actions are enabled or only output actions are enabled in each state, i.e., either $init(s) = I$ or $init(s) \subseteq O$ for all $s \in S$. Thus, traces of specification and implementation (deterministic) IOTSs may only differ on outputs and not on inputs.

Definition 2. Given the specification IOTS A, an implementation IOTS $B \in \mathfrak{I}$ that is not trace equivalent to A, and a test α, we say that α *detects* B if there exists a prefix of α that is a trace of the implementation IOTS B and not of A.

Given the specification IOTS A, the set $\mathfrak{I}$ of implementation IOTSs over the alphabet $I \cup O$, and a test α, $\mathfrak{I}_\alpha \subseteq \mathfrak{I}$ denotes the subset of implementations that are detected by α. The set $\mathfrak{I}_\alpha$ can be empty, it is the case when, for example, α is a trace of the specification.

Definition 3. A *test suite* is a finite set of tests. An implementation IOTS $B \in \mathfrak{I}$ that is not trace equivalent to A is said to be *detected* by a test suite if the test suite has a test that detects B.

If a test suite $TS = \{ \alpha_1, ..., \alpha_k \}$ and $\mathfrak{I}_{TS} \subseteq \mathfrak{I}$ denotes the subset of implementations that are detected by TS then $\mathfrak{I}_{TS} = \mathfrak{I}_{\alpha_1} \cup ... \cup \mathfrak{I}_{\alpha_k}$.

Given a test α over the alphabet $I \cup O$, we derive a tree IOTS $IOTS_\alpha = <T, I, O, \lambda, \varepsilon>$, where T is the prefix closure of α with the empty sequence ε as the initial state. Given a proper prefix β of α and symbol $a \in I \cup O$, $(\beta, a, \beta a) \in \lambda$ if βa is a prefix of α. By definition, state α is a deadlock state.

Given a test suite TS consisting of test cases $\alpha_1, ..., \alpha_k$ over the alphabet $I \cup O$, a tree IOTS $IOTS_{TS}$ is determined by first deriving the union of the IOTSs $IOTS_{\alpha_1}$, $IOTS_{\alpha_2}, ..., IOTS_{\alpha_k}$ [HoUI79] and then determinizing the obtained IOTS.

Definition 4. Given the specification IOTS A and a set $\mathfrak{I}' \subseteq \mathfrak{I}$ of implementation IOTSs over alphabets $I \cup O$, let TS be a test suite. The test suite TS is *exhaustive* in $\mathfrak{I}'$, if the test suite detects each $B \in \mathfrak{I}'$ that is not trace equivalent to A.

Given a test case $\alpha\gamma$ that is not reduced and $B \in \mathfrak{I}_{\alpha\gamma}$, in order to detect B we can use the shortest prefix α of $\alpha\gamma$ that is not a trace of the specification A. In other words, in order to detect all possible faulty implementations of the fault domain $\mathfrak{I}_{\alpha\gamma}$ it is sufficient to use the reduced test α, i.e., the following statement holds.

Proposition 2. Given the specification IOTS A, let α and $\alpha\gamma$ be test cases such that α is not a trace of the specification A. The set of implementation IOTSs that are detected by $\alpha\gamma$ coincides with the set of those implementations that are detected by α, i.e., $\mathfrak{I}_\alpha = \mathfrak{I}_{\alpha\gamma}$.

Given a test suite exhaustive in $\mathfrak{S}\,'$, we can reduce the length of this test suite by deleting every test that is a trace of the specification and replacing each remaining non-reduced test with its shortest prefix that is not a trace of the specification. According to Proposition 2, the resulting test suite is also exhaustive in $\mathfrak{S}\,'$.

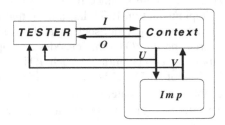

Fig. 2. Test architecture

3.2 Test Architecture

We consider the composition of IOTSs *Context* and *Emb* with the IOTS *TESTER*, and assume that during testing all actions can be observed (Figure 2). In this case, the closed system is the parallel composition *TESTER* ‖ *Context* ‖ *Emb* with the output set $I \cup O \cup U \cup V$.

As usual, we assume that the *Context* component is fault free and only an implementation of the embedded component may be faulty. Moreover, we assume that each possibly faulty implementation is a complete deterministic FSM with a restricted number of states represented as an IOTS and denote $\mathfrak{I}(Emb)$ the fault domain of *Emb*, i.e., $\mathfrak{I}(Emb)$ is the set of IOTSs that represent all possible *Emb* implementations. Thus, a fault domain of the system *MAX* ‖ *Context* ‖ *Emb* is $\mathfrak{I}(Con\text{-}Emb) = \{MAX \, \| \, Context \, \| \, Imp : Imp \in \mathfrak{I}(Emb)\}$. Given $Imp \in \mathfrak{I}(Emb)$, *Imp* is said to be a *conforming* (in the given context) implementation of *Emb* if IOTSs *MAX* ‖ *Context* ‖ *Imp* and *MAX* ‖ *Context* ‖ *Emb* are trace equivalent. Otherwise, *Imp* is a *nonconforming* implementation. Not every implementation of the embedded component that is not trace equivalent to *Emb* and thus, can be detected in isolation, is a nonconforming implementation in context [10]. As an example, consider the specification *Emb* and the faulty implementation Imp_1 shown in Figures 3a and 3b, respectively. The context IOTS is shown in Figure 4. The composition *MAX* ‖ *Context* ‖ Imp_1 is trace equivalent to the *MAX* ‖ *Context* ‖ *Emb*. Therefore, the fact that the implementation Imp_1 is not trace equivalent to *Emb* cannot be established within the given context.

According to the above test architecture, during the testing process a tester applies actions of the set *I* to the external input of *Context* and draws a conclusion whether an implementation *Imp* of the embedded component conforms to its specification by observing the outputs over the set $O \cup U \cup V$. Thus, traces of a tester are defined over the alphabet $I \cup O \cup U \cup V$. Since we are interested in the system of communicating IOTSs *Context* and *Imp* that work in a slow environment, the tester has also to be slow, i.e., the tester can apply the next symbol $i \in I$ only after it has obtained, from *Context*, an external output $o \in O$ to the previously applied input of the set *I*. We call such tester a *slow* tester and according to Proposition 1, a slow tester executes traces in the set $(I(UV)^*O)^*$. Each trace of a tester is called an *external* test (case). In fact, a tester is a tree IOTS derived from an external test suite. As usual, an *external* test suite is a finite set of external tests. Following Definitions 2 and 3, an external test detects each implementation system *MAX* ‖ *Context* ‖ *Imp* of the set $\mathfrak{I}(Con\text{-}Emb)$ if some prefix of the external test is a trace of *MAX* ‖ *Context* ‖ *Imp*, but not a trace of the composition *MAX* ‖ *Context* ‖ *Emb*. In this case, the tester detects a fault that makes *Imp* nonconforming. Otherwise, i.e., when the external test is a trace of *MAX* ‖ *Context* ‖ *Emb*, the implementation *Imp* has no faults that can be detected by this test.

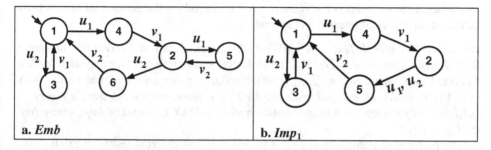

Fig. 3. Specification *Emb* and a faulty implementation *Imp*₁

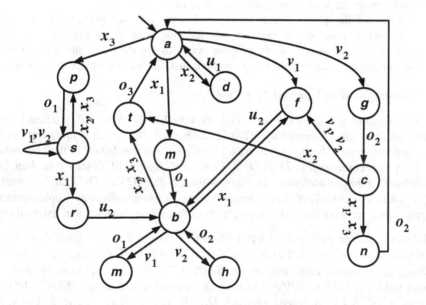

Fig. 4. Context IOTS

3.3 Problem Definition

Given the embedded component *Emb* over input alphabet U and output alphabet V, an *internal* test (case) is a trace over the alphabet $U \cup V$. Since the IOTS *Emb* has an

FSM behavior, an internal test is a non-empty sequence of the language $(UV)^*$. Correspondingly, an *internal* test suite is a finite set of internal tests.

When the implementation is tested through the context, the internal inputs of the embedded component are not directly controllable; except for the context of FIFO queues [4,11]. For other types of contexts, internal tests have to be translated to external tests.

Given an internal test *InTest*, let $\mathfrak{I}_{InTest}(Emb) \subseteq \mathfrak{I}(Emb)$ denote the set of possible faulty implementations of the embedded component *Emb* that can be detected by *InTest* when testing the IOTS *Emb* in isolation. Naturally, it makes sense to consider internal tests which detect at least one nonconforming implementation, i.e., tests which belong to the set $(UV)^*\backslash Tr(Emb)$. We first introduce the notion that relates fault detection capability of internal and external tests.

Definition 5. Given *InTest* $\in (UV)^*\backslash Tr(Emb)$, an external test *ExtTest* has the *same fault detection power* as *InTest* if *ExtTest* detects each implementation system *MAX* ‖ *Context* ‖ *Imp*, where *Imp* $\in \mathfrak{I}_{InTest}(Emb)$. Similarly, given an internal test suite *InTS* $\subseteq (UV)^*\backslash Tr(Emb)$, an external test suite *ExtTS* has the *same fault detection power* as *InTS*, if *ExtTS* detects each implementation system *MAX* ‖ *Context* ‖ *Imp*, where *Imp* $\in \mathfrak{I}_{InTS}(Emb)$.

The problem of translating *InTest* is to determine an external test (if it exists) with the same fault detection power, i.e., to determine an external test case that detects each IOTS *MAX* ‖ *Context* ‖ *Imp*, where *Imp* $\in \mathfrak{I}_{InTest}(Emb)$. The problem is called the *test translation* or the *fault propagation* problem [2,7].

In the rest of the paper, given an internal test suite, we propose a method of translating (when possible) it into an external one with the same fault detection power. Moreover, in Section 4, we propose methods for deriving internal test suites with the guaranteed fault coverage that can be translated within the given context.

3.4 Translation of an Internal Test Case

Given an internal test case *InTest* $\in (UV)^*\backslash Tr(Emb)$, let *Imp* $\in \mathfrak{I}_{InTest}(Emb)$ be an implementation that is detected by *InTest*, i.e., *Imp* has a trace that is not a trace of the embedded component *Emb*. Therefore, a tester, that induces *InTest* at the channels *U* and *V* in the composition *TESTER* ‖ *Context* ‖ *Imp*, will detect that *Imp* is a nonconforming implementation. In other words, if the IOTS (*TESTER* ‖ *Context* ‖ *Imp*)$\downarrow_{U \cup V}$ has a trace *InTest*, then a tester detects the nonconforming implementation *Imp*, and, thus, we have the following definition that relates internal and external tests.

Definition 6. Given *InTest* $\in (UV)^*\backslash Tr(Emb)$, an external test $\in \langle (I(UV)^*O)^* \rangle$ is a *translation* of *InTest*, denoted *Transl(InTest)*, if the IOTS (*IOTS*$_{Transl(InTest)}$ ‖ *Context* ‖ *IOTS*$_{InTest}$)$\downarrow_{U \cup V}$ is trace equivalent to the IOTS *IOTS*$_{InTest}$. Correspondingly, given an internal test suite *InTS* $\subseteq (UV)^*\backslash Tr(Emb)$, an external test suite $\subseteq \langle (I(UV)^*O)^* \rangle$ is a *translation* of *InTS*, denoted *Transl(InTS)*, if the (*IOTS*$_{Transl(InTS)}$ ‖ *Context* ‖ *IOTS*$_{InTS}$)$\downarrow_{U \cup V}$ is trace equivalent to the IOTS *IOTS*$_{InTS}$.

The following statement is implied immediately.

Proposition 3. Given *InTS* $\subseteq (UV)^*\backslash Tr(Emb)$, an external test suite *Transl(InTS)* detects each implementation system *MAX* ‖ *Context* ‖ *Imp*, *Imp* $\in \mathfrak{I}_{InTS}(Emb)$, i.e., *Transl(InTS)* has the same fault detection power as *InTS*.

Due to Definition 6, in order to determine a translation of a given internal test *InTest* we have to establish conditions under which the composition (*TESTER* ‖ *Context* ‖ *Imp*)$_{\downarrow U \cup V}$ has the trace *InTest*. According to the definition of the parallel composition, the following statement holds for traces of the composition (*TESTER* ‖ *Context* ‖ *Imp*)$_{\downarrow U \cup V}$.

Proposition 4. Given an internal test case *InTest*, the composition (*TESTER* ‖ *Context* ‖ *Imp*)$_{\downarrow U \cup V}$ has a trace *InTest* if and only if the IOTS *Imp* and the composition (*TESTER* ‖ *Context*)$_{\downarrow U \cup V}$ have the trace *InTest*. Correspondingly, given an internal test suite *InTS*, the set of traces of the composition (*TESTER* ‖ *Context* ‖ *Imp*)$_{\downarrow U \cup V}$ contains *InTS* if and only if the set of traces of the IOTS *Imp* and of the composition (*TESTER* ‖ *Context*)$_{\downarrow U \cup V}$ contains the set *InTS*.

Here we note that not each internal case can be translated, as the context may render it impossible. According to Proposition 4, the following sufficient and necessary conditions can be established for the translation of an internal test in the given context.

Let $IOTS_{InTest}^{Aug}$ denote the IOTS obtained from $IOTS_{InTest}$ by adding self-loops labeled with all $i \in I$ (input) and $o \in O$ (output) at every non-deadlock state.

Theorem 1. Given a context *Context* and internal test *InTest*, the test *InTest* can be translated in the context if and only if the IOTS (*MAX* ‖ *Context* ‖ $IOTS_{InTest}$)$_{\downarrow U \cup V}$ is trace equivalent to $IOTS_{InTest}$. Moreover, if the test *InTest* is reduced and can be translated in the context then the set of traces with the ($U \cup V$)-restriction *InTest* that take the IOTS *MAX* ‖ *Context* ‖ $IOTS_{InTest}^{Aug}$ into a deadlock state coincides with the set of all reduced translations of the test *InTest*.

In fact, the first statement of the theorem is a direct corollary to Proposition 4. In the second statement of the theorem, we use $IOTS_{InTest}^{Aug}$ instead of $IOTS_{InTest}$ in the composition, to force a tester to stop after the first unexpected output v of *InTest* is produced by an implementation. Thus, each trace with the ($U \cup V$)-restriction *InTest* that takes the IOTS *MAX* ‖ *Context* ‖ $IOTS_{InTest}^{Aug}$ into a deadlock state is a reduced external test if *InTest* is reduced. As the set of traces the IOTS *MAX* ‖ *Context* has each trace that can occur in the composition of the system *Context* ‖ *Imp*, *Imp* ∈ $\mathfrak{I}(Emb)$, with a slow tester, the set of all traces with the ($U \cup V$)-restriction *InTest* that take the IOTS *MAX* ‖ *Context* ‖ $IOTS_{InTest}^{Aug}$ into a deadlock state coincides with the set of all reduced translations of the test *InTest*.

According to Theorem 1, each trace with the ($U \cup V$)-restriction *InTest* that takes the IOTS *MAX* ‖ *Context* ‖ $IOTS_{InTest}^{Aug}$ into a deadlock state has the same fault detection power as the internal test *InTest* and is a translation of *InTest*.

In general, the IOTS *MAX* ‖ *Context* ‖ $IOTS_{InTest}^{Aug}$ has many traces that lead to a deadlock state and have the ($U \cup V$)-restriction *InTest*. Each such trace can be selected as a translation of *InTest*. However, we are interested in a shortest translation, i.e., in the translation that is a reduced external test. According to Theorem 1, in order to determine a shortest translation of *InTest* it is sufficient to find a shortest trace that takes $MAX \parallel Context \parallel IOTS_{InTest}^{Aug}$ to a deadlock state and has the $(U \cup V)$-restriction $InTest$. Therefore, the problem of finding a shortest translation $Transl(InTest)$ of $InTest$ can be solved by determining a shortest trace that takes the IOTS $MAX \parallel$ $Context \parallel IOTS_{InTest}^{Aug}$ to a deadlock state and has $InTest$ as its $(U \cup V)$-restriction.

In our working example, consider the internal test u_2v_2. By direct inspection (Figure 5), one can assure that the trace $x_1o_1x_1u_2v_2$ is a translation of u_2v_2. For the internal test u_2v_2, we have two shortest translations $x_1o_1x_1u_2v_2$ and $x_3o_1x_1u_2v_2$ (Figure 5).

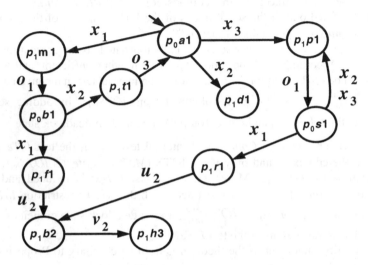

Fig. 5. IOTS $MAX \parallel Context \parallel IOTS_{u_2v_2}^{Aug}$

3.5   Translation of an Internal Test Suite

Since an internal test suite is a finite set, it can be translated by translating each of its test cases separately, as described in the previous subsection. To reduce the length of a resulting external test suite we have to select for each internal test $InTest$ a shortest translation of $InTest$.

However, all the tests of an internal test suite $InTS$ can be translated altogether. Given $IOTS_{InTS}$, we denote $IOTS_{InTS}^{Aug}$ the IOTS obtained from $IOTS_{InTS}$ by adding self-loops labeled with all $i \in I$ and $o \in O$ at every non-deadlock state. According to Theorem 1, the following statement holds.

Theorem 2. Given a context $Context$ and an internal test suite $InTS$, the test suite $InTS$ can be translated in the context, if and only if the IOTS $(MAX \parallel Context \parallel IOTS_{InTS})\downarrow_{U \cup V}$ is trace equivalent to $IOTS_{InTS}$. Moreover, if the test suite $InTS$ has only reduced tests and can be translated in the context then the set of traces with the $(U \cup V)$-restriction $InTS$ that take the IOTS $MAX \parallel Context \parallel IOTS_{InTS}^{Aug}$ into a deadlock state contains each set of reduced translations of the test suite $InTS$.

Due to Theorem 2, each subset of traces with the $(U \cup V)$-restriction $InTS$ that take the IOTS $MAX \parallel Context \parallel IOTS_{InTest}^{Aug}$ into a deadlock state has the same fault detection power as the internal test suite $InTS$ and is a translation of $InTS$.

Here we note the resulting translation of $InTS$ (i.e. an external test suite) $Transl(InTS)$ can have tests whose I-restrictions are prefixes of the same sequence over the alphabet I. According to our test architecture, for a tester it is sufficient to apply to the context longest I-restrictions of sequences of $Transl(InTS)$. As an example, consider the internal test suite $\{u_2v_2, u_2v_1u_2v_1\}$. One of its translations is an external test suite $\{x_2u_2v_2, x_2u_2v_1u_2v_1\}$. When executing the external test suite $\{x_2u_2v_2, x_2u_2v_1u_2v_1\}$ that is a translation of the internal test suite $\{u_2v_2, u_2v_1u_2v_1\}$ the tester has to apply only the external input x_2 to the context and observe the obtained outputs.

4 Exhaustive External Test Suites

When an internal test suite cannot be translated throughout the given context, there may still exist another internal test suite that detects the same set of faulty implementations of Emb and can be translated in the given context. Therefore, given a fault domain $\Im(Emb)$, we would like to derive an internal test suite for Emb that can be translated in the given context to obtain a translation exhaustive in the fault domain $\Im(Con\text{-}Emb)$.

As an example, consider a fault domain $\Im(Emb)$ of Emb (Figure 3a) that contains each IOTS with a behavior of a complete deterministic FSM with at most two states and an exhaustive test suite for Emb w.r.t. the fault domain $\Im(Emb)$. Such a test suite can be derived using the W-method [1,12] or its derivatives. The W-method provides an exhaustive test suite $E = \{u_2u_1, u_1u_1u_1, u_1u_2u_1\}$ as a set of input sequences over alphabet U. In order to transform this set into an internal test suite $InTS$ for the IOTS Emb we proceed as follows. For each sequence $u_1...u_k$ of the set E we determine a corresponding trace $u_1v_1...u_kv_k$ of the embedded component Emb. Then, we append each prefix $u_1v_1...u_j$, $j \le k$, of the trace $u_1v_1...u_kv_k$ with all possible wrong internal outputs $v' \in V\setminus\{v_j\}$ and include the resulting sequences into the internal test suite $InTS$. In our example, we obtain $InTS = \{u_2v_2, u_2v_1u_1v_2, u_1v_2, u_1v_1u_1v_1, u_1v_1u_1v_2u_1v_2, u_1v_1u_2v_1, u_1v_1u_2v_2u_1v_2\}$, the I-restriction of this set is exactly the set E.

By direct inspection, one can assure that this test suite cannot be translated through the given context. The reason is that, for example, an internal test case $u_1v_1u_1v_1$ is not in the set of traces of the IOTS $(MAX \parallel Context)_{\downarrow U \cup V}$ and thus, cannot be executed in the given context.

Therefore, to derive an exhaustive internal test suite w.r.t. the above fault domain $\Im(Emb)$ that can be translated into an exhaustive external test suite w.r.t. the fault domain $\Im(Con\text{-}Emb)$, we have to consider only the behavior of the embedded component Emb for the sequences that can be executed in the given context. To this end, we define an IOTS a so called observable equivalent of Emb, by removing from it the sequences that cannot be executed with the given context.

Definition 7. Given IOTSs MAX, $Context$ and Emb, the IOTS Eq_{Emb} is an *observable equivalent* of Emb if $Tr(Eq_{Emb}) = Tr(Emb) \cap Tr((MAX \parallel Context)_{\downarrow U \cup V})$.

Due to Definition 7 and Proposition 4, the observable equivalent Eq_{Emb} of an embedded component Emb can be derived as follows: $Eq_{Emb} = (MAX \parallel Context \parallel Emb)_{\downarrow U \cup V}$. For our working example, the observable equivalent is shown in Figure 6.

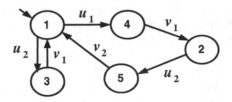

Fig. 6. Observable equivalent IOTS Eq_{Emb}

As a corollary to Theorem 2, the following statement holds.

Theorem 3. A reduced internal test $u_1v_1...u_kv_k \in (UV)^* \backslash Tr(Emb)$ can be translated in the given context if and only if the input sequence $u_1...u_k$ is a trace of the U-restriction of the IOTS Eq_{Emb}, i.e., $u_1...u_k \in (MAX \parallel Context \parallel Emb)_{\downarrow U}$.

If a reduced internal test $InTest$ can be translated in the given context then every external translation of $InTest$ has the same fault detection power as $InTest$ (Proposition 3). According to Theorem 3, a reduced internal test $u_1v_1...u_kv_k$ can be translated if the internal input sequence $u_1...u_k$ is a trace of the U-restriction of the IOTS Eq_{Emb}. Therefore, the following two statements hold as corollaries to Theorem 3.

Corollary 2. Given a reduced internal test $InTest$ such that the U-restriction of $InTest$ is a trace of the IOTS $(Eq_{Emb})_{\downarrow U}$, the composition of the context and a faulty implementation $Imp \in \mathfrak{I}(Emb)$ is detected by the external test $Transl(InTest)$ if and only if Imp is detected by $InTest$.

Corollary 3. Given an internal test suite $InTS$ with reduced tests such that the U-restriction of each test of $InTS$ is a trace of the IOTS $(Eq_{Emb})_{\downarrow U}$, the composition of the context and a faulty implementation $Imp \in \mathfrak{I}(Emb)$ is detected by the external test suite $Transl\text{-}InTS$ if and only if Imp is detected by $InTS$.

There is a special case when $Tr(Eq_{Emb})_{\downarrow U} = U^*$. This means that the context has a behavior of a complete FSM and any internal test case can be translated.

Corollary 4. Given the observable equivalent Eq_{Emb} of the embedded component, if $Tr(Eq_{Emb})_{\downarrow U} = U^*$ then each reduced internal test can be translated through the given context.

Here we note that the notion of the observable equivalent is close to the notion of the embedded equivalent in [10]. However, in that work, the observable equivalent is derived under the assumption that the internal channels are not observable; in fact, the construction refines a so-called conforming part of the embedded component Emb restricting it to alphabets of Emb.

According to Corollary 3, internal test suites are derived from the specification of the embedded component that has a behavior of a partial deterministic FSM. Then an internal test suite for the embedded component can be derived, using the State

Counting (SC) method in [8], exhaustive w.r.t. the fault model $<Spec, \leq, FD>$, where *Spec* is a partial FSM, $\leq$ is the quasi-equivalence relation, called weak conformance in [13], and *FD* is the set of all possible implementation FSMs with a restricted number of states.

Applied to the partial FSM that is encoded as the IOTS Eq_{Emb}, the SC-method returns a set E of internal (over the alphabet U) input sequences. In order to transform this set into an internal test suite *InTS* we again for each sequence $u_1...u_k$ of the set E, determine a corresponding trace $u_1v_1...u_kv_k$ of the embedded component *Emb*, append each prefix $u_1v_1...u_j$, $j \leq k$, of the trace $u_1v_1...u_kv_k$ with all possible wrong internal outputs $v' \in V\backslash\{v_j\}$ and include the resulting sequences into the internal test suite *InTS*.

Consider the observable equivalent IOTS Eq_{Emb} of *Emb* in Figure 6. The IOTS Eq_{Emb} has a behavior of a partial FSM with two states. If we consider the fault domain $\mathfrak{I}(Emb)$ of all IOTSs that have a behavior of a complete deterministic FSM with at most two states, then we can derive, using the SC-method or the method in [13], the test suite $\{u_1v_2, u_1v_1u_2v_1, u_1v_1u_2v_2u_2v_2, u_2v_2, u_2v_1u_2v_2\}$ exhaustive w.r.t. the fault domain $\mathfrak{I}(Emb)$ and quasi-equivalence relation. The corresponding external tests are $\{x_2u_1v_2, x_2u_1v_1u_2v_1, x_1o_1x_1u_2v_2, x_1o_1x_1u_2v_1o_1x_1u_2v_2\}$ and according to Theorem 3, this external test suite is exhaustive w.r.t. the fault domain $\mathfrak{I}(Con\text{-}Emb)$.

Another approach for test derivation from the embedded equivalent is mutant-based testing. A mutant may model certain suspected faults, which have to be tested for their presence. The approach is based on the enumeration of mutants of the embedded component *Emb* and finding external tests that kill these mutants. To this end, given a mutant $Imp \in \mathfrak{I}(Emb)$, we consider the IOTS $Imp \| Eq_{Emb}$. We first note that the observable equivalent Eq_{Emb} does not deadlock, since each IOTS *Context* and *Emb* has a behavior of a complete FSM. Secondly, given $Imp \in \mathfrak{I}(Emb)$, Imp is not trace equivalent to *Emb* if and only if the IOTS $Imp \| Emb$ deadlocks. If the IOTS $Imp \| Eq_{Emb}$ does not deadlock then the mutant IOTS Imp is a conforming implementation of *Emb*. Otherwise, each trace of Imp such that its U-restriction takes the IOTS $(Imp \| Eq_{Emb})\downarrow_U$ to a deadlock state is an internal test that detects a faulty implementation Imp and this internal test can be translated through the given context.

As an example, consider the faulty implementation Imp_1 (Figure 3b) of the embedded component *Emb* (Figure 3a). The composition $Imp_1 \| Eq_{Emb}$ is similar to the Eq_{Emb} in Figure 6; only state labels are renamed 11, 22, 33, 44, and 55. Since the composition $Imp_1 \| Emb$ does not deadlock, the faulty implementation Imp_1 cannot be detected through the given context, and thus Imp_1 is a conforming implementation (in the given context). As another example, consider the faulty implementation Imp_2 which is similar to Imp_1 of Fig. 3b except that the transition connecting states 5 and 1 has the label v_1 instead of v_2. The composition $Imp_2 \| Emb$ deadlocks after the trace $u_1v_1u_2$ and thus Imp_2 can be detected through the given context.

5 Conclusions

In this paper, we proposed an approach for translating internal tests derived for a component embedded within a modular system into external tests of the system. The system is represented as two complete deterministic communicating finite state machines, an embedded component machine to be tested and a context machine that represents the remaining part of the system. The context is assumed to be fault free and the interactions between the component machines are observable. Also, in this paper, we established necessary and sufficient conditions for an internal test (suite) to be translated in the given context. If a test cannot be translated, we demonstrated another test with the guaranteed fault detection power could be determined (if such a test exists) that can be translated in the given context. In our future work, we intend to generalize the fault translation approach elaborated in this paper for communicating finite state machines to input output transition systems.

References

1. T. S. Chow, "Test design modeled by finite-state machines", *IEEE Trans. SE*, vol. 4, no.3, pp. 178-187, 1978.
2. K. El-Fakih and N. Yevtushenko, "Fault propagation by equation solving", *Proc. of the IFIP 24th International Conference on Formal Techniques for Networked and Distributed Systems*, Madrid, Spain, LNCS 3235, pp. 185-198, 2004.
3. J. E. Hopcroft and J. D. Ullman, *Introduction to automata theory, languages, and computation*, Addison-Wesley, N.Y., 1979.
4. C. Jard, T. Jéron, L. Tanguy, and C. Viho, "Remote testing can be as powerful as local testing", *Proc. of the IFIP Joint Intl. Conf. Formal Description Techniques for Distributed Systems and Communication Protocols and Protocol Specification, Testing and Verification (FORTE XII / PSTV XIX)*, volume 156 of *IFIP Conference Proceedings*, Beijing, China, Oct. 5-8, Kluwer, pp. 25–40, 1999.
5. L. P. Lima, "A pragmatic method to generate test sequences for embedded systems", Ph.D. Thesis, *Institute National des Telecommunications*, Evry, France, 1998.
6. L. P Lima and A. R. Cavalli, "A pragmatic approach to generating test sequences for embedded systems", *Proc. of the 10th International Workshop on Testing of Communicating Systems*, pp: 125-140, 1997.
7. A. Petrenko and N. Yevtushenko, "Testing faults in embedded components", *Proc. of the 10th International Workshop on Testing of Communicating Systems*, pp. 272-287, 1997.
8. A. Petrenko and N. Yevtushenko, "Testing from partial deterministic FSM specifications", *IEEE Transactions on Computers*, vol. 54, no. 9, pp. 1154-1165, 2005.
9. A. Petrenko, N. Yevtushenko, and G. v. Bochmann, "Fault models for testing in context", *Proc. International Conference on Formal Techniques for Networked and Distributed Systems*, pp. 125-140, 1996.
10. A. Petrenko, N. Yevtushenko, G. v. Bochmann, and R. Dssouli, "Testing in context: framework and test derivation", *Computer communications*, Vol. 19, pp. 1236-1249, 1996.
11. J. Tretmans and L. Verhaard, "A queue model relating synchronous and asynchronous communication", In R. J. Linn, Jr. and M. Ü. Uyar, eds., *Proc. of the IFIP TC6/WG6.1 12th Intl. Symp. Protocol Specification, Testing and Verification*, volume C-8 of *IFIP Transactions*, Lake Buena Vista, Florida, USA, pp. 131–145, 1992.
12. M. P. Vasilevskii, "Failure diagnosis of automata", translated from Kibernetika, no.4, pp. 98-108, 1973.
13. M. Yannakakis and D. Lee, "Testing finite state machines", *Proc. of the 23rd Annual ACM Symposium on Theory of Computing*, New Orleans, Louisiana, pp. 476-485, 1995.

Using Distinguishing and UIO Sequences Together in a Checking Sequence

M. Cihan Yalcin and Husnu Yenigun

Faculty of Engineering and Natural Sciences, Sabanci University,
Tuzla 34956, Istanbul, Turkey

Abstract. If a finite state machine M does not have a distinguishing sequence, but has UIO sequences for its states, there are methods to produce a checking sequence for M. However, if M has a distinguishing sequence $\bar{D}$, then there are methods that make use of $\bar{D}$ to construct checking sequences that are much shorter than the ones that would be constructed by using only the UIO sequences for M. The methods to applied when a distinguishing sequence exists, only make use of the distinguishing sequences. In this paper we show that, even if M has a distinguishing sequence $\bar{D}$, the UIO sequences can still be used together with $\bar{D}$ to construct shorter checking sequences.

1 Introduction

Finite state machines (FSM) have been successfully used to model the externally observable behavior of systems [1]. Based on the FSM model M of a system under test (SUT) N, a test sequence can be constructed to check if N is implemented correctly [2,3].

Such a test sequence, which will be called a checking sequence, is a sequence of inputs such that, if N produces the expected outputs then this information provides sufficient evidence to conclude that N is a correct implementation of M. Of course, such a checking sequence cannot be found in general. Two important assumptions are made on N in practice. First assumption is that N is deterministic and does not change during the experiments. The second assumption is that N has at most the same number of states as M. Although the latter assumption seems to be restrictive, this assumption provides a basis to construct a checking sequence. Based on the methods that can generate checking sequences under this assumption, it is possible to extend these methods to generate checking sequences when this assumption is relaxed and N is assumed to have at most $n + \Delta$ states for some constant Δ, where n is the number of states in M (e.g. see [4]).

Basically, a checking sequence consists of parts that challenge N to provide evidence for the correct implementation of every transition in M. To do this, the checking sequence brings N to a state, applies an input at that state (to see if it would produce the correct output), and then it applies a sequence of inputs to recognize the state reached. As we will explain, bringing N to a certain state is also based on recognizing states, which can only be performed by observing distinct outputs produced to the same input sequence by different states.

M.Ü. Uyar, A.Y. Duale, and M.A. Fecko (Eds.): TestCom 2006, LNCS 3964, pp. 259–273, 2006.

Recognizing states can be based on distinguishing sequences [3], a characterization set [3] or unique input-output (UIO) sequences [5]. It is known that a distinguishing sequence may not exist for every minimal FSM [6], and that determining the existence of a distinguishing sequence for an FSM is PSPACE-complete [7]. However, if M has a distinguishing sequence, there are methods already available in the literature (e.g. [3, 8, 9]) to produce a checking sequence in which distinguishing sequences are used to recognize the states. It is quite easy to understand why a distinguishing sequence $\bar{D}$ can be used to recognize a state, since all the states in M produces a different output sequence to the same input sequence $\bar{D}$.

If an FSM M does not have a distinguishing sequence, it is still possible to construct a checking sequence for M. For example in [5] and in [10], it is shown how a checking sequence can be constructed by using UIO sequences, which are sequences that may exist even when a distinguishing sequence is not available. However, the authors of [11] show that, the original method proposed in [5] is not sufficient, and they propose the UIOv method to fix the problems of the method given in [5]. Since the UIO sequences of the states are not necessarily the same, although the response of a state to is UIO $\bar{U}$ is unique in the specification, we have to make sure that no other state produces the same response to $\bar{U}$ in N. As this must be guaranteed for the UIO sequences of all the states, checking sequences based on UIO sequences tend to be longer. Hence the UIOv and the other UIO based methods are considered only when a distinguishing sequence does not exist.

In this paper we propose that, even if there exists a distinguishing sequence for an FSM M, UIO sequences for the states of M (which are guaranteed to exist since M is known to have a distinguishing sequence) can also be used to construct a checking sequence in conjunction with the distinguishing sequence. We explain a method to show how to construct such a checking sequence. We also give an example for which the length of the checking sequence based on the distinguishing sequence and UIO sequences is less than the length of the checking sequence based on the distinguishing sequence only.

The rest of the paper is organized as follows. Section 2 introduces the concepts used in constructing checking sequences. In Section 3, an existing method to construct checking sequences based on distinguishing sequences is given. Section 4 explains the conditions under which a UIO sequence can be used to recognize states in a checking sequence. In Section 5, we give a modification of the method in Section 3 that constructs checking sequences in which UIO sequences are also used for state recognition. Finally, Section 6 concludes the paper and provides future research directions on the topic.

2   Preliminaries

We directly adopt the formalism and the notation for finite state machines from [12] and include it below for completeness. A deterministic FSM M is defined by a tuple $(S, s_1, X, Y, \delta, \lambda)$ where

- S is a finite set of *states*,
- $s_1 \in S$ is the *initial state*,
- X is the finite *input alphabet*,
- Y is the finite *output alphabet*,
- $\delta : S \times X \to S$ is the *next state function*, and
- $\lambda : S \times X \to Y$ is the *output function*.

Throughout the paper, we use barred symbols (e.g. $\bar{x}, \bar{P}, \ldots$) to denote sequences, and juxtaposition to denote concatenation. The next state function δ and the output function λ can be extended to sequences in a straightforward manner as, for an input symbol $a \in X$, a sequence of inputs $\bar{x} \in X^*$, and a state $s \in S$,

$$\delta(s, a\bar{x}) = \delta(\delta(s, a), \bar{x}) \text{ and } \lambda(s, a\bar{x}) = \lambda(s, a)\lambda(\delta(s, a), \bar{x})$$

The number of states of M is denoted n and the states of M are enumerated, giving $S = \{s_1, s_2, \ldots, s_n\}$. An FSM is *completely specified* if the functions λ and δ are total.

An FSM, that will be denoted M_0 throughout this paper, is described in Figure 1. Here, $S = \{s_1, s_2, s_3\}$, $X = \{a, b\}$ and $Y = \{0, 1\}$.

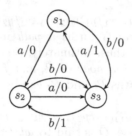

Fig. 1. The FSM M_0

In an FSM M, $s_i \in S$ and $s_j \in S$, $s_i \neq s_j$, are *equivalent* if, $\forall \bar{x} \in X^*$, $\lambda(s_i, \bar{x}) = \lambda(s_j, \bar{x})$. If $\exists \bar{x} \in X^*$ such that $\lambda(s_i, \bar{x}) \neq \lambda(s_j, \bar{x})$ then $\bar{x}$ is said to *distinguish* s_i and s_j. An FSM M is said to be *minimal* if none of its states are equivalent.

A *distinguishing sequence* for an FSM M is an input sequence $\bar{D}$ for which each state of M produces a distinct output. More formally, for all $s_i, s_j \in S$ if $s_i \neq s_j$ then $\lambda(s_i, \bar{D}) \neq \lambda(s_j, \bar{D})$. Thus, for example, M_0 in Figure 1 has distinguishing sequence aa.

A *unique input output sequence* (a UIO sequence, or simply a UIO) for a state s_i of an FSM M is an input sequence $\bar{U}_i$ which distinguishes s_i from the other states. More formally, $\bar{U}_i$ is a UIO for s_i if for all $s_j \in S$, if $s_j \neq s_i$, then $\lambda(s_i, \bar{U}_i) \neq \lambda(s_j, \bar{U}_i)$. Thus, for example, s_3 of M_0 has UIO $\bar{U}_3 = b$.

It is known that some FSMs do not have a distinguishing sequence, and some states do not have UIO sequences. However, when we consider a machine $M = (S, s_1, X, Y, \delta, \lambda)$ with a distinguishing sequence $\bar{D}$ (let $\bar{D}$ be a shortest such sequence), and a state $s_i \in S$ with a UIO sequence $\bar{U}_i$ (let $\bar{U}_i$ be a shortest such sequence), we can easily observe the following fact: $\bar{D}$ distinguishes between all pairs of states (s_i and s_j, $\forall s_i, s_j \in S$), whereas $\bar{U}_i$ distinguishes only between certain pairs of states (s_i and s_j, $\forall s_j \in S$). Hence, $\bar{U}_i$ must be at most as long as $\bar{D}$. In fact, any distinguishing sequence is also a UIO sequence for all the states by definition.

For example, for the state s_3 in M_0 of Figure 1, $\bar{U}_3 = b$ is shorter than the distinguishing sequence $\bar{D} = aa$. However, for the states s_1 and s_2, shortest UIO sequences are of length 2, which is the same as the length of the distinguishing sequence.

Therefore, when we do have a distinguishing sequence for an FSM M, we may be able to find shorter UIO sequences for the states of M. It is this observation that will allow us to form shorter checking sequences, as explained in the rest of the paper.

An FSM M can be represented by a directed graph (*digraph*) $G = (V, E)$ where a set of vertices V represents the set S of states of M, and a set of directed edges E represents all transitions of M. Each edge $e = (v_j, v_k, x/y) \in E$ represents a transition $t = (s_j, s_k, x/y)$ of M from state s_j to state s_k with input x and output y where $s_j, s_k \in S$, $x \in X$, and $y \in Y$ such that $\delta(s_j, x) = s_k$, $\lambda(s_j, x) = y$.

A sequence $\bar{P} = (n_1, n_2, x_1/y_1)(n_2, n_3, x_2/y_2)\ldots(n_{k-1}, n_k, x_{k-1}/y_{k-1})$ of pairwise adjacent edges from G forms a *path* in which each *node* n_i represents a vertex from V and thus, ultimately, a state from S. Here $initial(\bar{P})$ denotes n_1, which is the *initial node* of $\bar{P}$, and $final(\bar{P})$ denotes n_k, which is the *final node* of $\bar{P}$. Two paths $\bar{P}_1$ and $\bar{P}_2$ can be concatenated as $\bar{P}_1\bar{P}_2$ only if $final(\bar{P}_1) = initial(\bar{P}_2)$.

The sequence $\bar{Q} = (x_1/y_1)(x_2/y_2)\ldots(x_{k-1}/y_{k-1})$ is the *label* of $\bar{P}$ and is denoted $label(\bar{P})$. In this case, $\bar{Q}$ is said to *label* the path $\bar{P}$. $\bar{Q}$ is said to be a *transfer sequence* from n_1 to n_k. The path $\bar{P}$ can be represented by the tuple $(n_1, n_k, \bar{Q})$ or by the tuple $(n_1, n_k, \bar{x}/\bar{y})$ in which $\bar{x} = x_1 x_2 \ldots x_{k-1}$ is the *input portion* of $\bar{Q}$ and $\bar{y} = y_1 y_2 \ldots y_{k-1}$ is the *output portion* of $\bar{Q}$.

A *tour* is a path whose initial and final nodes are the same. Given a tour $\bar{\Gamma} = e_1 e_2 \ldots e_k$, $\bar{P} = e_j e_{j+1} \ldots e_k e_1 e_2 \ldots e_{j-1}$ is a path formed by *starting* $\bar{\Gamma}$ with edge e_j, and hence by *ending* $\bar{\Gamma}$ with edge e_{j-1}. An *Euler Tour* is a tour that contains each edge exactly once. A set E' of edges from G is *acyclic* if no tour can be formed using the edges in E'.

A digraph is *strongly connected* if for any ordered pair of vertices (v_i, v_j) there is a path from v_i to v_j. An FSM is *strongly connected* if the digraph that represents it is strongly connected. It will be assumed that any FSM considered in this paper is deterministic, minimal, completely specified, and strongly connected.

Given an FSM M, let $\Phi(M)$ be the set of FSMs each of which has at most n states and the same input and output alphabets as M. Let N be an FSM of $\Phi(M)$. N is *isomorphic* to M if there is a one-to-one and onto function f on the state sets of M and N such that for any state transition $(s_i, s_j, x/y)$ of M, $(f(s_i), f(s_j), x/y)$ is a transition of N. A *checking sequence* of M is an input sequence starting at the initial state s_1 of M that distinguishes M from any N of $\Phi(M)$ that is not isomorphic to M. In the context of testing, this means that in response to this input sequence, any faulty implementation N from $\Phi(M)$ will produce an output sequence different from the expected output, thereby indicating the presence of a fault/faults. As stated earlier, a crucial part of testing the correct implementation of each transition of M in N from $\Phi(M)$ is recognizing the starting and terminating states of the transition which lead to the notions of state recognition and transition verification used in algorithms for constructing checking sequences (for example, [9,13]).

3   An Existing Approach

In this section, we will present an existing approach for generating checking sequences. The approach is based on distinguishing sequences only, and directly imported from [12] for completeness. After understanding the components (and their purpose) that are put together to form a checking sequence by this approach, it will be easier to understand how we can use UIO sequences instead of some of these components, that will hopefully make the generated checking sequences shorter. In fact, the algorithm for generating a checking sequence that will be proposed in this paper is a modification on the algorithm of [12], which was first given in [13].

3.1   Basics

The checking sequence $\bar{C}$ will be a sequence of inputs to be applied to SUT N, that will identify whether N is a correct implementation of M or not, i.e. whether N is isomorphic to M or not. Suppose that we trace $\bar{C}$ on the digraph $G = (V, E)$ representing M. Since M is deterministic, the trace will correspond to a unique path $\bar{P} = (n_1, n_2, x_1/y_1)(n_2, n_3, x_2/y_2) \ldots (n_{k-1}, n_k, x_{k-1}/y_{k-1})$. Below we will refer to the checking sequence $\bar{C}$ as the input portion of the input/output sequence $\bar{Q}$ which is the label of the path $\bar{P}$.

$\bar{P}$ can also be viewed as the application of $\bar{C}$ to N. In this view, the nodes $n_1, n_2, \ldots, n_k$ (or equivalently the states of N visited during this application) are not known. A checking sequence $\bar{C}$, or equivalently $\bar{P}$, should be designed in such a way that, the inputs and the corresponding outputs should provide sufficient evidence to let us identify these unknown states that are visited during the application of $\bar{C}$ to N.

If M has a distinguishing sequence $\bar{D}$, then $\bar{D}$ can be used in $\bar{C}$ to help to identify the states. Let us call $\bar{T}_i = \bar{D}/\lambda(s_i, \bar{D})\bar{B}_i$ as a T–sequence, where $\bar{B}_i = \bar{I}_i/\lambda(\delta(s_i, \bar{D}), \bar{I}_i)$ for a possibly empty transfer sequence $\bar{I}_i$. For example, for FSM M_0 in Figure 1, if we take $\bar{I}_1, \bar{I}_2$ and $\bar{I}_3$ as empty sequences, $\bar{T}_1 = aa/00$, $\bar{T}_2 = aa/01$, $\bar{T}_3 = aa/10$.

Inference Rule IR1: Let $\bar{R}_i = (n_p, n_q, \bar{T}_i)$ be a subpath in $\bar{P}$. Since the response of N to $\bar{D}$ at n_p is $\lambda(s_i, \bar{D})$, this unknown state n_p of N at step p, has some relation to the state s_i of M. Of course, this does not guarantee that n_p

is equivalent to the state s_i under the light of this evidence only. N may be a faulty implementation of M, yet it may still have a state that produces the same output $\lambda(s_i, \bar{D})$ to $\bar{D}$. Therefore we only say that, if n_p produces the same output to $\bar{D}$ as s_i, then n_p is *recognized* as state s_i of M in $\bar{Q}$.

Based on the assumption that N does not change during the experiments, the following inference rule can also be used.

Inference Rule IR2: If $\bar{P}_1 = (n_p, n_q, \bar{x}/\bar{y})$ and $\bar{P}_2 = (n_r, n_s, \bar{x}/\bar{y})$ are two subpaths of $\bar{P}$ such that n_p and n_r are recognized as state s_i of M and n_q is recognized as state s_j of M, then n_s is said to be recognized (in $\bar{Q}$) as state s_j of M. Intuitively, this rule says that if $\bar{P}_1$ and $\bar{P}_2$ are labeled by the same input/output sequence and their starting vertices are both recognized as the same state s_i of M, then their terminating vertices correspond to the same state s_j of M.

For N to be a correct implementation of M, first of all, for each state s_i of M, N must have a state which is recognized as s_i. If P has subpaths $\bar{R}_i = (n_p, n_q, \bar{T}_i)$ for all $i \in \{1, 2, \ldots, n\}$, then it will check existence of the corresponding states in N. If N does not produce the expected outputs, then N is a faulty implementation of M. However, if N produces the expected outputs, then for each state s_i in M, N must have at least one state corresponding to (recognized as) s_i. When combined with the assumption that N has at most n states, this will form a one–to–one correspondence between the states of M and the states of N.

As explained in the paragraph above, for each $\bar{T}_i$, $\bar{P}$ will have at least one subpath $\bar{R}_i = (n_p, n_q, \bar{T}_i)$. Based on IR1, $initial(\bar{R}_i)$ will be recognized as s_i. Note that, if there exists another subpath $\bar{R}'_i = (n'_p, n'_q, \bar{T}_i)$, $initial(\bar{R}'_i)$ will again be recognized as s_i. In other words, for every subpath with the label $\bar{T}_i$, the initial node of the subpath will be recognized as s_i. We will abuse the notation and let $initial(\bar{T}_i)$ denote the state s_i. Since, N is deterministic and does not change during experiments, we can also argue that for any subpath $\bar{R}_i$ with the label $\bar{T}_i$, $final(\bar{R}_i)$ will be recognized as the same state s_j, where $s_j = \delta(s_i, \bar{D}\bar{I}_i)$. We will use $final(\bar{T}_i)$ to denote this state s_j. Below we explain how $final(\bar{R}_i)$ can be recognized as well.

In order to recognize $final(\bar{R}_i)$, $\bar{P}$ will include subpaths with the labels as explained below. Let α'-set $A = \{\bar{\alpha}'_1, \bar{\alpha}'_2, \ldots \bar{\alpha}'_q\}$ be a set of input/output sequences such that $\bar{\alpha}'_k$ ($1 \leq k \leq q$) is the sequence $\bar{T}_{k_1} \bar{T}_{k_2} \ldots \bar{T}_{k_{r_k}}$, for some $1 \leq k_1, k_2, \ldots, k_{r_k} \leq n$, such that $\forall i \in \{1, 2, \ldots r_k - 1\}$, $initial(\bar{T}_{k_{i+1}}) = final(\bar{T}_{k_i})$. Each $\bar{\alpha}'_k$ is called an α'-sequence, and an α'-set A satisfies the following condition [13]: For all $i \in \{1, 2, \ldots, n\}$, there exists a $j \in \{1, 2, \ldots, n\}$ and a $k \in \{1, 2, \ldots, q\}$, such that $\bar{T}_i \bar{T}_j$ is a subsequence of $\bar{\alpha}'_k$. For example $\{\bar{T}_1 \bar{T}_3, \bar{T}_3 \bar{T}_2, \bar{T}_2 \bar{T}_1\}$ is an α'-set for FSM M_0 given in Figure 1.

Lemma 1. Let $T = \{\bar{T}_1, \bar{T}_2, \ldots, \bar{T}_n\}$ be a T-set, and $A = \{\bar{\alpha}'_1, \bar{\alpha}'_2, \ldots, \bar{\alpha}'_q\}$ be an α'-set based on T. If $\bar{Q} = label(\bar{P})$ includes all $\bar{\alpha}'_k$, $1 \leq k \leq q$, as a subsequence then:

1. For all $k \in \{1, 2, \ldots q\}$, if $(n_p, n_q, \bar{\alpha}'_k)$ is a subpath in $\bar{P}$, then n_p is recognized.
2. For all $i \in \{1, 2, \ldots, n\}$, $\bar{T}_i$ is a subsequence in $\bar{Q}$.

3. For all $i \in \{1, 2, \ldots, n\}$, if $(n_r, n_s, \bar{T}_i)$ is a subsequence in $\bar{P}$, then n_s is recognized in $\bar{P}$.
4. For all $k \in \{1, 2, \ldots q\}$, if $(n_p, n_q, \bar{\alpha}'_k)$ is a subpath in $\bar{P}$, then n_q is recognized.

Proof. 1. Since $\bar{\alpha}'_k$ starts with a $\bar{T}_i$ that has a prefix $\bar{D}/\lambda(s_i, \bar{D})$, n_p is recognized as s_i in $\bar{Q}$ (IR1).
2. Since for each $\bar{T}_i$, there exists a $\bar{T}_j$ such that $\bar{T}_i \bar{T}_j$ is a subsequence of some $\bar{\alpha}'_k$, which in turn is a subsequence in $\bar{Q}$, $\bar{T}_i$ is a subsequence in $\bar{Q}$.
3. There exists a $\bar{T}_j$ such that $\bar{T}_i \bar{T}_j$ is a subsequence of some $\bar{\alpha}'_k$, which in turn is a subsequence in $\bar{Q}$. In other words, there exists a subpath $(n_p, n_t, \bar{T}_i \bar{T}_j)$ in $\bar{P}$. After dividing this path into two as $(n_p, n_q, \bar{T}_i)(n_q, n_t, \bar{T}_j)$, it is easy to see that, n_p and n_q are recognized as states s_i and s_j respectively. But then, we can use IR2 on $(n_p, n_q, \bar{T}_i)$ and $(n_r, n_s, \bar{T}_i)$ to deduce that n_s is recognized as s_j.
4. Since $\bar{\alpha}'_k$ ends with a $\bar{T}_i$, based on the discussion given in (3) above, n_q is recognized.   □

Different α' sets can be found for a given set of T–sequences $\{\bar{T}_1, \bar{T}_2, \ldots, \bar{T}_n\}$. For example $\{\bar{T}_3 \bar{T}_2 \bar{T}_1 \bar{T}_3\}$ and $\{\bar{T}_1 \bar{T}_3 \bar{T}_2 \bar{T}_1\}$ are also α'–sets for M_0 of Figure 1. Since $\bar{C}$ will have the input portion of α'–sequences as subsequences, it may be desirable to minimize the total length of α'–sequences. Note that, this is just a heuristic to minimize the length of $\bar{C}$. In [14] authors explain how to find a set of α'–sequences with a minimal total length from a given set of T–sequences.

Besides these components to recognize the states in N, a checking sequence will also have components to check if the transitions are implemented correctly. We say that the transition $(s_i, s_j, x/y)$ of M is *verified* in $\bar{Q} = label(\bar{P})$ if $(n_p, n_q, x/y)$ is a subpath of $\bar{P}$, n_p is recognized as s_i and n_q is recognized as s_j. n_p will have to be recognized using IR2. n_q can be recognized using IR1, by applying a $\bar{T}_i$. Since α'–sequences start with $\bar{T}_i$'s, they can also be used to recognize the end state of the transitions [13].

In the next section we will explain a method to generate a checking sequence, which is based on Theorem 1.

Theorem 1. *(Theorem 1, [9]) Let $\bar{Q}$ be the label of a path $\bar{P}$ on G representing an FSM M that starts at s_1. If every transition of M is verified in $\bar{Q}$, then the input portion of $\bar{Q}$ is a checking sequence of M.*

3.2 Checking Sequence Construction

In [13], the following method is explained to produce a checking sequence. Given $G = (V, E)$ corresponding to an FSM M, a T–sequence set $\mathcal{T} = \{\bar{T}_1, \bar{T}_2, \ldots, \bar{T}_n\}$, and an α'–set $A = \{\bar{\alpha}'_1, \bar{\alpha}'_2, \ldots, \bar{\alpha}'_q\}$, first another digraph $G' = (V', E')$ is produced by augmenting the digraph G as follows (Figure 2 is the digraph G' corresponding to the digraph G of FSM M_0 given in Figure 1):

a) $V' = V \cup U'$ where $U' = \{v' : v \in V\}$, i.e. for each vertex v in G, there are two copies of v in G'. In Figure 2, the nodes on the left are the nodes in V, and the nodes on the right are the nodes in U'.

b) $E' = E_C \cup E_T \cup E_{\alpha'} \cup E''$ where

   i) $E_C = \{(v'_i, v_j, x/y) : (v_i, v_j, x/y) \in E\}$. The solid edges leaving the nodes on the right in Figure 2 are the edges in E_C.

   ii) $E_T = \{(v_i, v'_j, \bar{T}_i) : \bar{T}_i \in \mathcal{T}, s_i = initial(\bar{T}_i), s_j = final(\bar{T}_i)\}$. For example, since $initial(\bar{T}_1) = s_1$ and $final(\bar{T}_1) = s_2$, there is an edge $(v_1, v'_2, \bar{T}_1)$ in Figure 2.

   iii) $E_{\alpha'} = \{(v_i, v'_j, \bar{\alpha}'_k) : \bar{\alpha}'_k \in A, \bar{\alpha}'_k = \bar{T}_i \ldots \bar{T}_l, initial(\bar{T}_i) = s_i, final(\bar{T}_l) = s_j\}$. For example, in Figure 2 we consider a singleton α'–set $A = \{\bar{\alpha}'_1 = \bar{T}_1\bar{T}_3\bar{T}_2\bar{T}_1\}$. There is an edge $(v_1, v'_3, \bar{\alpha}'_1)$ in Figure 2 since $initial(\bar{T}_1) = s_1$ (the first T–sequence in $\bar{\alpha}'_1$ is $\bar{T}_1$), and $final(\bar{T}_1) = s_3$ (the last T–sequence in $\bar{\alpha}'_1$ is $\bar{T}_1$).

   iv) $E'' \subseteq \{(v'_i, v'_j, x/y) : (v_i, v_j, x/y) \in E\}$. E'' is a subset of the copies of the edges in E placed between the corresponding nodes in U'. E'' is selected in such a way that, $G'' = (U', E'')$ does not have a tour and G' is strongly connected.

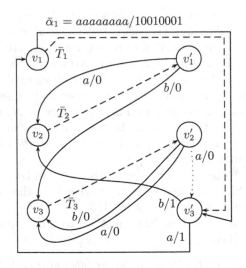

$\bar{\alpha}_1 = aaaaaaaa/10010001$

Fig. 2. G' for M_0

We would like to highlight the followings about G':

- The edges in E_C represent the transitions to be verified.
- On a path in G', an edge in E_C will have to be followed by an edge in E_T or $E_{\alpha'}$. Since an α'–sequence also starts with a T–sequence, this means that a transition will always be followed by a T–sequence, hence the end state of the transition will be recognized.
- On a path in G', the nodes in U' will be recognized. If a node v' in U' is reached by using an edge in E_T or an edge $E_{\alpha'}$, it is easy to show that v' is recognized since the final states of T–sequences and α'–sequences are recognized as explained previously in Lemma 1. As long as $G'' = (U', E'')$

is acyclic, it is also guaranteed that v' will be recognized if it is reached by using an edge in E'' (please see the proof of Theorem 2 in [9] for the sketch of a proof of this claim).
- Based on the previous claim, the initial states of the transitions will also be recognized in a path $\bar{P}$ in G', since the edges in E_C representing the transitions always have their initial nodes in U'.

Suppose that we form a path $\bar{P}$ in G' that starts from and ends at v_1 such that, it includes all the edges in $E_{\alpha'}$ (so that the states are recognized), and it also includes all the edges in E_C (so that the transitions are verified). On the basis of Theorem 1, it is argued in [13] that the input portion of the label of such a path $\bar{P}$ which is followed by $\bar{D}$ is a checking sequence of M.

In fact, since we would like to keep the length of the checking sequence small, an optimization is used to find a short path. The approach given in [13] forms a minimal symmetric augmentation G^* of the digraph induced by $E_{\alpha'} \cup E_C$ by adding replications of edges from E'. If G^*, with its isolated vertices removed, is connected, then G^* has an Euler tour. Otherwise, a heuristic such as the one given in [9] is applied to make G^* connected and an Euler tour of this new digraph is formed to find a path from v_1 to v_1.

$$(v_1, v_3', \bar{\alpha}_1')(v_3', v_2, b/1)(v_2, v_1', \bar{T}_2)(v_1', v_2, a/0)(v_2, v_1', \bar{T}_2)(v_1', v_3, b/0)(v_3, v_2', \bar{T}_3)$$
$$(v_2', v_3, a/0)(v_3, v_2', \bar{T}_3)(v_2', v_3, b/0)(v_3, v_2', \bar{T}_3)(v_2', v_3', a/0)(v_3', v_1, a/1)$$

Fig. 3. An tour in G'

The checking sequence constructed based on the tour given in Figure 3 would be the label of the path of Figure 3 followed by $\bar{D}$. Hence the length of the checking sequence is 27.

4   Using UIO Sequences for State Recognition

The method explained in Section 3 uses a distinguishing sequence to recognize the end state of a transition $(s_i, s_j, x/y)$ by applying $\bar{D}$ after the execution of the transition, and by observing the output $\lambda(s_j, \bar{D})$ which is unique among all the states. The purpose of an edge $(v_i, v_j', \bar{T}_i)$ in G' is twofold: (i) it recognizes the final state of a transition, and (ii) it also recognizes the final state of itself (see Lemma 1). In other words, when the input portion of $\bar{T}_i$ is applied to SUT N and the expected output is observed, we do not only recognize the state before the application, but we also recognize the state that is reached after the application of the input part of $\bar{T}_i$. This is obviously based on the fact that, the input portion of all the α'–sequences are also applied and the expected outputs are observed from N.

A UIO sequence $\bar{U}_j$ for a state s_j also provides a similar information. In other words, to recognize the end state of a transition $(s_i, s_j, x/y)$, one can apply $\bar{U}_j$ after the execution of the transition, and observe the output $\lambda(s_j, \bar{U}_j)$ which is also unique among all the states. Since $\bar{U}_j$ will be at most as long as $\bar{D}$, using UIO sequences instead of distinguishing sequences may shorten the overall checking sequence.

However, for a UIO sequence $\bar{U}_j$ for a state s_j, suppose that $\bar{P}$ contains $(n_p, n_q, \bar{U}_j/\lambda(s_j, \bar{U}_j))$ as a subpath. (i) Can we conclude that n_p must be recognized as s_j? (ii) Can we conclude that n_q must be recognized as $\delta(s_j, \bar{U}_j)$? Below we explain under what conditions both of these questions can be answered positively.

For a sequence $\bar{x} \in X^*$, let $symb(\bar{x}) \subseteq X$ denote the set of input symbols that appear in $\bar{x}$. For example, if $\bar{x} = aba$, then $symb(\bar{x}) = \{a, b\}$.

Theorem 2. *Let $\bar{Q}$ be the label of a path $\bar{P}$ in $G = (V, E)$ corresponding to an FSM M, and $\bar{U}_j$ be a UIO for a state s_j in M. Assume that $\forall x \in symb(\bar{U}_j)$ and for all states s in M, the transition $(s, \delta(s, x), x/\lambda(s, x))$ is verified in $\bar{Q}$. If $(n_p, n_q, \bar{U}_j/\lambda(s_j, \bar{U}_j))$ is a subpath of $\bar{P}$, then n_p is recognized as s_j and n_q is recognized as $\delta(s_j, \bar{U}_j)$.*

We will need the following result to prove Theorem 2.

Lemma 2. *Let $\bar{Q}$ be the label of a path $\bar{P}$ in $G = (V, E)$ corresponding to an FSM M, and $\bar{x}' \in X^*$ be an input sequence. Assume that $\forall x \in symb(\bar{x}')$ and for all states s in M, the transition $(s, \delta(s, x), x/\lambda(s, x))$ is verified in $\bar{Q}$. If $(n_r, n_s, \bar{x}'/\lambda(s', \bar{x}'))$ is a subpath of $\bar{P}$ and n_r is recognized as s', then n_s is recognized as $\delta(s', \bar{x}')$.*

Proof. The proof is based on induction on the length of $\bar{x}'$. When the length of $\bar{x}'$ is 1, i.e. when $\bar{x}' = a$ for some $a \in X$, we have $\bar{P}_1 = (n_r, n_s, a/\lambda(s', a))$ as a subpath in $\bar{P}$. Since $\forall x \in symb(\bar{x}') = \{a\}$ and for all states s in M, the transition $(s, \delta(s, x), x/\lambda(s, x))$ is verified in $\bar{Q}$, there must exist a subpath $\bar{P}_2 = (n_p, n_q, a/\lambda(s', a))$ in $\bar{P}$ such that n_p is recognized as s', and n_q is recognized as $\delta(s', a)$. Using $\bar{P}_1$ and $\bar{P}_2$ and the inference rule IR2, we can deduce that n_s is recognized as $\delta(s', a)$.

For the inductive step, assume that $\bar{x}' = a\bar{x}''$, in other words we have a subpath $\bar{P}_1 = (n_r, n_s, a\bar{x}''/\lambda(s', a\bar{x}''))$, or equivalently by dividing $\bar{P}_1$ into two, we have the subpaths $\bar{P}_{11} = (n_r, n_t, a/\lambda(s', a))$, $\bar{P}_{12} = (n_t, n_s, \bar{x}''/\lambda(\delta(s', a), \bar{x}''))$. Based on the discussion given in the base step of the proof, n_t is recognized as $\delta(s', a)$. This completes the proof, since n_t is recognized, and $\bar{x}''$ is shorter than $\bar{x}'$. □

We can now go back to the proof of Theorem 2:

Proof (of Theorem 2). We know that the transitions of all the states for all the input symbols in $\bar{U}_j$ are implemented correctly. Since $\bar{U}_j$ is a UIO sequence for s_j, this means that only the state that should be recognized as state s_j in N produces the output $\lambda(s_j, \bar{U}_j)$ to $\bar{U}_j$. Hence, for the subpath $(n_p, n_q, \bar{U}_j/\lambda(s_j, \bar{U}_j))$ of $\bar{P}$, n_p must be recognized as s_j.

When n_p is recognized, we can use Lemma 2 to show that n_q is also recognized. □

What Theorem 2 suggests is that, when it is guaranteed that the transitions of the states for the input symbols that appear in a UIO sequence $\bar{U}_j$ are verified, then $\bar{U}_j/\lambda(s_j, \bar{U}_j)$ can be used in a checking sequence exactly in the same way and for the same purpose as the T–sequence $\bar{T}_j$. Based on this observation, we will propose a modification on the method given in Section 3.2 for constructing checking sequences.

5   Modified Method for Checking Sequence Construction

The modification will actually be quite intuitive, and very simple for a reader who understands the purposes of the components of the digraph G' given in Section 3.2.

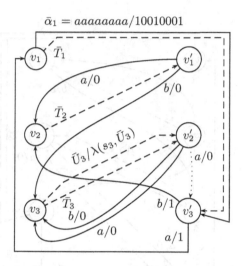

$$\bar{\alpha}_1 = aaaaaaaa/10010001$$

Fig. 4. The first (unseccuessful) attempt for the modification

Let us explain the modified method on our running example first. We will provide the method formally later. Consider M_0 in Figure 1, and the digraph G' for M_0 given in Figure 2, and let us focus on the edge $(v_2', v_3, a/0)$. In G', this edge will have to followed by the edge $(v_3, v_2', \bar{T}_3)$, which would both recognize v_3 as s_3, and also recognize v_2' as s_2.

The state s_3 in M_0 has the UIO $\bar{U}_3 = b$. Based on the discussions given Section 4, we can add outgoing edge to $(v_3, v_2', \bar{U}_3/\lambda(s_3, \bar{U}_3))$ in G', since $\bar{U}_3/\lambda(s_3, \bar{U}_3)$ can also be used in a similar way as $\bar{T}_3$ is used in G' (Figure 4).

However, we also require that the input symbols that appear in the UIO sequences that are used to recognize states to be verified. We have to avoid verifying an edge depending on the correctness of itself. In other words, there are some transitions with the input b whose final states are s_3. Namely the edges $(v_1', v_3, b/0)$ and $(v_2', v_3, b/0)$ in Figure 4. The verification of the corresponding transitions of these edges will have to be performed in the conventional way. In other words, we will need to force to use the edge with the label $\bar{T}_3$ when these two edges are used to reach v_3, to guarantee that b transitions of s_2 and s_3 are verified.

This can be achieved by having two copies of v_3 in G'. One copy of v_3 will be the usual v_3 that already exists in G', and have the outgoing edge with label $\bar{T}_3$. The other copy of v_3 (say v_3^U) will have an outgoing edge with the label $\bar{U}_3/\lambda(s_3, \bar{U}_3)$. Note that, having this edge as the only outgoing edge of v_3^U would force $\bar{U}_3$ to be used to recognize the node v_3^U. However, if we also add an edge (v_3^U, v_3, ϵ), this would introduce the possibility and the flexibility of using $\bar{T}_3$ (and any α'-sequence originating from v_3 if there were any) to recognize the node v_3^U. The final digraph G' that will be used for our example is given in Figure 5.

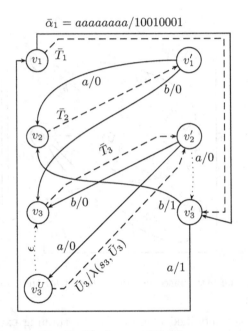

Fig. 5. G' after modification

We now explain the modified method more formally. Given $G = (V, E)$ corresponding to an FSM M, a T-sequence set $\mathcal{T} = \{\bar{T}_1, \bar{T}_2, \ldots, \bar{T}_n\}$, and an α'-set $A = \{\bar{\alpha}'_1, \bar{\alpha}'_2, \ldots, \bar{\alpha}'_q\}$, we will again generate a digraph $G' = (V', E')$ by augmenting G. Assume that we are also given a set of UIO sequences for some of the states to recognize these states. Let $\mathcal{U} = \{\bar{U}_{i_1}, \bar{U}_{i_2}, \ldots, \bar{U}_{i_k}\}$ be such a set of UIO sequences. Suppose that the UIO sequence $\bar{U}_{i_j} \in \mathcal{U}$ is a UIO sequence for the state s_{i_j}. Let $symb(\mathcal{U}) = symb(\bar{U}_{i_1}) \cup symb(\bar{U}_{i_2}) \cup \cdots \cup symb(\bar{U}_{i_k})$ below.

a) $V' = V \cup V^U \cup U'$ where
   i) $U' = \{v' : v \in V\}$. For each $v \in V$, we have a copy of v in U'.
   ii) $V^U = \{v_j^u : v_j \in V, j \in \{i_1, i_2, \ldots, i_k\}\}$
      If $\mathcal{U}$ includes a UIO sequence $\bar{U}_j$ for the state s_j, then for the corresponding node $v_j \in V$, we create a copy v_j^U in V^U.

b) $E' = E_C \cup E_T \cup E_U \cup E_\epsilon \cup E_{\alpha'} \cup E''$ where
   i) $E_C = \{(v_i', v_j^U, x/y) : (v_i, v_j, x/y) \in E, x \notin symb(\mathcal{U}), j \in \{i_1, i_2, \ldots, i_k\}\} \cup \{(v_i', v_j, x/y) : (v_i, v_j, x/y) \in E, (x \in symb(\mathcal{U})$ or $j \notin \{i_1, i_2, \ldots, i_k\})\}$.
      E_C will again correspond to the transitions to be verified. However, we have now two different types of edges in E_C. If the input symbol of the transition is not one of the input symbols in $symb(\mathcal{U})$ (i.e. it does not appear in any of the UIO sequences provided), and there exists a UIO sequence $\bar{U}_j \in \mathcal{U}$ for the recognition of final state s_j of the transition, then the edge is connected to the node v_j^U. Otherwise, the edge will be connected to the node v_j.
   ii) $E_T = \{(v_i, v_j', \bar{T}_i) : \bar{T}_i \in \mathcal{T}, s_i = initial(\bar{T}_i), s_j = final(\bar{T}_i)\}$. There is no change in this component.
   iii) $E_U = \{(v_i^U, v_j', \bar{U}_i/\lambda(s_i, \bar{U}_i)) : \bar{U}_i \in \mathcal{U}, s_j = \delta(s_i, \bar{U}_i)\}$. If $\bar{U}_i \in \mathcal{U}$ is a UIO sequence for a state s_i, then we place the outgoing edge from v_i^U for the UIO recognition, hence it has the label $\bar{U}_i/\lambda(s_i, \bar{U}_i)$.
   iv) $E_\epsilon = \{(v_i^U, v_i', \epsilon) : \bar{U}_i \in \mathcal{U}\}$. If $\bar{U}_i \in \mathcal{U}$ is a UIO sequence for a state s_i, then we insert an ϵ edge from v_i^U to v_i for increased flexibility of using $\bar{T}_i$ from v_i (or an α'–sequence outgoing from v_i, if exists) for recognizing the end state of an edge in E_C that ends in v_i^U.
   v) $E_{\alpha'} = \{(v_i, v_j', \bar{\alpha}_k') : \bar{\alpha}_k' \in A, \bar{\alpha}_k' = \bar{T}_i \ldots \bar{T}_l, initial(\bar{T}_i) = s_i, final(\bar{T}_l) = s_j\}$. There is no change in this component.
   vi) $E'' \subseteq \{(v_i', v_j', x/y) : (v_i, v_j, x/y) \in E\}$. E'' is again a subset of the copies of the edges in E placed between the corresponding nodes in U'. E'' is selected in such a way that, $G'' = (U', E'')$ does not have a tour and G' is strongly connected.

As in the case of the previous method, a tour is found in G' that includes all the edges in $E_{\alpha'} \cup E_C$. Figure 6 shows a tour in G' given in Figure 5. The tour includes the necessary edges, and hence can be used to form a checking sequence as explained below.

$$(v_1, v_3', \bar{\alpha}_1')(v_3', v_2, b/1)(v_2, v_1', \bar{T}_2)(v_1', v_2, a/0)(v_2, v_1', \bar{T}_2)(v_1', v_3, b/0)(v_3, v_2', \bar{T}_3)$$
$$(v_2', v_3^U, a/0)(v_3^U, v_2', \bar{U}_3/\lambda(s_3, \bar{U}_3))(v_2', v_3, b/0)(v_3, v_2', \bar{T}_3)(v_2', v_3', a/0)(v_3', v_1, a/1)$$

Fig. 6. An tour in the modified G'

The checking sequence constructed based on the tour given in Figure 6 would be the label of the path of Figure 6 followed by $\bar{D}$. Hence the length of the checking sequence is 26. The length of the new checking sequence is 1 less than the length of the checking sequence produced by the previous method.

6 Conclusion and Future Work

We have shown that, for a FSM M with a distinguishing sequence, UIO sequences for states can also be used to recognize states in a checking sequence. Existing methods in the literature use only distinguishing sequences to recognize states in a checking sequence when M has a distinguishing sequence. However, when an FSM M has a distinguishing sequence, the states of M may have shorter UIO sequences. Therefore using UIO sequences instead of distinguishing sequences may result in shorter checking sequences. We have given an example of such a case, where the length of the checking sequence is reduced.

We have also shown how a checking sequence that uses UIO sequences for state recognition can be constructed by modifying an already existing checking sequence construction technique, which is based on using distinguishing sequences only for state recognition.

It is assumed that we are given a set of UIO sequences to be used for state recognition. Further research is required to compute a set of UIO sequences for an FSM M, that will help shortening the length of a checking sequence. Intuitively, if for a state s_j, there is a large number of transitions incoming into the state s_j, and if we can find a UIO $\bar{U}_j$ for s_j such that a small number of different input symbols appear in $\bar{U}_j$, then heuristically, using $\bar{U}_j$ for recognizing s_j seems to be promising to reduce the length of the checking sequence.

This paper shows that it is possible to decrease the length of a checking sequence using the method proposed. However, an experimental study would also be useful to understand the magnitude of a typical reduction.

References

1. Tanenbaum, A.S.: Computer Networks. 3rd edn. Prentice Hall International Editions, Prentice Hall (1996)
2. Gill, A.: Introduction to the Theory of Finite–State Machines. McGraw–Hill, New York (1962)
3. Hennie, F.C.: Fault–detecting experiments for sequential circuits. In: Proceedings of Fifth Annual Symposium on Switching Circuit Theory and Logical Design, Princeton, New Jersey (1964) 95–110
4. Lee, D., Yannakakis, M.: Principles and methods of testing finite–state machines – a survey. Proceedings of the IEEE 84(8) (1996) 1089–1123
5. Sabnani, K., Dahbura, A.: A protocol test generation procedure. Computer Networks 15 (1988) 285–297
6. Kohavi, Z.: Switching and Finite Automata Theory. McGraw–Hill, New York (1978)
7. Lee, D., Yannakakis, M.: Testing finite state machines: state identification and verification. IEEE Trans. Computers 43(3) (1994) 306–320
8. Gonenc, G.: A method for the design of fault detection experiments. IEEE Transactions on Computers 19 (1970) 551–558
9. Ural, H., Wu, X., Zhang, F.: On minimizing the lengths of checking sequences. IEEE Transactions on Computers 46(1) (1997) 93–99

10. Aho, A., Dahbura, A., Lee, D., Uyar, M.: An optimization technique for protocol conformance test generation based on UIO sequences and rural chinese postman tours. IEEE Transactions on Communications **39**(11) (1991) 1604–1615
11. Chan, W., Vuong, C., Otp, M.: An improved protocol test generation procedure based on UIOS. ACM SIGCOMM Computer Communication Review **19**(4) (1989) 283–294
12. Tekle, K.T., Ural, H., Yalcin, M.C., Yenigun, H.: Generalizing redundancy elimination in checking sequences. In: 20th International Symposium on Information and Computer Sciences (ISCIS). Volume 3733 of Lecture Notes in Computer Science., Istanbul, Turkey (2005) 915–926
13. Hierons, R.M., Ural, H.: Reduced length checking sequences. IEEE Transactions on Computers **51**(9) (2002) 1111–1117
14. Hierons, R.M., Ural, H.: Optimizing the length of checking sequences. IEEE Transactions on Computers (2004) accepted for publication.

Reducing the Lengths of Checking Sequences by Overlapping

Hasan Ural and Fan Zhang

School of Information Technology and Engineering, University of Ottawa
ural@site.uottawa.ca, fzhang@site.uottawa.ca

Abstract. There are two main shortcomings in the existing models for generating checking sequences based on distinguishing sequences. First, these models require a priori selection of state recognition sequences (called α-sequences) which may not be the best selection for yielding substantial reduction in the length of checking sequences. Second, they do not take advantage of overlapping to further reduce the length of checking sequences. This paper proposes an optimization model that tackles these shortcomings to reduce the lengths of checking sequences beyond what is achieved by the existing models by replacing the state recognition sequences with a set of basic sequences called α-elements and by making use of overlapping.

1 Introduction

To ensure the correct functioning of implementations of a *Finite State Machine* (FSM) M, a *fault detection experiment* can be formed [14]: Such an experiment consists of applying an input sequence (derived from M) to an implementation N of M, observing the actual output sequence produced by N in response to the application of the input sequence, and comparing the actual output sequence to the expected output sequence. The applied input sequence is called a *checking sequence* which determines whether N is a correct or faulty implementation of M [8, 10].

A checking sequence of M is constructed in such a way that the output sequence produced by N in response to the application of the checking sequence provides sufficient information to verify that every state transition of M is implemented correctly by N. That is, in order to verify the implementation of a transition from state a to state b under input x, 1) N is transferred to the state recognized as state a of M; 2) the output produced by N in response to x is checked to be as specified in M (to detect an *output fault*); and 3) the state reached by N after the application of x is recognized as state b of M (to detect a *transfer fault*). Hence, a crucial part of testing the correct implementation of each transition is recognizing the starting and terminating states of the transition which can be achieved by a distinguishing sequence [8], a characterization set [8] or a unique input-output (UIO) sequence [6]. A *distinguishing sequence* for M is an input sequence for which each state of M produces a distinct output sequence. It is known that a distinguishing sequence may not exist for every minimal FSM [14], and that determining the existence of a distinguishing sequence for an FSM is PSPACE-complete [15]. However, based on distinguishing sequences, various methods have been proposed for FSM based testing (e.g., [4, 9-11, 16, 17]).

M.Ü. Uyar, A.Y. Duale, and M.A. Fecko (Eds.): TestCom 2006, LNCS 3964, pp. 274–288, 2006.

Recent methods for constructing reduced length checking sequences based on distinguishing sequences utilize optimization models. In these models, a distinguishing sequence for M is used to form both α-sequences and test segments [11, 16]. The α-sequences, which consist of consecutive applications of the distinguishing sequence for M, are formed to ensure that each state of M is also a distinct state of N; the test segments, which consist of the application of the input triggering the corresponding transition and the distinguishing sequence for M, are formed to verify that every state transition of M is implemented correctly by N. The α-sequences collectively confirm that if N produces the corresponding distinct output sequence for each state of M, then the distinguishing sequence for M is also a distinguishing sequence for N, that is, the distinguishing sequence used in the formation of the α-sequences defines a bijection between states of M and N. Thus, when a path P of the directed graph G representing M is formed such that the input sequence that induces P on G covers each α-sequence and each test segment, that input sequence is a checking sequence of M.

In these models, however, there are two main shortcomings. These models require *a priori* selection of a set of α-sequences which may not guarantee a substantial reduction in the length of a resulting checking sequence. Also, these models connect the α-sequences and test segments to form a checking sequence and thus do not take advantage of potential overlapping among the α-sequences and test segments that could be used to further reduce the lengths of checking sequences.

This paper proposes a novel optimization model that tackles these shortcomings in generating the minimal-length checking sequences: The proposed model does not require selection of α-sequences in advance. It employs a set of α-elements where there is an α-element for each state of M which consists of the application of the distinguishing sequence for M twice. The set of α-elements are then used for the same purpose as the α-sequences in the earlier models. The proposed model does not simply connect the α-sequences and test segments to form a checking sequence. It facilitates the use of overlapping among the α-elements and test segments to further reduce the lengths of resulting checking sequences.

In the remainder of the paper, the proposed model is presented after some preliminary definitions. An example is used to illustrate the model and the steps of its construction. It is then proven that the proposed model constructs a checking sequence. The extensions and the potential uses of the model are discussed in the concluding remarks.

2   Preliminaries

A deterministic and completely specified *FSM* (*finite state machine*) is a quintuple $M = (S, X, Y, \delta, \lambda)$, where $S = \{s_1, s_2, ..., s_n\}$ is a finite set of states with $n = |S|$ and $s_1 \in S$ as the *initial state*, X is a finite set of inputs, Y is a finite set of outputs, δ is a state transition function that maps $S \times X$ to S, and λ is an output function that maps $S \times X$ to Y. M is *minimal* if, for any different states $s_i, s_j \in S$, there is an input sequence $I \in X^*$ such that $\lambda(s_i, I) \neq \lambda(s_j, I)$. M can be represented by a directed graph (digraph) $G = (V, E)$ (Figure 1) where a set of vertices $V = \{v_1, v_2, ..., v_n\}$ represents the set of states of

M and a set of directed edges $E=\{(v_j, v_k; x/y): v_j, v_k \in V\}$ represents all specified transitions of M. More specifically, each edge $e = (v_j, v_k; x/y) \in E$ represents a state transition $t = (s_j, s_k; x/y)$ of M from state s_j to s_k with input $x \in X$ and output $y \in Y$, and the (input/output) pair x/y is the *label* of e.

A *path* $P = (n_1, n_2; x_1/y_1)(n_2, n_3; x_2/y_2) \dots (n_{r-1}, n_r; x_{r-1}/y_{r-1})$, $r > 1$, of $G = (V, E)$ is a finite sequence of adjacent (not necessarily distinct) edges in E, where each node n_i, $1 \le i \le r$, represents a vertex of V; n_1 and n_r are called *start* and *end* of P, and the input/output sequence $(x_1/y_1)(x_2/y_2) \dots (x_{r-1}/y_{r-1})$ is called *label* of P, denoted *label*(P). P is represented by $(n_1, n_r; I/O)$, where I/O is called the *transfer sequence* T from n_1 to n_r, $I = "x_1 x_2 \dots x_{r-1}"$ is called the *input portion* of I/O, $O = "y_1 y_2 \dots y_{r-1}"$ is called the *output portion* of I/O. In this case, I (or I/O) is said to *induce* P at n_1. The *length* (or *cost*) *of an input sequence* I (or *input/output sequence* I/O) is its number of inputs, denoted $|I|$ (or $|I/O|$). The *length* (*or cost*) *of a path* $P = (n_1, n_r; I/O)$ is the length (or cost) of the input sequence I, denoted $|P|$. A sequence $i_1 i_2 \dots i_k$ is a *subsequence* of $x_1 x_2 \dots x_m$ if there exists a Δ, $0 \le \Delta \le m-k$, such that for all j, $1 \le j \le k$, $i_j = x_{j+\Delta}$. Subsequence $i_1 i_2 \dots i_k$ is a *prefix* of $x_1 x_2 \dots x_m$ if $i_1 = x_1$.

A digraph $G = (V, E)$ is *strongly connected* if, for every pair of vertices v_j and v_k, there exists a path from v_j to v_k. A *Rural Postman* (*RP*) *path* from vertex v_i to vertex v_j over a subset of edges E' in $G = (V, E)$ is a path which starts at v_i, ends at v_j, and includes all edges of E'; the *Rural Chinese Postman* (*RCP*) *Problem* is to find an RP path of minimum cost i.e., an RCP path, which is the optimization model we will formulate. Algorithms for solving the RCP problem and its special cases important to testing can be found in [1, 16], which are left outside of the scope of this paper.

Let $M = (S, X, Y, \delta, \lambda)$ denote a completely specified, minimal, and deterministic FSM, which is represented by a strongly connected digraph $G = (V, E)$. Given an FSM M, let $\Phi(M)$ be the set of FSMs each of which has at most n states and the same input and output sets as M. Let N be an FSM of $\Phi(M)$. N is *isomorphic* to M if there is a one-to-one and onto function f on the state sets of M and N such that for any state transition $(s_i, s_j; x/y)$ of M, $(f(s_i), f(s_j); x/y)$ is a transition of N. A *checking sequence* of M is an input sequence starting at the initial state s_1 of M that distinguishes M from any N of $\Phi(M)$ that is not isomorphic to M i.e., the output sequence produced by any such N of $\Phi(M)$ is different from the output sequence produced by M. In the context of testing, this means that in response to this input sequence, any faulty implementation N from $\Phi(M)$ will produce an output sequence different from the expected output, thereby indicating the presence of a fault(s). As stated earlier, a crucial part of testing the correct implementation of each transition of M in N from $\Phi(M)$ is recognizing the starting and terminating states of the transition which lead to the notions of state recognition and transition verification used in algorithms for constructing checking sequences (for example, [11], [16]). These notions are defined below in terms of a given distinguishing sequence D for FSM M.

A *distinguishing sequence* (DS) of M is an input sequence D such that the output sequence produced by M in response to D is different for each state of M (i.e., $\forall s_i, s_j \in S, s_i \ne s_j, \lambda(s_i, D) \ne \lambda(s_j, D)$). A distinguishing sequence for FSM M_0 is shown in Table 1. Based on this definition, the concepts of *state recognition* and *transition verification* can be defined as follows. Let an IO-sequence Q be the label of a path P = $e_1e_2 \ldots e_r$ of G starting at v_1, where $e_j = (n_j, n_{j+1}; x_j/y_j)$ for all j, $1 \le j \le r$, i.e., $Q = (x_1/y_1) \ldots (x_r/y_r)$. Then the following defines state recognition and transition verification as in [16].

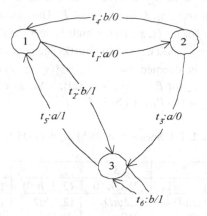

Fig. 1. FSM M_0

Table 1. A DS D = "*aa*" and responses of each state of FSM M_0

Start state	End state	D = "*aa*"
1	3	00
2	1	01
3	2	10

Recognition of a node n_i of P (in Q) as some state of M is defined recurrently and is associated with a nonnegative number *depth(n_i)*: Let depth(n_i) = ∞ initially;

- 1) A node n_i of P is *d-recognized* (*in Q*) as some state s of M if n_i is the start of a subpath of P whose label is $D/\lambda(s, D)$; depth(n_i) $\leftarrow$ 0.
- 2) Suppose that $(n_q, n_i; T)$ and $(n_j, n_k; T)$ are subpaths of P such that nodes n_q and n_j are *d*-recognized as state s of M, and n_k is *d*-recognized as state s' of M but n_i is not *d*-recognized. Then, node n_i is *t-recognized* as state s' of M; depth(n_i) $\leftarrow$ 1.
- 3) Suppose that $(n_q, n_i; T)$ and $(n_j, n_k; T)$ are subpaths of P such that n_q and n_j are either *d*-recognized or *t*-recognized as state s of M, and n_k is either *d*-recognized or *t*-recognized as state s' of M. Then, node n_i is *t-recognized* as state s' of M; depth(n_i) $\leftarrow$ min{depth(n_i), 1 + max{depth(n_q), depth(n_j), depth(n_k)}}.
- A node of P is said to be *recognized* if it is either *d*-recognized or *t*-recognized as some state of M.

A transition $t = (s, s'; x/y)$ of M is *verified* (in Q) if there is an edge $(n_i, n_{i+1}; x_i/y_i)$ of P such that nodes n_i and n_{i+1} are recognized as states s and s' of M, and $x_i/y_i = x/y$.

Verification of the transitions of M leads to forming checking sequences as shown in Theorem 1 below, which forms the foundation for our proposed model.

Theorem 1: [16] Let Q be the label of a path P of G (for FSM M) such that every transition is verified in Q. Then the input portion of Q is a checking sequence of M.

In the proposed model, recognition of nodes n_i and n_{i+1} of edge $(n_i, n_{i+1}; x_i/y_i)$ of P (of G for FSM M), which corresponds to the transition $t = (s, s'; x/y)$ of M, will be achieved as follows. The node n_{i+1} will be d-recognized in Q as state s' of M, hence there must be a subpath $(n_{i+1}, n_k; D/\lambda(s', D))$ of P. This subpath of P will be used to construct a *test segment* for $t = (s, s'; x/y)$, which is denoted $t' = (n_i, n_{i+1}; x_i/y_i)(n_{i+1}, n_k; D/\lambda(s', D))$ (or $t' = (n_i, n_k; x_iD/\lambda(s, x_iD))$ in short). The collection of test segments for all transitions of M will be denoted P_C, i.e., $P_C = \{(s_i, s_j; x/y)(s_j, \delta(s_j, D); D/\lambda(s_j, D))$: for all $t = (s_i, s_j; x/y)$ of $M\}$ (or $P_C = \{(s_i, \delta(s_i, xD); xD/\lambda(s_i, xD))$: for all $t = (s_i, s_j; x/y)$ of $M\}$ in short). Table 2 shows P_C for FSM M_0.

Table 2. Test Segments of FSM M_0 with DS $D =$ "aa"

k	1	2	3	4	5	6
t_k	$(1, 2; a/0)$	$(1, 3; b/1)$	$(2, 3; a/0)$	$(2, 1; b/0)$	$(3, 1; a/1)$	$(3, 3; b/1)$
t_k'	$(1,1;aD$ $/001)$	$(1,2;bD$ $/110)$	$(2,2;aD$ $/010)$	$(2,3;bD$ $/000)$	$(3,3;aD/$ $100)$	$(3,2;bD$ $/110)$

The node n_i of P will be t-recognized in Q as state s of M, hence there must be subpaths $(n_j, n_k; T)$ and $(n_q, n_i; T)$ of P such that n_j and n_q are either d-recognized or t-recognized as state s of M, and n_k is either d-recognized or t-recognized as state s' of M. These subpaths will be formed by using what is called α-*elements*. A set of α-*elements for M* is a set of paths $\{\alpha_i = (s_i, \delta(s_i, D); D/\lambda(s_i, D))(\delta(s_i, D), \delta(\delta(s_i, D), D); D/\lambda(\delta(s_i, D), D))$: $i = 1, \ldots, n\}$ (or $\{\alpha_i = (s_i, \delta(s_i, DD); DD/\lambda(s_i, DD))$: $i = 1, \ldots, n\}$ in short), denoted by P_α. For example, Table 3 shows P_α for FSM M_0 with $D =$ "aa".

Proposition 1: Let Q be the label of a path P of G (for FSM M with a distinguishing sequence D) such that Q contains n subsequences of the form $DD/\lambda(s_i, DD)$, $i = 1, \ldots, n$. If Q induces a path in N of $\Phi(M)$ then D is also a distinguishing sequence for N and defines a bijection from the states of M to the states of N.

Proof: Since D is a distinguishing sequence for M, each of these subsequences of the form $D/\lambda(s_i, D)$, which is a prefix of $DD/\lambda(s_i, DD)$, $i = 1, \ldots, n$, is unique. If Q induces a path of N from $\Phi(M)$ then, since N has at most n states, D must also be a distinguishing sequence for N. This says that if n different responses to D are observed in N, then D defines a one-to-one correspondence between the states of M and N. In this case, we say that the uniqueness of the response of each of the n states of N to D is verified and hence N has n distinct states [13].   □

Proposition 2: Let Q be the label of a path P of G (for FSM M with a distinguishing sequence D) such that each α-element $\alpha_i = (s_i, \delta(s_i, DD); DD/\lambda(s_i, DD))$, $i = 1, \ldots, n$, is a subpath P of G. Then, for each $(s_i, \delta(s_i, D); D/\lambda(s_i, D))$, $1 \le i \le n$, appearing in P as a subpath $(n_j, n_k; D/\lambda(s_i, D))$,
1. the start node n_j of $(n_j, n_k; D/\lambda(s_i, D))$ is d-recognized
2. the end node n_k of $(n_j, n_k; D/\lambda(s_i, D))$ is t-(or d-)recognized

Proof: Part 1) is a direct consequence of the definition state recognition. Part 2) can easily be shown as follows. The α-element $\alpha_i = (s_i, \delta(s_i, DD); DD/\lambda(s_i, DD))$ appears in P as a subpath $(n_q, n_r; D/\lambda(s_i, D))(n_r, n_v; D/\lambda(\delta(s_i, D), D))$. As n_q, n_r and n_j are d-recognized, the end node n_k of $(n_j, n_k; D/\lambda(s_i, D))$ must be t-recognized as $\delta(s_i, D)$ if it is not d-recognized, by the definition of state recognition. $\square$

Table 3. α-elements for FSM M_0 (with D = "aa")

	Start state s_i	End state	label(α_i) = $DD/\lambda(s_i, DD)$
α_1	1	2	$aaaa/0010$
α_2	2	3	$aaaa/0100$
α_3	3	1	$aaaa/1001$

3   The Optimization Model

We wish to pose the following optimization problem: Given an FSM M (represented by a digraph $G = (V, E)$) and DS D for M, generate a minimum-length checking sequence of M starting at the initial state s_1 through composing an RCP path P of G which starts at v_1 and contains every element of $P_\alpha \cup P_C$. As in the earlier models for constructing reduced length checking sequences based on distinguishing sequences, it will be shown that since this RCP path P of G contains every element of $P_\alpha \cup P_C$, it establishes that all states of M are recognized and all transitions of M are verified.

In order to reduce the overall length of the resulting checking sequence, we will take advantage of the overlapping among elements of $P_\alpha \cup P_C$ in generating a minimum-length checking sequence in our model as follows: Let P_1 and P_2 denote two paths of G. If P_1 has a suffix R that is a prefix of P_2, namely, $P_1 = R_1 R$ and $P_2 = RR_2$ for some paths R_1 and R_2 of G, we say that P_1 *overlaps* P_2 by R. In this case, a new path $P_{1,2}$ of G can be formed by overlapping P_1 and P_2 by R, namely, $P_{12} = R_1 RR_2$, with $|P_{12}| = |P_1| + |P_2| - |R|$. Furthermore, if label($P_2$) has D as the prefix of its input portion, we call overlap of this type D-*overlap* by R. This definition offers a way to check if P_1 D-overlap P_2 or not by first checking if D is the prefix of the input portion of label(P_2) and then identifying the maximal overlapping portion R.

D-*overlap of a sequence of paths* (of G) $P_1, P_2, ..., P_k$, where $k > 2$, can be defined inductively as follows: If D-overlapping of the sequence $P_1, ..., P_{k-1}$ forms a new path $P_{1,k-1}$ and if this $P_{1,k-1}$ D-overlaps P_k forming a path $P_{1,k}$, then D-overlapping of the sequence $P_1, P_2, ..., P_k$ forms $P_{1,k}$.

The proposed algorithm for the solution of the optimization problem augments $G = (V, E)$ to form a digraph $G^* = (V^*, E^*)$ and then formulates the construction of a minimum-length checking sequence for M starting at the initial state s_1 as finding an RCP path P of G^* which starts at v_1 and contains every element of $P_\alpha \cup P_C$.

The proposed algorithm is given as follows:

Initially, $G^* = (V^*, E^*) \leftarrow G = (V, E)$

1. For every $\tau = (s_i, s_j; I_\tau/O_\tau)$ that is either an α-element or a test segment,
    a)   add to V^* two new vertices $s'_{i\tau}$, $s''_{j\tau}$ for the start and end states of τ, resp.
    b)   add to E^*

- an edge $(s'_{i\tau}, s''_{j\tau}; I_\tau/O_\tau)$ with cost $|I_\tau|$
- an edge $(s''_{j\tau}, s_j)$ with cost 0
- an edge $(s_i, s'_{i\tau})$ with cost 0

2. a) Add to V^* an artificial node s_1^* (representing the initial state s_1)
   b) Add to E^* an edge $(s_1^*, s'_{1\tau})$ with cost 0, for each of those $\tau = (s_1, s_j; I_\tau/O_\tau) \in P_\alpha \cup P_C$ such that D is a prefix of I_τ

3. For any pair of two different $\tau = (s_i, s_j; I_\tau/O_\tau)$ and $\mu = (s_k, s_r; I_\mu/O_\mu)$, each being either an α-element or a test segment, such that τ D-overlaps μ by R (which can be determined by first checking if D is the prefix of I_μ and if yes then identifying the maximal overlapping portion R of τ and μ), add to E^* an edge $(s''_{j\tau}, s'_{k\mu})$ with (**negative**) cost $-|R|$ (which reflects the effect of D-overlapping).

4. Find an RCP path P of G^* starting from s_1^* traversing at least once the edges representing the α-elements and the test segments. Use the input portion of label(P) as a checking sequence of M.

   More specifically, the algorithm is as follows:
   Construct $G^* = (V^*, E^*)$ whose vertex-set and edge-set are
   $V^* = V \cup V' \cup V'' \cup \{ s_1^* \}$ and $E^* = E \cup E_0 \cup E_\alpha \cup E_C \cup E' \cup E'' \cup E_D$
   from $G = (V, E)$ representing a given FSM M, where

   $V' = \{s'_{i\tau}:$ for all $\tau = (s_i, s_j; I_\tau/O_\tau) \in P_\alpha \cup P_C\}$,
   $V'' = \{s''_{j\tau}:$ for all $\tau = (s_i, s_j; I_\tau/O_\tau) \in P_\alpha \cup P_C\}$,
   $E_0 = \{(s_1^*, s'_{1\tau}; \varepsilon)$ with cost 0: for all $\tau = (s_1, s_j; I_\tau/O_\tau) \in P_\alpha \cup P_C$
       such that D is a prefix of $I_\tau\}$,
   $E_\alpha = \{(s'_{i\tau}, s''_{j\tau}; I_\tau/O_\tau)$ with cost $|I_\tau|:$ for all $\tau = (s_i, s_j; I_\tau/O_\tau) \in P_\alpha\}$,
   $E_C = \{(s'_{i\tau}, s''_{j\tau}; I_\tau/O_\tau)$ with cost $|I_\tau|:$ for all $\tau = (s_i, s_j; I_\tau/O_\tau) \in P_C\}$,
   $E' = \{(s_i, s'_{i\tau}; \varepsilon)$ with cost 0: for all $\tau = (s_i, s_j; I_\tau/O_\tau) \in P_\alpha \cup P_C\}$,
   $E'' = \{(s''_{j\tau}, s_j; \varepsilon)$ with cost 0: for all $\tau = (s_i, s_j; I_\tau/O_\tau) \in P_\alpha \cup P_C\}$, and
   $E_D = \{(s''_{j\tau}, s'_{k\mu}; \varepsilon)$ with cost $-|R|:$ for all $\tau = (s_i, s_j; I_\tau/O_\tau), \mu = (s_k, s_r; I_\mu/O_\mu) \in P_\alpha \cup P_C$ such that τ D-overlaps μ by some $R\}$.

   Find an RCP path P of G^* that starts at $s'_{1\tau}$ and contains all edges of $E_\alpha \cup E_C$.
   The input portion of label(P) is a checking sequence of M.

Example: For FSM M_0 with $D = $ "aa", Table 4 shows D-overlapping between pairs of elements of $P_\alpha \cup P_C$ and the resulting negative cost from each overlapping. More specifically, it shows all pairs $\tau, \mu \in P_\alpha \cup P_C$ such that τ D-overlaps μ by R with negative cost $-|R|$. Figure 2 shows an example of the result of application of the proposed algorithm to G (for M_0), using only part of D-overlapping in Table 4 (so that Figure 2 does not become too complicated to follow). In Figure 2, thick lines represent edges of $E_\alpha \cup E_C$ and for simplicity, s'_i, s''_j are used for $s'_{i\tau}$ and $s''_{j\tau}$. Note that in Tables 4-6, we dropped output portion of the paths for ease of presentation. An RCP path P (starting at vertex 1^* and ending at vertex $2''$) is found as

$$P = (1^*,1'; \varepsilon)t_1'\text{-}\alpha_1\text{-}t_3'\text{-}\alpha_2\text{-}t_5'\text{-}\alpha_3 (1'',1; \varepsilon)(1,1'; \varepsilon)t_2'(2'',2; \varepsilon)(2,2'; \varepsilon)t_4'(3'',3; \varepsilon)(3,3'; \varepsilon)t_6'$$

where each hyphen sign indicates an occurrence of D-overlapping. Its corresponding input sequence (without overlapping) is:

"$\varepsilon aaa\text{-}aaaa\text{-}aaa\text{-}aaaa\text{-}aaa\text{-}aaaa\varepsilon\varepsilon baa\varepsilon\varepsilon baa\varepsilon\varepsilon baa$".

Table 4. Overlapping among $P_\alpha \cup P_C$ for FSM M_0

$\tau = P_1$	$\mu = P_2$	R	$-|R|$	$E \in E_D$	Inuse*
$\alpha_1 = (1, 2, aaaa)$	$\alpha_2 = (2, 3; aaaa)$	$(2, 2; aaa)$	-3	$(2_\tau'', 2_\mu')$	y
α_1	$\alpha_3 = (3, 1; aaaa)$	$(3, 2; aa)$	-2	$(2_\tau'', 3_\mu')$	y
α_1	$t'_1 = (1, 1; aaa)$	$(1, 2; a)$	-1	$(2_\tau'', 1_\mu')$	
α_1	$t'_3 = (2, 2; aaa)$	$(2, 2; aaa)$	-3	$(2_\tau'', 2_\mu')$	y
α_1	$t'_5 = (3, 3; aaa)$	$(3, 2; aa)$	-2	$(2_\tau'', 3_\mu')$	
$\alpha_2 = (2, 3; aaaa)$	$\alpha_1 = (1, 2, aaaa)$	$(1, 3; aa)$	-2	$(3_\tau'', 1_\mu')$	y
α_2	$\alpha_3 = (3, 1; aaaa)$	$(3, 3; aaa)$	-3	$(3_\tau'', 3_\mu')$	y
α_2	$t'_1 = (1, 1; aaa)$	$(1, 3; aa)$	-2	$(3_\tau'', 1_\mu')$	
α_2	$t'_3 = (2, 2; aaa)$	$(2, 3; a)$	-1	$(3_\tau'', 2_\mu')$	
α_2	$t'_5 = (3, 3; aaa)$	$(3, 3; aaa)$	-3	$(3_\tau'', 3_\mu')$	y
$\alpha_3 = (3, 1; aaaa)$	$\alpha_1 = (1, 2, aaaa)$	$(1, 1; aaa)$	-3	$(1_\tau'', 1_\mu')$	y
α_3	$\alpha_2 = (2, 3; aaaa)$	$(2, 1; aa)$	-2	$(1_\tau'', 2_\mu')$	y
α_3	$t'_1 = (1, 1; aaa)$	$(1, 1; aaa)$	-3	$(1_\tau'', 1_\mu')$	y
α_3	$t'_3 = (2, 2; aaa)$	$(2, 1; aa)$	-2	$(1_\tau'', 2_\mu')$	
α_3	$t'_5 = (3, 3; aaa)$	$(3, 1; a)$	-1	$(1_\tau'', 3_\mu')$	
$t'_1 = (1, 1; aaa)$	$\alpha_1 = (1, 2, aaaa)$	$(1, 1; aaa)$	-3	$(1_\tau'', 1_\mu')$	y
t'_1	$\alpha_2 = (2, 3; aaaa)$	$(2, 1; aa)$	-2	$(1_\tau'', 2_\mu')$	y
t'_1	$\alpha_3 = (3, 1; aaaa)$	$(3, 1; a)$	-1	$(1_\tau'', 3_\mu')$	
t'_1	$t'_3 = (2, 2; aaa)$	$(2, 1; aa)$	-2	$(1_\tau'', 2_\mu')$	y
t'_1	$t'_5 = (3, 3; aaa)$	$(3, 1; a)$	-1	$(1_\tau'', 3_\mu')$	
$t'_2 = (1, 2; baa)$	$\alpha_1 = (1, 2, aaaa)$	$(1, 2; a)$	-1	$(2_\tau'', 1_\mu')$	
t'_2	$\alpha_3 = (3, 1; aaaa)$	$(3, 2; aa)$	-2	$(2_\tau'', 3_\mu')$	y
t'_2	$t'_1 = (1, 1; aaa)$	$(1, 2; a)$	-1	$(2_\tau'', 1_\mu')$	
t'_2	$t'_5 = (3, 3; aaa)$	$(3, 2; aa)$	-2	$(2_\tau'', 3_\mu')$	y
$t'_3 = (2, 2; aaa)$	$\alpha_1 = (1, 2, aaaa)$	$(1, 2; a)$	-1	$(2_\tau'', 1_\mu')$	
t'_3	$\alpha_2 = (2, 3; aaaa)$	$(2, 2; aaa)$	-3	$(2_\tau'', 2_\mu')$	y
t'_3	$\alpha_3 = (3, 1; aaaa)$	$(3, 2; aa)$	-2	$(2_\tau'', 3_\mu')$	y
t'_3	$t'_1 = (1, 1; aaa)$	$(1, 2; a)$	-1	$(2_\tau'', 1_\mu')$	
t'_3	$t'_5 = (3, 3; aaa)$	$(3, 2; aa)$	-2	$(2_\tau'', 3_\mu')$	y
$t'_4 = (2, 3; baa)$	$\alpha_1 = (1, 2, aaaa)$	$(1, 3; aa)$	-2	$(3_\tau'', 1_\mu')$	y
t'_4	$\alpha_2 = (2, 3; aaaa)$	$(2, 3; a)$	-1	$(3_\tau'', 2_\mu')$	
t'_4	$t'_1 = (1, 1; aaa)$	$(1, 3; aa)$	-2	$(3_\tau'', 1_\mu')$	y
t'_4	$t'_3 = (2, 2; aaa)$	$(2, 3; a)$	-1	$(3_\tau'', 2_\mu')$	
$t'_5 = (3, 3; aaa)$	$\alpha_1 = (1, 2, aaaa)$	$(1, 3; aa)$	-2	$(3_\tau'', 1_\mu')$	
t'_5	$\alpha_2 = (2, 3; aaaa)$	$(2, 3; a)$	-1	$(3_\tau'', 2_\mu')$	
t'_5	$\alpha_3 = (3, 1; aaaa)$	$(3, 3; aaa)$	-3	$(3_\tau'', 3_\mu')$	y
t'_5	$t'_1 = (1, 1; aaa)$	$(1, 3; aa)$	-2	$(3_\tau'', 1_\mu')$	y
t'_5	$t'_3 = (2, 2; aaa)$	$(2, 3; a)$	-1	$(3_\tau'', 2_\mu')$	
$t'_6 = (3, 2; baa)$	$\alpha_1 = (1, 2, aaaa)$	$(1, 2; a)$	-1	$(2_\tau'', 1_\mu')$	
t'_6	$\alpha_3 = (3, 1; aaaa)$	$(3, 2; aa)$	-2	$(2_\tau'', 3_\mu')$	y
t'_6	$t'_1 = (1, 1; aaa)$	$(1, 2; a)$	-1	$(2_\tau'', 1_\mu')$	
t'_6	$t'_5 = (3, 3; aaa)$	$(3, 2; aa)$	-2	$(2_\tau'', 3_\mu')$	y

The checking sequence obtained from *P* with *D*-overlapping starting at state 1 is

"εaaa-~~aaaa~~-~~aaa~~-~~aaaa~~-~~aaa~~-~~aaaa~~a baa baa baa"

whose length is 15.

For the same example, a reduced length checking sequence was found to have length 32 in [11].

The last column with "y" indicates the overlapping is used in Figure 2 for generating an RCP path. The ones with blank space indicated the overlapping is not considered in Figure 2 as we do not want the Figure too complicated.

Note that *D*-overlapping between two elements $P_\alpha \cup P_C$ is shown explicitly in the optimization model whereas *D*-overlapping among a sequence of elements are formed

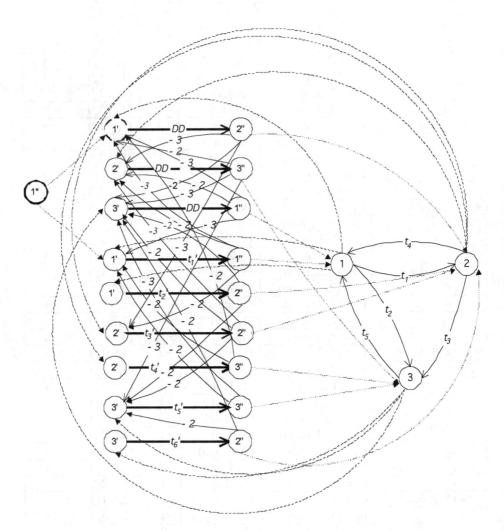

Fig. 2. Optimization Model for FSM M_0

automatically in the process of finding an RCP path P and can be identified in P. To prove the correctness of the proposed algorithm, let Π denote a set paths (of G) obtained from D-overlapping among elements (paths) of $P_\alpha \cup P_C$, such that every element of $P_\alpha \cup P_C$ is a subpath of a path of Π. In G^*, such a Π is naturally formed to consist of the maximal paths generated from D-overlapping among elements of $P_\alpha \cup P_C$ and those elements of $P_\alpha \cup P_C$ not contained in any D-overlapping path. Let E_π be a set of edges, whose end vertices are in V of G, that represent the paths of Π.

Theorem 2: Let P be an RCP path of $G_0 = (V, E \cup E_\pi)$ such that P starts at v_1 and contains every edge of E_π and D is the prefix of the input portion of its label Q. Let E_1 denote the set of edges of P after excluding E_π, i.e., $E_1 = E(P) \cap E$. If $G_1 = (V, E_1)$ does not contain a cycle, then the input portion of Q forms a checking sequence of M.

Proof: Suppose that $P = (n_1, n_2; L_1)\ldots (n_r, n_{r+1}; L_r)$ and its label $Q = L_1 L_2 \ldots L_r$, where each $(n_j, n_{j+1}; L_j)$, $1 \le j \le r$, is an edge (of G_0) representing either a single edge or a subpath of $G = (V, E)$. First we claim that every n_j of P, $1 \le j \le r$, is recognized (in Q).

Suppose the claim is not true. Since $G_1 = (V, E_1)$ does not contain a cycle, it is well known [2] that the vertices of V can be assigned an order "$\propto$" such that $u \propto v$ if there exists a path from u to v in G_1. Let n_i be a node corresponding to the smallest member of V (with respect to "$\propto$") such that n_i is not recognized. Note that $i > 1$ as n_1 is d-recognized. We consider $(n_{i-1}, n_i; L_{i-1})$ of P and derive a contradiction for each of all possible cases below.

If $(n_{i-1}, n_i; L_{i-1})$ corresponds to an edge of E_π, then n_i corresponds to the end of either an α-sequence or a test segment, which must be t-recognized, a contradiction. If $(n_{i-1}, n_i; L_{i-1}) \in E_1$, i.e., $(n_{i-1}, n_i; L_{i-1}) = (u, v; x/y) \in E$, then n_{i-1} is recognized as $u \propto n_i$. Note that P contains the test segment for this edge of E, say $(n_j, n_{j+1}; L_j)$ where $L_j = (xD_v)/\lambda(u, xD_v)$. As n_j corresponds to $u \propto n_i$, n_j is recognized. Also the node adjacent to n_j in the subpath $(n_j, n_{j+1}; L_j)$, $\delta(n_j, x)$, is d-recognized. Thus, n_i is t-recognized as $\delta(u, x)$ of M, another contradiction. Therefore, every n_j is recognized.

For every transition t of M, its test segment t' is contained in a subpath P_i (of P) represented by an edge $(n_{i-1}, n_i; L_{i-1})$ of P. If the start state of t' is n_{i-1}, from the argument above, n_{i-1} recognized; otherwise, t' is contained through D-overlapping, its start state is d-recognized. Hence, t' is verified in Q, and by Theorem 1, the input portion of Q is a checking sequence of M.   □

Correctness of the proposed algorithm is a direct consequence of Theorem 2. Notice that E_π is formed naturally in the process of solving for an RCP path P of G^*. The RCP path P of G^* can be viewed as a path of $G_0 = (V, E \cup E_\pi)$, by mapping the nodes of P into the corresponding vertices of G. Thus, as long as the premise of Theorem 2 holds, the correctness of the proposed algorithm is guaranteed.

Up to this point, we have presented the proposed optimization model with a simplification for ease of presentation. This simplification is in the formation of α-elements, that is, instead of using a more general form DT_iDT_j, where T_i, T_j are transfer sequences, we used DD (equivalently, assumed $T_i = T_j = \varepsilon$). In the previous models [11, 16] for constructing reduced-length checking sequences, a given set of transfer sequences $T_i = I_i/O_i$ starting at state $\delta(s_i, D)$, $i = 1, \ldots, n$, is used for two main purposes. First, it is used to redefine a set of test segments as $P_C = \{(s_i, s_j; x/y)(s_j, \lambda(s_j, DI_j); DI_j/\lambda(s_j, DI_j))$: for all $t = (s_i, s_j; x/y)$ of $M\}$, so that every state can be reached by at least one of these test segments. Second, it is used to increase the flexibility of the models to obtain a possible further reduction in the lengths of checking sequences.

Now we present a generalization of the optimization model that incorporates a given set of transfer sequences $T_i = I_i/O_i$ starting at state $\delta(s_i, D)$, $i = 1, \ldots, n$ through an adjustment of α-elements and test segments as follows.

Let $P_\alpha = \{(s_i, s_j; DI_i/\lambda(s_i, DI_i))(s_j, \delta(s_j, DI_j); DI_j/\lambda(s_j, DI_j))$: $i = 1, \ldots, n\}$ and $P_C = \{(s_i, s_j; x/y)(s_j, \lambda(s_j, DI_j); DI_j/\lambda(s_j, DI_j))$: for all $t = (s_i, s_j; x/y)$ of $M\}$.

This adjustment amounts to replacing subsequences of the form $DD/\lambda(s_i, DD)$ with subsequences of the form $DI_iDI_j/\lambda(s_i, DI_iDI_j)$, $i = 1, \ldots, n$, $1 \le j \le n$ as the labels of α-elements; and replacing subsequences of the form $xD/\lambda(s_i, xD)$ with subsequences of the form $xDI_j/\lambda(s_i, xDI_j)$, $i = 1, \ldots, n$, $1 \le j \le n$ as the labels of test segments. As such, the adjustment does not alter the validity of the Propositions 1 and 2, and Theorem 2: Their proofs are similar to the proofs of those given for the optimization model presented in the previous section. With these new P_α and P_C, we can apply the proposed algorithm to solve the same optimization problem as the one given earlier.

Example: For FSM M_0, $D = $ "aa", and a given set of transfer sequences $T_1 = a/1$, $T_2 = T_3 = \varepsilon$, the set P_α of α-elements and the set P_C of test segments are listed in Table 5 and Table 6 below. Figure 3 shows the general optimization model for FSM M_0 with the given T_i, $i = 1, 2, 3$. (Note that t_1' and t_3' are prefixes of α-elements, and thus, they are eliminated from the model.) We obtain the optimal solution to the general model as an RCP path P of G^* starting at 1^* (and ending at state $2''$), which is:

$$P = (1^*, 1'; \varepsilon)t_1'\text{-}\alpha_1\text{-}\alpha_2\text{-}\alpha_3\text{-}t_3'\text{-}t_5'(1'', 1; \varepsilon)(1, 1'; \varepsilon)t_2'(2'', 2; \varepsilon)(2, 2'; \varepsilon)t_4'(1'', 1; \varepsilon)t_2(3, 3'; \varepsilon)t_6'$$

where each hyphen sign indicates an occurrence of D-overlapping. Its corresponding input sequence (without overlapping) is:

$$\varepsilon aaa\text{-}aaaaaa\text{-}aaaaa\text{-}aaaa\text{-}aaa\text{-}aaa\varepsilon\varepsilon baa\varepsilon\varepsilon baa\varepsilon b\varepsilon baa$$

The checking sequence obtained from P with D-overlapping starting at state 1 is

$$aaa\text{-}\cancel{aaa}aaa\text{-}\cancel{aaaaa}\text{-}\cancel{aaaa}\text{-}\cancel{aaa}\text{-}\cancel{aaa} \; baa \; baa \; b \; baa$$

whose length is: $3+3+1+1+3+3+1+3 = 18$.

Table 5. α-elements for FSM M_0 (with $D = $ "aa", $T_1 = a/1$, $T_2 = T_3 = \varepsilon$)

start s_i	$\lambda(s_i, DI_i)$	$s_j = \delta(s_i, DI_i)$	$\lambda(s_j, DI_j)$	end $\delta(s_j, DI_j)$
1	001	1	001	1
2	01	1	001	1
3	10	2	01	1

Table 6. Test Segments of FSM M_0 (with $D = $ "aa", $T_1 = a/1$, $T_2 = T_3 = \varepsilon$)

k	1	2	3	4	5	6
t_k	(1,2; a/0)	(1,3; b/1)	(2,3; a/0)	(2,1; b/0)	(3,1; a/1)	(3,3; b/1)
t_k'	(1,1;aD/ 001)	(1,2;bD/ 110)	(2,2;aD/ 010)	(2,1;bDa/ 0001)	(3,1;aDa/ 1001)	(3,2;bD/ 110)

This result is longer than the checking sequence produced by the optimization model presented earlier where all $T_i = \varepsilon$, $i = 1, 2, 3$. Indeed, an experimental study reported in [12] confirms the intuitive hypothesis that using empty transfer sequences results in shorter checking sequences based on distinguishing sequences.

Theorem 3. Given an FSM M represented by graph G, a DS D for M and a set of TS $T_i = I_i/O_i$ starting at state $\delta(s_i, D)$, $i = 1, ..., n$, the minimum-length checking sequence constructed by our general model is at least as short as the ones constructed by the previous optimization models of [11, 16].

Proof: In the previous models [11, 16], a set of I/O sequences of M is first generated, where each I/O sequence, called an α-*sequence* α_k in [16], α'-*sequence* α_k' in [11], is the label of a path P_k (of G, $k = 1,..., q$) that is formed by concatenating some of the

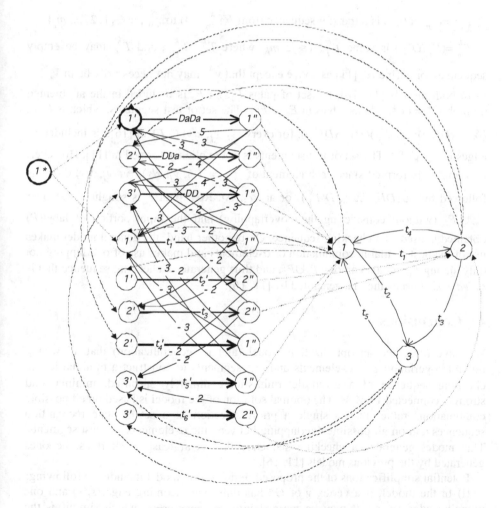

Fig. 3. The General Model for FSM M_0

paths $\{(s_i, \delta(s_i, DI_i); D/\lambda(s_i, DI_i)): i = 1,\ldots, n\}$, such that, by including all $P_1, \ldots, P_q$ in a path P of G, the following requirements are satisfied:

- For each $D/\lambda(s_i, D)T_i$ in label(P), $1 \le i \le n$, its start node is d-recognized;
- For each $D/\lambda(s_i, D)T_i$ in label(P), $1 \le i \le n$, its end node is t-(or d-)recognized.

α_k (or α'_k) = label(P_k), $1 \le k \le q$ which is formed as follows: Let $V=\{v_1, v_2, \ldots, v_n\}$ and let $V_1, V_2, \ldots, V_q$, $q \ge 1$, be subsets of V, i.e., $V_k \subseteq V$, $1 \le k \le q$, whose union is V. Without loss of generality, assume that $V_k = \{v_1^k, v_2^k, \ldots, v_{m_k}^k\}$, $1 \le k \le q$, and for V_k,

define α_k [16] as: $\alpha_k = D/\lambda(v_1^k, D)T_1^k \, D/\lambda(v_2^k, D)T_2^k \ldots D/\lambda(v_{m_k}^k, D)T_{m_k}^k \, D/\lambda(v_w^k, D)T_w^k$

where $T_j^k = (I_j^k/O_j^k)$ is a transfer sequence from $\delta(v_j^k, D)$ to v_{j+1}^k for $j = 1, 2, \ldots, m_{k-1}$

$T_{m_k}^k = (I_{m_k}^k/O_{m_k}^k)$ is a transfer sequence from $\delta(v_{m_k}^k, D)$ to v_w^k, $w \in \{1, 2, \ldots, m_k\}$

$T_w^k = (I_w^k/O_w^k)$ is some T_j^k, $1 \le j \le m_k$, where T_j^k, $T_{m_k}^k$, and T_w^k may be empty

sequences; or define α'_k [11] as above except that v_w^k may not necessarily be in V_k.

In both models [11, 16], the set of paths $P_1, \ldots, P_q$ is included in the augmented digraph $G^* = (V^*, E^*)$ as edges in $E_\alpha \subset E^*$. The set of test segments, which is $\{(v_i, (\delta(v_i, xDI_j^k)); (xDI_j^k)/\lambda(v_i, xDI_j^k)):$ for every $(v_i, v_j; x/y) \in E\}$ in [16], is included as edges in $E_C \subset E^*$. The set of test segments is not explicitly formed in [11]. However, it is implicitly formed since each element of $E_c = \{(v_i, v_j; x/y): (v_i, v_j; x/y) \in E\}$ is followed by a $(DI_j^k)/\lambda(v_j, D\,I_j^k))$ or an α'_k. An RCP path P is sought in G^* over $E_\alpha \cup E_C$ (without considering their overlapping) and the input portion of label(P) obtained is used as a checking sequence. On the other hand, the general model makes use of the best available combination of α-elements and makes use of overlapping not only among P_α but also among $P_\alpha \cup P_C$, and thus generates a checking sequence that is *at least as short* as the one generated by [11, 16].

4   Conclusions

We have presented an optimization model (and its generalization) that allow any possible overlapping of α-elements and test segments to construct a minimal-length checking sequence of a given deterministic, completely specified, minimal and strongly connected FSM M. The optimal solution of the model is based on all possible combinations rather than a single a priori selection of a set of state recognition sequences and on all possible overlapping between the α-elements and test segments. This model generates a checking sequence that is at least as short as the ones generated by the previous models [11, 16].

Potential simplifications of the proposed optimization model include the following:

(i) In the model, if a vertex v of G^* has only one incoming edge (s, v) and one outgoing edge (v, s'), it may be merged into another vertex, which simplifies the model. This is particularly useful when a test segment is not involved in D-

overlapping. This simplification is clearly applicable to both the proposed model and its generalization.

(ii) Further simplification may also be achieved by eliminating some test segments from the model as they are part of the α-elements. The elimination of test segments that are related to the transitions traversed as the last transition of a path induced by D in an α-sequence, formed by concatenated $DI_i/\lambda(s_i, DI_i)$'s where $I_i = \varepsilon$, has been proposed in [3]. Such elimination can be incorporated into the model as follows: First, identify the test segments that are covered by α-elements and do not include them in P_C. Then, find an RCP P in the simplified model that contains all α-elements and those remaining test segments, and use the input portion of label(P) as a checking sequence. Since the generalization of the proposed model utilizes $DI_i/\lambda(s_i, DI_i)$'s where $I_i \neq \varepsilon$, incorporation of this simplification requires further study.

It must be noted that the RCP problem is NP-complete [7]. However, for some systems, the given FSM M has a reset feature, i.e. there is input r such that $\delta(s_i, r) = s_1$ for every state s_i. With this reset feature, an optimal solution of our proposed model can be found in polynomial-time and is guaranteed to be a checking sequence for M as follows. For each transition t of form $(s_i, s_1; r/\lambda(s_i, r))$, i.e., t is triggered by r, its test segment $t' = (s_i, \delta(s_1, DI_1); (rDI_1)/\lambda(s_i, rDI_1))$ is added to the graph G^* with both vertices in V (as no overlapping is involved). These edges consist of a connected spanning subgraph of G. In this case, the problem of finding an RP path with minimum cost is reduced to a min-cost flow problem as in [1, 16], which can be solved in polynomial-time [5].

In the generalized model, a given set of transfer sequences $\{T_i: i = 1, ..., n\}$ is used together with a DS in forming α-elements and test segments. Although it was shown experimentally that empty TS (i.e., $T_i = \varepsilon$, $i = 1, ..., n$) leads to shorter checking sequences [12], the best selection of such a set is unknown and worth further study.

References

1. A.V. Aho, A.T. Dahbura, D. Lee, and M.U. Uyar, "An optimization technique for protocolconformance test sequence generation based on UIO sequences and rural Chinese postmantours", *IEEE Trans. on Comm.* vol.39, pp.1604-1615, 1991.
2. J.A.Bondy and U.S.R. Murty, *Graph Theory with Applications*, New York: Elsevier North Holland, Inc. 1976.
3. J. Chen, R.M. Hierons, H. Ural and H. Yenigun, "Eliminating redundant tests in a checkingsequence", *Proc. of IFIP TestCom* 2005, May 2005, pp.146-158.
4. T. Chow, "Testing software design modeled by finite-state machines", *IEEE Trans. Software Eng.*, vol.SE-4, pp.178-187, 1978.
5. W.J. Cook, W.H. Cunningham, W.R. Pulleyblank and A. Schrijver, *Combinatorial Optimization*, John Wiley and Sons, New York, 1998.
6. A.T. Dahbura, K.K. Sabnani, and M.U. Uyar, "Formal methods for generating protocol conformance test sequences", *Proc. of IEEE*, vol.78, pp.1317-1325, 1990.
7. M.R. Garey and D.S. Johnson, *Computers and Intractability*, W.H. Freeman and Company, New York, 1979.
8. A. Gill, *Introduction to the Theory of Finite-State Machines*, NY: McGraw-Hill, 1962.
9. G. Gonenc, "A method for the design of fault detection experiments", *IEEE Trans. on Computer*, vol.19, pp.551-558, June 1970.

10. F.C. Hennie, "Fault detecting experiments for sequential circuits", *Proc. 5th. Symp. Switching Circuit Theory and Logical Design*, pp.95-110, Princeton, N.J.,1964.
11. R.M. Hierons and H. Ural, "Reduced length checking sequences", *IEEE Trans. on Computers,* vol.51(9), pp.1111-1117, 2002.
12. R. Hieron and H. Ural, "Optimizing the length of checking sequences", *submitted to IEEE Trans on Computers*, 2004.
13. K. Inan and H. Ural, "Efficient checking sequences for testing finite state machines",*Information and Software Technology*, vol.41, pp.799-812, 1999.
14. Z. Kohavi, *Switching and Finite State Automata Theory*, McGraw-Hill, 1978.
15. D. Lee and M. Yannakakis, "Testing finite state machines: state identification and verification", *IEEE Trans. on Computers,* vol.43, pp.306-320, 1994.
16. H. Ural, X. Wu and F. Zhang, "On minimizing the length of checking sequence", *IEEE Trans. on Computers*, vol.46, pp.93-99, 1997.
17. M.P. Vasilevskii, "Failure diagnosis of automata", *Kibernetika*, vol.4, pp.98-108, 1973.

Test Case Minimization for Real-Time Systems Using Timed Bound Traces[*]

Ismaïl Berrada[1], Richard Castanet[1], Patrick Félix[1], and Aziz Salah[2]

[1] LaBRI - CNRS - UMR 5800 Université Bordeaux 1,
33405 Talence cedex, France
{berrada, castanet, felix}@labri.fr
[2] Département d'Informatique,
Université du Québec à Montréal,
201, avenue du Président-Kennedy, Montreal,
Quebec H2X 3Y7, Canada
aziz.salah@uqam.ca

Abstract. Real-Time systems (RTS for short) are those systems whose behavior is time dependent. Reliability and safety are of paramount importance in designing and building RTS because a failure of an RTS puts the public and/or the environment at risk. For the purpose of effective error reporting and testing, this paper considers the trace inclusion problem for RTS: given a path ρ (resp. ρ') of length n of a timed automaton A (resp. B), find whether the set of timed traces of ρ of length n are included in the set of timed traces of ρ' of length n such that A is known but not B. We assume that the traces of ρ' are only defined by a decision procedure.

The proposed solution is based on the identification of a set of timed bound traces. The latter gives a finite representation of the trace space of a path. The number of these timed bounds varies between 1 and $2 \times (n+1)$. The trace inclusion problem is then reduced to the inclusion of timed bound traces. The paper shows also how these results can be used to reduce the number of test cases for an RTS.

Keywords: Timed Input Output Automata, Trace Inclusion, Black-Box Testing, Conformance Testing.

1 Introduction

Nowadays, real-time systems (RTS for short) span various domains of our daily life such as telephone systems, patient monitoring systems, and air traffic control. All these systems are time sensitive because their behavior does not only depend on the logical result of the computation but also on the time at which the inputs and outputs are observed. It is well-known to RTS research community that the misbehavior of an RTS is generally due to the violation of time

---

[*] This research has been supported by the French RNTL project AVERROES and the Marie Curie RTN TAROT (MCRTN 505121).

M.Ü. Uyar, A.Y. Duale, and M.A. Fecko (Eds.): TestCom 2006, LNCS 3964, pp. 289–305, 2006.

constraints. Such malfunctioning may have catastrophic consequences on both human lives and the environment. Therefore, it is very necessary to make sure that the implementation of an RTS is error-free before its deployment.

Two formal techniques, namely verification and testing, are usually used to detect errors in RTS systems. Verification aims at checking that a specification or a model of the system respects some functional and timing requirements. However, testing deals with the implementation of the system, usually referred to as Implementation Under Test or IUT for short, and checks its conformance to the specification of the system in three steps. First of all, test cases are generated according to some coverage criteria. Then, those test cases are executed against the IUT and its reactions are logged. Finally, the verdict is concluded by analyzing the reactions of the IUT: if the behavior of the IUT during test cases doesn't conform to its specification, the IUT is said faulty.

In this paper, we study the following problem:

Trace Inclusion Problem. Consider a path ρ (resp. ρ') of length $n \in \mathbb{N}$ of a timed automaton A (resp. B). How to show that $TTrace(\rho) \subseteq TTrace(\rho')$ such that:

- ρ is known: the different constraints and clock updates of ρ are given.
- ρ' is unknown: only the set $TTrace(\rho')$ is given (the different constraints and clock updates of ρ' are unknown).

with $TTrace(\rho)$ (resp. $TTrace(\rho')$) are the timed traces of ρ (resp. ρ') of length n[1].

Our motivation for studying this problem is testing. The testing research community distinguishes between three main testing strategies: black-box testing, white-box testing, and grey-box testing. Those testing strategies differ from each other on the way the test cases are generated. In the case of black-box testing of RTS, the code of IUT is unknown and only its timed traces are given. Black-box testing consists then of deriving test cases based solely on the specification of the IUT. The use of so called conformance relations give formal characterizations of conditions under which an IUT can be considered as conformant to its specification. Checking a conformance relation can be reduced, in general, to the trace inclusion problem between the implementation and the specification. By studying this problem, the paper gives the necessary and sufficient conditions to check a conformance relation based on trace inclusion. These conditions can be then used to reduce the number of test cases considered for testing an RTS.

The main contribution of this paper is the proposition of a solution to the trace inclusion problem. The proposed solution is based on the identification of the timed bound traces of a path. The latter considers only the behaviors of the RTS on the constraint bounds. Their number varies between 1 and $2 \times (n+1)$, where n is the length of the path. The proof of the existence of those traces 1) considers the constraint polyhedron corresponding to the set of constraints that each timed

---

[1] A formal definition of $TTrace()$ is given in section 3.

trace of the path has to satisfy and 2) uses some graph transformations that preserve the positivity of the graph cycles.

As a second contribution, the paper proposes an approach to reduce the number of test cases considered while testing RTS. The proposed approach is based on the use of the simulation graph introduced by Tripakis [19]. The fact that the trace inclusion problem can be solved by the inclusion of timed bound traces, provides a method to reduce the number of test cases.

The rest of this paper is structured as follows. Section 2 introduces the theoretical background of the paper. Section 3 presents the model of timed automata and its corresponding notations. Section 4 corresponds to the core of this paper and shows how to generate timed traces from a path. Section 5, based on the result of Section 4, outlines a method for minimizing the number of test cases considered while testing RTS. Section 6 presents the related work. Finally, we conclude and draw some perspectives in Section 7.

2 Background

Through-out this paper, we write $\mathbb{R}$, $\mathbb{R}^0$, $\mathbb{N}$ for the sets of reals, nonnegative reals and naturals, respectively. $+\infty$ (resp. $-\infty$) is the positive infinity (resp. negative) such that: $t \in \mathbb{R}$, $-\infty \leq t \leq +\infty$, $t + (+\infty) = (+\infty) + t = +\infty$ and $t + (-\infty) = (-\infty) + t = -\infty$. $\overline{\mathbb{R}}$ is the set $\mathbb{R} \cup \{+\infty, -\infty\}$. For a set P, 2^P is the powerset of P and for a given order on P, $min(P)$ is the smallest element of P. Logical "and" and "or" are written $\wedge$ and $\vee$, respectively.

2.1 Timed Event and Timed Sequence

Let Σ be a finite set of symbols. As usual, Σ^* will denote the set of finite sequences and $\epsilon \in \Sigma^*$ the empty sequence. τ will denote an action not in Σ and Σ_τ the set $\Sigma \cup \{\tau\}$. Let σ be a sequence and $X \subseteq \Sigma$. Then, $\sigma_{|X}$ is the sequence obtained by erasing from σ all symbols not in X (projection on X).

A timed event over Σ is a pair $u = (a, d)$ such that $a \in \Sigma$ and $d \in \mathbb{R}^0$. If a is interpreted to denote an event occurrence then d is interpreted as the timestamp of the occurrence of a. A timed sequence $\sigma = (a_1, d_1)...(a_n, d_n)$ over Σ is a member of $(\Sigma \times \mathbb{R}^0)^*$ such that the sequence of timestamps is monotonically increasing. For example, $\sigma = (a_1, 3)(a_2, 5)$ is a timed sequence, however $\sigma' = (a_1, 3)(a_2, 2)$ is not. The set of timed sequences over Σ is noted $TS(\Sigma)$. Note that, when $X \subseteq \Sigma$, the projection of a timed sequence σ over X is obtained by erasing from σ all symbols such that the associated event is not in X.

2.2 Valuations and Polyhedra

Valuations. Let V be a finite set of variables ranged over $\mathbb{R}^0$. A valuation ν over V is a function $\nu : V \mapsto \mathbb{R}^0$ that assigns to each variable a real value. $\mathcal{V}(V)$ will denote the set of all valuations over V. Let $X \subseteq V$, $d \in \mathbb{R}$ and $\nu \in \mathcal{V}(V)$.

Then $\nu[X := 0]$ is the valuation defined by $\nu[X := 0](x) = \nu(x)$ if $x \notin X$ and $\nu[X := 0](x) = 0$ otherwise. Intuitively, $\nu[X := 0]$ assigns to each variable in X the value 0 and leaves the rest of variables unchanged. $\nu + d$ is a valuation such that for all $x \in V$, $(\nu + d)(x) = \nu(x) + d$. Intuitively, $\nu + d$ is obtained from ν by advancing all variables by d.

c-Closure [19]. Let $c \in \mathbb{N}$. Two valuations ν and ν' over V are called c-equivalent if:

- for any $x \in V$, either $\nu(x) = \nu'(x)$ or ($\nu(x) > c$ and $\nu'(x) > c$).
- for any pair $(x, y) \in V^2$, either $\nu(x) - \nu(y) = \nu'(x) - \nu'(y)$ or ($|\nu(x) - \nu(y)| > c$ and $|\nu'(x) - \nu'(y)| > c$).

Polyhedra. An atomic constraint over V is an expression of the form $x \bowtie n$ or $x - y \bowtie m$ where $(x, y) \in V^2$, $\bowtie \in \{\leq, \geq\}$ and $(n, m) \in \mathbb{N}^2$. The set of formulas that are finite conjunctions of atomic constraints (resp. of constraints of the form $x \bowtie n$) will be denoted by $\Phi(V)$ (resp. $\Phi_I(V)$). Elements of $\Phi(V)$ are called *polyhedra*. We write **true** for $\bigwedge_{\forall x \in V} x \geq 0$ and **zero** for $\bigwedge_{\forall x \in V}(x \leq 0 \wedge x \geq 0)$. Let $\nu \in \mathcal{V}(V)$ and $Z \in \Phi(V)$. Then ν satisfies Z, noted $\nu \in Z$, if ν satisfies all constraints of Z. Z is bounded iff there is $d \in \mathbb{N}$ such that for all $\nu \in Z$, $\nu + d \notin Z$.

Given a polyhedron Z, the c-closure of Z, noted $close(Z, c)$, is the greatest polyhedron Z' such that $Z \subseteq Z'$, and for all $\nu' \in Z'$ there exists $\nu \in Z$ such that ν and ν' are c-equivalent.

Operations on Polyhedra. We define the operations $Z[X := 0]$ and $Z^\nearrow$ of forward clock reset and forward time elapse of a polyhedron Z, respectively, as follows ($X \subseteq V$):

$$Z[X := 0] = \{\nu[X := 0] \mid \nu \in Z\} \qquad Z^\nearrow = \{\nu + d \mid \nu \in Z, d \in \mathbb{R}^{\geq 0}\}$$

3   Timed Automata

A clock is a variable that allows to record the passage of time. It is ranged over $\mathbb{R}^{\geq 0}$, and the only assignment allowed is clock reset of the form $x := 0$.

Timed Automata [1]. A timed automaton (TA) A over Σ is a tuple $A = (L, l_0, \Sigma, C, I, \rightarrow)$ such that:

- L is a finite set of locations,
- l_0 is the initial location,
- Σ is an alphabet of actions,
- C is a finite set of clocks,
- $I : L \mapsto \Phi_I(C)$ is a mapping that assigns invariants to locations, and
- $\rightarrow \subseteq L \times \Phi(C) \times \Sigma_\tau \times 2^C \times L$ is the set of edges. An edge has a source, a label, a guard, a set of clocks to be reset with this edge, and a target.

The labels in Σ represent the observable interactions of A; the special label $\tau \notin \Sigma$ represents an unobservable, internal action. A transition $t = (l, Z, a, r, l') \in \rightarrow$ is noted by $l \xrightarrow{Z,a,r} l'$. $\mathcal{TA}(\Sigma)$ denotes the set of all TAs over Σ.

Semantics. The semantics of a TA A is defined by associating a labeled transition system (LTS) $S(A) = (S, s_0, \Gamma, \rightarrow_A)$. A state of $S(A)$ is a couple $(l, \nu) \in S$ such that l is a location of A and ν is valuation over C such that ν satisfies the invariant $I(l)$. The initial state s_0 of $S(A)$ is (l_0, ν) where $\nu \in$ **zero**. Labels of Γ are included in $\Sigma_\tau \cup \{\epsilon(d) \mid d \in \mathbb{R}\}$ such that $\{\epsilon(d) \mid d \in \mathbb{R}\}$ corresponds to the elapse of time (Waiting d units of time is noted $\epsilon(d)$). There are two types of transitions in $S(A)$:

- *State change due to elapse of time*: for a state (l, ν) and $d \in \mathbb{R}^{\geq 0}$ $(l, \nu) \xrightarrow{\epsilon(d)}_A$ $(l, \nu + d)$ if for all $0 \leq d' \leq d$, $\nu + d' \in I(l)$ (a timed transition).
- *State change due to a location-edge*: for a state (l, ν) and an edge (l, Z, a, r, l'), $(l, \nu) \xrightarrow{a}_A (l', \nu[r := 0])$ if $\nu \in Z$ and $\nu[r := 0] \in I(l')$ (a discrete transition).

Runs. Let $A = (L, l_0, \Sigma, C, I, \rightarrow) \in \mathcal{TA}(\Sigma)$ and $\sigma = (a_1, d_1)...(a_n, d_n) \in TS(\Sigma_\tau)$. A run r of A over σ, denoted by $(\bar{l}, \bar{\nu})$, is a finite sequence of the form:

$$r : (l_0, \nu_0) \xrightarrow{(a_1, d_1)} (l_1, \nu_1) \ ... \ (l_{n-1}, \nu_{n-1}) \xrightarrow{(a_n, d_n)} (l_n, \nu_n)$$

with $l_i \in L$, and $\nu_i \in V(C)$, for all $i \in [0, n]$, satisfying the following requirements:

1. Initiation: for all $x \in C$, $\nu_0(x) = 0$.
2. Consecution: for all $i \in [1, n]$, there is an edge $t_i = (l_{i-1}, Z_i, a_i, r_i, l_i)$ of A, such that:
   - $\nu_{i-1} + (d_i - d_{i-1}) \in Z_i$.
   - ν_i equals to $(\nu_{i-1} + (d_i - d_{i-1}))[r_i := 0]$.
   - $\nu_{i-1} + d \in I(l_{i-1})$ holds for all $0 \leq d \leq d_i - d_{i-1}$.

Intuitively, at the initial location l_0, the values of clocks are defined to be zero. When the transition t_{i+1} from state l_i to l_{i+1} occurs, we use the value $\nu_i + (d_{i+1} - d_i)$ to check the clock constraints, however, at time d_{i+1}, the value of clocks that are reset in t_{i+1} is defined to be 0. By convention, d_0 is equal to 0.

Example 1. Consider the TA A of the Fig.1 and the timed sequence $(a, 2)(b, 3.7)$. The run corresponding to this sequence is given below. A clock interpretation is represented by listing the values $[x, y]$.

$$(l_1, [0, 0]) \xrightarrow{(a, 2)} (l_2, [2, 0]) \xrightarrow{(b, 3.7)} (l_3, [3.7, 1.7]). \qquad \square$$

The set of timed sequences of A, noted $Run(A)$, is defined by:

$$Run(A) = \{\sigma \mid A \text{ has a run over } \sigma \in TS(\Sigma_\tau)\}.$$

The set of timed traces of A, noted $TTrace(A)$, is defined by:

$$TTrace(A) = \{\sigma \mid \exists \sigma' \in Run(A), \sigma'_{|\Sigma} = \sigma\}.$$

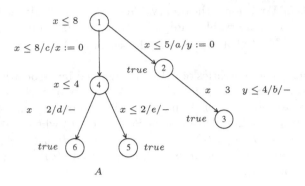

Fig. 1. Timed automata

Finally, for a path ρ of A of length n (i.e. a suite of n transitions of A), we use $TTrace(\rho)$ to denote the set of timed traces of length n of the automaton A_ρ induced by ρ [2].

4   Timed Bound Traces of a Path

The goal of this section is to provide an approach to extract timed traces from a given path. As we will see in the next section, these traces can be used to test RTS.

The idea behind our approach is as follows: to a path ρ, we can associate a constraint polyhedron Z_ρ defining the set of constraints to be satisfied by each trace of $TTrace(\rho)$. From this polyhedron, we identify some timed traces of ρ called the *timed bound traces* (TBT). These latter give a finite representation of the trace space of ρ. The proof of the existence of TBT is based on some transformations on the constraint graph associated to a polyhedron, and can be found in Annex B.

For the rest of this paper, $\rho = t_1 \cdots t_n$ will denote a path of a TA $A = (L, l_0, \Sigma, C, I, \rightarrow)$ such that $t_i = (l_{i-1}, Z_i, a_i, r_i, l_i)$, for all $i \in [1, n]$. $V = \{v_1, v_2, ..., v_n\}$ will denote a set of variables ranged over $\mathbb{R}^{\geq 0}$, and $V_0 = V \cup \{v_0\}$ the set V extended with a fictive variable v_0 which is always equals to 0. We will confound elements of $\Phi(V)$ with elements of $\Phi(V_0)$ and a valuation over V with a valuation over V_0.

4.1   Constraint Polyhedron

Let $\sigma = (a_1, d_1) \ldots (a_n, d_n) \in TTrace(\rho)$. According to the definition of $TTrace(\rho)$, the different instants $(d_i)_{i \in [1,n]}$ satisfy a set of constraints related to the transitions of ρ. So, we can associate to ρ a constraint polyhedron Z_ρ

---

[2] A_ρ has the same states and transitions as ρ.

over variables $V_0 = \{v_0, v_1, \cdots, v_n\}$ such that: $\sigma \in TTrace(\rho)$ iff the valuation $\nu \in \mathcal{V}(V_0)$ defined by $\nu(v_i) = d_i$, for all $\forall i \in [0, n]$, is in Z_ρ.

In order to define the constraint polyhedron, we need some additional notations. For a clock $x \in C$ and $i \in [1, n]$, $\rho_i = t_1...t_i$ will denote the path composed of the first i transitions of ρ. $last_i^x$ will denote the index of the transition where the clock x has been reset most recently before i. Recall that all clocks are reset at the initial location l_0, and thus $last_i^x = 0$ if x was not reset in ρ_i. Z_i will denote the constraint polyhedron associated to ρ_i. The construction of Z_i is done by induction: for $i \in [1, n]$,

1. $Z_0 = true$
2. Z_i is obtained from Z_{i-1} as follows:
    - $Z_i := Z_{i-1} \wedge v_{i-1} \leq v_i$
    - If the guard of t_i has a term of the form $x \bowtie k$ then $Z_i := Z_i \wedge v_i - v_j \bowtie k$, where $j = last_i^x$.
    - If the guard of t_i has a term of the form $x - y \bowtie k$ then $Z_i := Z_i \wedge v_p - v_q \bowtie k$, where $q = last_i^x$ and $p = last_i^y$.
    - If the invariant of l_{i-1} has a term of the form $x \bowtie k$ then $Z_i := Z_i \wedge v_i - v_j \bowtie k$, where $j = last_i^x$.

Next, Z_n will be noted Z_ρ.

Proposition 1. $\sigma = (a_1, d_1) \cdots (a_n, d_n) \in TTrace(\rho)$ iff there is a valuation $\nu \in Z_\rho$ such that $\nu(v_i) = d_i$, for all $\forall i \in [0, n]$. $\square$

Proof. See Annex A. $\square$

Example 2. Consider the path ρ of the automaton of Fig.1, defined by:

$$(l_1, x \leq 8) \xrightarrow{x \leq 5 / a / y := 0} (l_2, true) \xrightarrow{x \; 3 \; y \leq 4 / b / -} (l_3, true).$$

Recall that v_0 is equal to zero all time. Then,

$$Z_0 = true, \quad Z_1 = v_0 \leq v_1 \wedge v_1 - v_0 \leq 5 \wedge v_1 - v_0 \leq 8$$

$$Z_\rho = Z_2 = v_0 \leq v_1 \wedge v_1 - v_0 \leq 5 \wedge v_1 - v_0 \leq 8 \wedge v_1 \leq v_2 \wedge v_2 - v_0 \geq 3 \wedge v_2 - v_1 \leq 4$$

By consequence, $\sigma = (a, d_1).(b, d_2) \in TTrace(\rho)$ iff the valuation ν defined by $\nu(v_1) = d_1$, $\nu(v_2) = d_2$ is in Z_ρ. $\square$

Convention. Without losing the generality and for simplicity reasons, we assume that Z_ρ can be written (syntactically) as:

$$Z_\rho = \bigwedge_{v_i, v_j \in V_0, v_i = v_j} (v_i - v_j \leq l_{ij}), \quad l_{ij} \in \mathbb{R}.$$

In fact, a constraint of the form $v_i \leq c$ can be written as $v_i - v_0 \leq c$ (v_0 is equal to 0) and $v_i \leq c \wedge v_i \leq c'$ can be written as $v_i - v_0 \leq min(c, c')$. Furthermore, if v_i does not have a upper bound in Z_ρ, then we can add the constraint $v_i - v_0 \leq +\infty$. These remarks hold for a constraint of the form $v_i - v_j \leq c$.

Definition 1. $Z_\rho = \bigwedge_{v_i, v_j \in V_0, v_i = v_j} (v_i - v_j \leq l_{ij}) \neq \emptyset$ *is said in its canonical form if for all* $i \in [0, n]$, $j \in [0, n]$, *there exists a valuation* $\nu \in Z_\rho$ *such that :*
$$\nu(v_i) - \nu(v_j) = l_{ij} \qquad \square$$

Definition 2. *The canonical form of* Z_ρ, *noted* $cf(Z_\rho)$, *is the greatest canonical polyhedron included in* Z_ρ. $\qquad \square$

Note that $cf(Z_\rho)$ and Z_ρ represent the same space portion and $cf(Z_\rho) = Z_\rho$ if Z_ρ is in its canonical form.

4.2 Main Results

Theorem 1. *Let* ρ *be a path and* $cf(Z_\rho) = \bigwedge_{v_i, v_j \in V_0, v_i = v_j} (v_i - v_j \leq l_{ij})$ *be the canonical form of its constraint polyhedron. Assume that* Z_ρ *is bounded and not empty* $(Z_\rho \neq \emptyset)$. *Then, for all* $k \in [0, n]$:

1. *There is a valuation* $\nu_k^M(Z_\rho)$ *of* Z_ρ *such that: for all* $i \in [0, n]$, $i \neq k$,
   $\nu_k^M(Z_\rho)(v_i) - \nu_k^M(Z_\rho)(v_k) = l_{ik}$.
2. *There is a valuation* $\nu_k^m(Z_\rho)$ *of* Z_ρ *such that: for all* $i \in [0, n]$, $i \neq k$,
   $\nu_k^m(Z_\rho)(v_k) - \nu_k^m(Z_\rho)(v_i) = l_{ki}$. $\qquad \square$

Intuitively, if Z_ρ is bounded and nonempty, then for each variable $v_k \in V_0$, there is a valuation $\nu_k^M(Z_\rho)$ (resp. $\nu_k^m(Z_\rho)$) which reaches the bounds $(l_{ik})_{k=i, i \in [0,n]}$ (resp. $(l_{ki})_{k=i, i \in [0,n]}$) of $cf(Z_\rho)$ constraints, where v_k is a right (resp. left) member. We have assumed that Z_ρ is bounded to ensure the existence of $\nu_k^M(Z)$. The valuations $\nu_k^m(Z)$ exist even Z_ρ is not bounded because variables of V_0 are ranged over $\mathbb{R}$ [0].

Proof. See Annex B. $\qquad \square$

Computation of $\nu_k^M(Z_\rho)$ and $\nu_k^m(Z_\rho)$. Theorem 1 establishes the existence of valuations $(\nu_k^M(Z_\rho))_{k \in [0,n]}$ and $\nu_k^m(Z_\rho)_{k \in [0,n]}$, and their unicity. Having in mind that $v_0 = 0$, a direct application of this theorem gives: for all $k \in [0, n]$,

1. $\nu_k^M(Z_\rho)$ is the valuation defined by:
   - If $k = 0$ then $\nu_k^M(Z_\rho)(v_i) = l_{i0}$.
   - Else $\nu_k^M(Z_\rho)(v_i) = -l_{0k} + l_{ik}$ and $\nu_k^M(Z_\rho)(v_k) = -l_{0k}$
   for all $i \in [1, n]$, $i \neq k$.
2. $\nu_k^m(Z_\rho)$ is the valuation defined by:
   - If $k = 0$ then $\nu_k^m(Z_\rho)(v_i) = -l_{0i}$.
   - Else $\nu_k^m(Z_\rho)(v_i) = l_{k0} - l_{ki}$ and $\nu_k^m(Z_\rho)(v_k) = l_{k0}$
   for all $i \in [1, n]$, $i \neq k$.

Example 3. Let $Z_\rho = 0 \leq v_1 \wedge v_1 \leq v_2 \wedge v_1 \leq 5 \wedge v_2 \geq 3 \wedge v_2 - v_1 \leq 4$ be the constraint polyhedron of the example 2. Z_ρ is bounded. Its canonical form is defined by: $cf(Z_\rho) = (v_2 - v_0 \leq 5) \wedge (v_0 - v_2 \leq 0) \wedge (v_1 - v_0 \leq 9) \wedge (v_0 - v_1 \leq -3) \wedge (v_1 - v_2 \leq 4) \wedge (v_2 - v_1 \leq 2)$. Then,

$$\nu_0^M(Z_\rho) = \begin{pmatrix} 9 \\ 5 \end{pmatrix} \begin{matrix} v_1 \\ v_2 \end{matrix} \quad \nu_1^M(Z_\rho) = \begin{pmatrix} 3 \\ 5 \end{pmatrix} \quad \nu_2^M(Z_\rho) = \begin{pmatrix} 4 \\ 0 \end{pmatrix}$$

$$\nu_0^m(Z_\rho) = \begin{pmatrix} 3 \\ 0 \end{pmatrix} \quad \nu_1^m(Z_\rho) = \begin{pmatrix} 9 \\ 5 \end{pmatrix} \quad \nu_2^m(Z_\rho) = \begin{pmatrix} 3 \\ 5 \end{pmatrix} \qquad \square$$

Now, consider the two suites of timed sequences $(\sigma^{Mk})_{k \in [0,n]}$ and $(\sigma^{mk})_{k \in [0,n]}$ defined by:

- $\sigma^{Mk} = (a_1, \nu_k^M(Z_\rho)(v_1)) \cdots (a_n, \nu_k^M(Z_\rho)(v_n))$.
- $\sigma^{mk} = (a_1, \nu_k^m(Z_\rho)(v_1)) \cdots (a_n, \nu_k^m(Z_\rho)(v_n))$.

Note that, for all $k \in [0,n]$, $\sigma^{Mk} \in TTrace(\rho)$ and $\sigma^{mk} \in TTrace(\rho)$ (according to proposition 1).

Definition 3. *The timed sequences $(\sigma^{Mk})_{k \in [0,n]}$ and $(\sigma^{mk})_{k \in [0,n]}$ are called the timed bound traces (TBT) associated to ρ.* $\qquad \square$

Timed bound traces give a finite representation of the trace space of a path. According to Theorem 1, the number of TBT is $2 \times (n+1)$ (n is the length of the path). However, this number varies between 1 and $2 \times (n+1)$ and depends on the number of clock resets used in the path. Thus, a path without clock resets has at most 2 TBT. The complexity of computing σ^{Mk} or σ^{mk} from $cf(Z_\rho)$ is $O(n)$. The computation of the canonical form of a polyhedron depends on the data structures used. The algorithm given in [8] allows to compute this form and to test if a polyhedron is empty. Its complexity is $O(n^3)$.

4.3 Trace Inclusion

Consider a path ρ (resp. ρ') of length $n \in \mathbb{N}$ of a timed automaton A (resp. B). To show that $TTrace(\rho) \subseteq TTrace(\rho')$ is equivalent to show that $cf(Z_\rho) \subseteq cf(Z_{\rho'})$. We consider here the case where Z_ρ is known and only the set $TTrace(\rho')$ is known. We assume that Z_ρ (resp. $Z_{\rho'}$) is bounded and not empty.

Corollary 1. $TTrace(\rho) \subseteq TTrace(\rho')$ *iff* $\sigma^{Mk} \in TTrace(\rho')$ *and* $\sigma^{mk} \in TTrace(\rho')$, *for all* $k \in [0,n]$. $\qquad \square$

Intuitively, the corollary gives the necessary and sufficient conditions to show that $TTrace(\rho) \subseteq TTrace(\rho')$. In fact, it is sufficient to show that timed bound traces of $TTrace(\rho)$ are also timed traces of $TTrace(\rho')$.

Proof. See Annex C. $\qquad \square$

5 Application: Testing

A test case (test for short) is an experience performed on the IUT by the tester. In the case of RTS, there are different types of tests, depending on the capabilities of the tester to observe and react to event. Analog-clock tests [9, 13] can measure precisely the real-time delay between observed actions. Digital-clock tests can only count how many "ticks" of a finite-granularity clock have occurred between two actions. Analog-clock testers can measure real-time precisely, but they are difficult (if not impossible) to implement for real-time IUT. Digital-clock testers have access to a periodic clock/counter and are implementable for any IUT. However, they can announce a "Pass" verdict when it is "Fail" (the reception of an event "a" after 2.7 units of time and the same reception after 2.8 units of time will look the same for a digital-clock tester). The use of a digital-clock tester does not mean the discretization of time, the specification is still dense-time but the capabilities of the tester are discrete-time. In this paper, we consider digital-clock testers. Furthermore, we will consider static tests, i.e. the response of the digital-clock tester is the same and known in advance.

5.1 Simulation Graph [19]

Tripakis defines a number of different abstractions for timed automata and study the properties they preserve. These abstractions are based on the simulation graph, which is built by forward reachability and preserves all linear properties. In the simulation graph, the passage of time is hidden and only the discrete-state changes can be observed.

Let A be a TA, $S = (l, Z)$ be a symbolic state (i.e. a location l of A and a polyhedron Z), and $t = (l, Z', a, r, l')$ be a transition of A. Then,

$$\mathbf{post}_c(S, t) = (l', close(((Z \cap Z')[r := 0]) , c))$$

Intuitively, $post_c()$ contains all states (and their c-closure) that can be reached from states in S by taking transition t and letting some time pass. Given the initial location l_0 of A, the simulation graph $S(A, c)$ (c is a natural constant greater than the closure of A) is generated using a depth-first search starting from $S_0 = (l_0, \mathbf{zero})$ and generating for each vertex $S = (l, Z)$ in the stack, the successors $S' = post_c(S, t)$, for each transition $t = (l, Z_1, a, r, l')$ of source l in A. The exploration of the branch leading to S_i is stopped if: either $S_i = \emptyset$ or there is a previously generated vertex $S_i \subset S'$. Otherwise, S_i is added to the set of vertexes and $S \xrightarrow{a} S_i$ to the set of edges of the simulation graph. It has been shown in [19] that $S(A, c)$ is finite and there is a run of A from l_0 to l_f if in the simulation graph there is a vertex $S = (l_f, -)$. Moreover, for each path $S_0 = (l_0, Z_0) \xrightarrow{a_1} S_1 = (l_1, Z_1)... \xrightarrow{a_n} S_n = (l_n, Z_n)$ in the simulation graph, there is a run $r = (\bar{l}, \bar{\nu})$ of A such that $\nu_i \in Z_i$, for all $i \in [0, n]$, and vice versa.

5.2 Digital-Clock Test Derivation

Our goal here is not to provide a complete method to derive digital-clock tests, but only to give the broad lines of an approach to build statically digital-clock tests. The reader can found in [3] a complete algorithm to derive tests for digital-clock/analog-clock testers.

For generating tests, our approach uses the simulation graph. In fact, as we have said, $S(A, c)$ gives a finite representation of the reachable state space; each path of $S(A, c)$ has a run of A and each run of A is inscribed in path of $S(A, c)$. Classical methods for untimed systems can be applied, in general, to derive a set of paths from the graph $S(A, c)$. Let $ATS_M(A)$ be a set of paths derived from $S(A, c)$ with a method M, and with respect to a given coverage criterion (states, transitions,...). Element of $ATS_M(A)$ can not be used directly to test a given implementation of A because they are abstract.

Each path $\rho \in ATS_M(A)$ defines a set of timed traces $TTrace(\rho)$. Corollary 1 has a great influence on test cases considered for the path ρ. In fact, according to this corollary, the number of distinct tests required for the trace inclusion is between 1 and $2 \times (n + 1)$ test cases corresponding to the timed bound traces of ρ. Here, we assume that the time is bounded for each $\rho \in ATS_M(A)$ because testing is a finite experience. When a path $\rho \in ATS_M(A)$ is not bounded, we can choose a natural constant MAX to limit the time of observations.

Thus, the approach that we introduce derives abstract paths form the simulation graph; for each abstract path derived, between 1 and $2 \times (n+1)$ test cases are generated corresponding to TBT of this path. These latter are then decorated by the different verdicts. Our approach does not suffer from the explosion problem, since we use only tests that meet the timed bound traces.

6   Related Work

Regarding works in analyzing RTS, [2] have studied the problem of timestamp generation. The solution proposed consists in computing one timed trace corresponding to the minimal accumulate delay run. The approach of Tripakis for generating timed diagnostics presented in [19, 20] was based on a symbolic analysis. The solution proposed uses the simulation graph to generate abstract paths. For each abstract path, the authors chose randomly the instant of firing the transitions. In [14], the authors show the existence of timed diagnostics associated to a symbolic path, but do not provide a method to compute them. In [10], the authors propose to use the verification tool Uppaal to generate the optimal timed trace corresponding to a state. In [16], the authors propose several algorithms to compute the minimal timed diagnostic that reach a given state.

Regarding works on testing, [13] propose a method to derive analog/digital-clock test cases. The approach proposed was based on a symbolic analysis. However, the proposed method for digital tests considers "ticks" of clocks as an observable event. As a consequence of this choice, is the presence of long chains of ticks in the test cases generated as reported in [13]. The authors propose then a heuristic to compact chains of ticks, but this heuristic does not give always minimal tests and it is not trivial.

An extension of test theory for Mealy machines in the case of dense RTS was proposed by Springintveld et al. [18]. The authors suggested to perform a kind of discretization of the region graph model. Another work generating test sequences for a discretized deterministic timed automaton is given by En-Nouaary et al. in [7]. The authors propose to build a grid automaton from the region graph, and use a Wp method for the generation assuring a good coverage of the initial specification, but the number of generated test cases can be large. In [5], an implicit clock is used, the time is discrete and the proposed model is a temporized transition system. In [12], the authors have chosen as model temporized automata with discrete time. The model is transformed into automaton without time, but with two special events on clocks: set and expire. In [6], the system specification is based on a constraint graph. From a fault model, the authors define test criteria and generate test cases according to the test criteria. Since constraint graph is used as a model, only delays can be expressed between two successive events, and the coverage of faults cannot be complete. In [15], the generation of test cases is produced from logic formula (time is expressed by using two constructors: future and past). A unique clock is used and the temporal domain is discrete. [11] propose a generation method based on must/may traceability. The authors propose to test first, the correctness of the implementation of states and transitions. For that, they transform the specification into a FSM, and use the UIOv-method to derive test cases. [17] use symbolic analysis for event-recording automata inspired by the Uppaal model-checker.

All of these methods successfully generate timed test cases but most of them suffer from an exorbitant number of test cases. The solution that we have proposed was based on the use of timed bound traces and does not suffer from these problems.

7   Discussion

In this paper, we have studied the trace inclusion problem of RTS. Our solution was based on the identification of the timed bound traces (TBT) corresponding to a given path. The trace inclusion problem is then reduced to the inclusion of TBT. As an application, the paper showed how to use these results to reduce the number of test cases for an RTS. The idea behind our approach was the use of the simulation graph to derive abstract paths and the generation of a finite set of test cases from each abstract path corresponding to the timed bound traces.

To our knowledge, the identification of TBT, and the solution proposed for trace inclusion problem are new results. Furthermore, our approach for generating tests, does not suffer from an exorbitant number of test cases because we consider only test cases corresponding to TBT.

To have a complete coverage of the timed trace space of the specification while testing (according to corollary 1), the assumption of the event determinism of the specification is required. This model is quite restrictive, and the generalization will benefit many RTS. Especially, the determinism assumption may be broken by the on-the-fly determinization techniques. Of course, for the class of event-recording automata (ERA), the determinism assumption is not a limitation since this class of timed automata can be determinized.

Finally, timed bounds traces can be used to report counterexamples during timing verification: once the verification tool determines the sequence of transitions that leads to a violation of a safety property, the timed bound traces provide greater diagnostic feedback. In this case, the TBT are called timed bound diagnostics.

References

1. R. Alur and D. Dill. A theory of timed automata, *Theoretical Computer Science*, 126:183-235, 1994.
2. R. Alur, R. Kurshan and M. Viswanathan. Membership problems for timed and hybrid automata. 19th IEEE Real-Time Systems Symposium, 1998.
3. Ismail Berrada. Modélisation, Analyse et Test des Systèmes Communicants à Contraintes Temporelles : Vers une Approche Ouverte du Test. *Phd thesis, Université Bordeaux 1*, Bordeaux, France, 14 December, 2005.
4. Laura Brandán and Ed Brinksma. A test generation framework for quiescent real-time systems. *Proceedings of the 4rd International Workshop on Formal Approaches to Testing of Software, FATES2004*, Linz, Austria September 21, 2004.
5. Rachel Cardell-Oliver. Conformance testing of real-time systems with timed automata specifications. *Formal Aspects of Computing*, 12(5):350-371, 2000.
6. Duncan Clarke and Insup Lee. Automatic test generation for the analysis of a real-time system: case study. In *3rd IEEE Real-Time Technology and Applications Symposium*,
7. A. En-Nouaary, R. Dssouli, F. Khenedek and A. Elqortobi. Timed test cases generation based on state characterization technique. *In 19th IEEE Real Time Systems Symposium (RTSS'98)*, Madrid, Spain, 1998.
8. Robert W. Floyd. Algorithm 97 (shortest path). *Communications of the ACM*,18(3):165-172, 1964.
9. T. Henzinger, Z. Manna and A. Pnueli. What good are digital clocks?. *ICALP'92*, LNCS 623, 1992.
10. Anders Hessel, Kim G. Larsen, Brian Nielson, Paul Pettersson and Arne Skou. Time-optimal real-time test case generation using Uppaal. In *FATES2003*, Montreal, Quebec, Canada, October, LNCS 2931, pp. 118-135, Springer.
11. T. Higashino, A. Nakata, K. Taniguchi and A. Cavalli. Generating test cases for a timed i/o automaton model. *TESTCOM99*, Budapest, Hungary, September 1999.
12. A. Koumsi, M. Akalay, R. Dssouli, A. En-Nouaary, L. Granger. An approach for testing real time protocols, *TESTCOM*, Ottawa, Canada, 2000.
13. M. Krichen and S. Tripakis. Black-box conformance testing for real-time systems. *In SPIN 2004*, Spring-Verlag Heidelberg, pp. 109-126, 2004.
14. Kim G. Larsen, Paul Pettersson and Wang Yi. Diagnostic model-checking for real-time systems. *In Proc. of WVCHS III*, number 1066 in LNCS, pp. 575-586. Springer-Verlag, October 1995.
15. Dino Mandrioli, Sandro Morasca and Angelo Morzenti. Generating test cases for real-time systems from logic specifications. *ACM Transactions on Computer Systems*, 13(4):365-398, 1995.
16. P. Niebert, S. Tripakis and S. Yovine. Minimum-time reachability for timed automata. *In Mediterranean Conference on Control and Automation*, 2000.
17. B. Neilson ans A. Skou. Automated test generation for timed automata. *TACAS'01*, LNCS 2031, Springer 2001.
18. Jan Springintveld, Frits Vaandrager and Pedro R. D'Argenio. Testing timed automata. *Theoretical Computer Science*, 252(1-2):225-257, March 2001.
19. Stavros Tripakis. The formal analysis of timed systems in practice. PhD thesis, Université Joseph Fourier, Grenoble, 1998.
20. Stavros Tripakis. Timed diagnostics for reachability properties. *In Tools and Algorithms for the Construction and Analysis of Systems, TACAS'99*, Amsterdam, Holland, 1999.

Annex A: Proof of Proposition 1

Proof. For all $i \in [0, n]$, v_i represents the instant of firing t_i according to a global clock. Thus, the suite $(v_i)_{i \in [0,n]}$ is monotonically increasing. By convention, we assume the existence of a transition t_0 where all clocks are reset at instant $v_0 = 0$. Initially, Z_0 is equal to *true*. At step $i \in [1, n]$, if t_i has a term over x of the form $x \bowtie k$, then the actual value of x corresponds to the time elapsed since the last reset of x. Thus, the value of x is exactly $v_i - v_j$ where $j = last_x^i$. The constraint $v_i - v_j \bowtie k$ is then added to Z_i. If t_i has a term of the form $x - y \bowtie k$ then, the constraint $v_p - v_q \bowtie k$ is added to Z_ρ, where $q = last_i^x$ et $p = last_i^y$. In fact, the time elapsed since the last reset of x (resp. y) in transition t_q (resp. t_p) is equal to $v_i - v_q$ (resp. $v_i - v_p$). Thus, $x - y = (v_i - v_q) - (v_i - v_p) = v_p - v_q$. Finally, the same approach is applied to the invariant of a location.   □

Annex B: Proof of Theorem 1

In subsection 4.1, we have showed that we can associate to ρ a constraint polyhedron Z_ρ. In order to proof the main theorem of subsection 4.2, we need to define the constraint graph G_ρ associated to the polyhedron Z_ρ and some transformations on G_ρ. Before that, let us recall some graph notions.

Graph Notations

Graphs. A directed labeled graph (DLG for short) G is a triple (V, E, w), where

- V is a finite set of elements $\{v_1, v_2, \cdots, v_k\}$ called vertexes,
- E is the set of couples of distinct elements of the cartesian product $V \times V$ called edges ($E = \{(v_i, v_j) | v_i, v_j \in V \wedge v_i \neq v_j\}$),
- $w_G : E \mapsto \overline{\mathbb{R}}$ is a function that assigns to each edge a weight.

The couple $(v_i, v_j) \in E$, noted $v_i \rightarrow v_j$, represents the edge of source v_i and target v_j. Note that G is a complete graph.

Paths. Let $G = (V, E, w)$ be a DLG. A path p is a sequence of edges $e_1.e_2...e_n$ (e_i is an edge). A path of length n is a path of n edges. The weight of p, noted $w(p)$, is defined by: $w(p) = \sum_{i \in [1,n]} w(e_i)$. Let $e = v_i \rightarrow v_j$ be an edge. Then, $path(e)$ is the set of paths of source v_i and target v_j. A cycle with root v_i is path from v_i to itself. An elementary cycle (e-cycle for short) is a cycle that does not visit a vertex twice, except from the root vertex. The graph G is said:

- nonnegative if the weight of each cycle of G is nonnegative. Formally, for all cycle c, $w(c) \geq 0$.
- minimal if the weight of each edge e is less than or equal to the weight of each path of $path(e)$. Formally, for all $e \in E$, for all $p \in path(e)$, $w(e) \leq w(p)$.

Next, we will use the term graph to denote a DLG.

Graph-Theoretic Formulation

Let $Z_\rho = \bigwedge_{v_i, v_j \in V_0, v_i = v_j} (v_i - v_j \leq l_{ij})$ be the constraint polyhedron associated to ρ. The constraint graph $G_\rho = (V_0, E, w)$ associated to Z_ρ is the graph defined by (Recall that $V_0 = \{v_0, v_1, ..., v_n\}$):

$$w(v_j \rightarrow v_i) = l_{ij} \wedge v_j \rightarrow v_i \in E \iff v_i - v_j \leq l_{ij} \text{ is a term of } Z_\rho.$$

Proposition 2. Z_ρ *is not empty iff* G_ρ *is nonnegative.* □

Intuitively, the set $TTrace(\rho)$ is not empty iff the constraint graph G_ρ does not contain negative cycles. The proof of the theorem can be found in [8].

Definition 4. Z_ρ *is said in its canonical form if its constraint graph* G_ρ *is minimal.* □

This definition is equivalent to the definition 1 (subsection 4.1). Next, we will introduce three transformations that keep the positivity and/or the minimality of the transformed graph. To save space we omitted the proof of the next lemmas, but they are based on the comparison of the weights of e-cycles, and can be found in [3]. Let $G = (V_0, E, w_G)$ be the constraint graph of Z_ρ.

Transformation $m()$. The function $m()$ associates to $G = (V_0, E, w_G)$ the graph $G' = (V_0, E, w_{G'})$ such that: for each edge $v_p \rightarrow v_q \in E$,

$$w_{G'}(v_p \rightarrow v_q) = min(\{w_G(p) \mid p \in path(v_p \rightarrow v_q)\}).$$

Intuitively, the weight of $e = v_p \rightarrow v_q$, in G', is equal to the minimal weight, in G, of all paths of source v_p and target v_q. This weight is either reached by a path, i.e. there is $p \in path(e)$ such that $w_{G'}(e) = w_G(p)$, or $w_{G'}(e) = -\infty$ when $\{w_G(p) \mid p \in path(e)\}$ is not bounded. Note that, G' is a minimal graph.

Proposition 3. G *is a nonnegative graph iff* $m(G)$ *is not.* □

Intuitively, the transformation $m()$ preserves the positivity of cycles.

Definition 5. *Let* $m(G_\rho) = (V_0, E, w_m)$ *be the minimal graph of* G_ρ. *The canonical form of* Z_ρ, *noted* $cf(Z_\rho)$, *is the polyhedron defined by:*

$$cf(Z) = \bigwedge_{v_i, v_j \in V_0, v_i = v_j} (v_i - v_j \leq l_{ij}) \text{ such that } v_j \rightarrow v_i \in E, w_m(v_j \rightarrow v_i) = l_{ij}.$$

□

This definition is equivalent to definition 2 (subsection 4.1) and gives a method to compute the canonical form of a polyhedron.

Transformation $R_{i \to *}()$. Let $i \in [0, n]$. The function $R_{i\ *}()$ associates to $G = (V_0, E, w_G)$ the graph $G' = (V_0, E, w_{G'})$ such that: for each edge $v_p \to v_q \in E$,

$$w_{G'}(v_p \to v_q) = \begin{cases} -w_G(v_i \to v_p) & if\ q = i \\ w_G(v_p \to v_q) & otherwise \end{cases}$$

Intuitively, if $v_p \to v_q$ is not an incoming edge of the vertex v_i then, this edge keeps the same weight in G and G'. Otherwise, the weight of $v_p \to v_q$ is replaced, in G', by the opposite weight of the outgoing edge $v_i \to v_q$ of v_i. The next lemma establishes some properties of this transformation related to the minimality and the positivity of the transformed graph.

Lemma 1. *Let G be a nonnegative graph and $i \in [0, n]$. Consider the graph $G' = m(R_{i\ *}(G))$. Then,*

1. *$R_{i\ *}(G)$ is a nonnegative graph.*
2. *If G is minimal then, for all edges $v_p \to v_q \in E$:*

$$w_{G'}(v_p \to v_q) = \begin{cases} w_G(v_i \to v_q) & if\ p = i \\ -w_G(v_i \to v_p) & if\ q = i \\ -w_G(v_i \to v_p) + w_G(v_i \to v_q) & otherwise \end{cases} \qquad \square$$

Intuitively, the transformation $R_{i\ *}()$ preserves the positivity of cycles. When G is minimal and nonnegative, the second point of the lemma gives a method to compute the minimal graph associated to $R_{i\ *}(G)$ using the weights of G.

Transformation $R_{* \to i}()$. This transformation is similar to $R_{i\ *}()$. The transformed graph $G' = (V_0, E, w_{G'})$ is defined by: for each edge $v_p \to v_q \in E$,

$$w_{G'}(v_p \to v_q) = \begin{cases} -w_G(v_q \to v_i) & if\ p = i \\ w_G(v_p \to v_q) & otherwise \end{cases}$$

Intuitively, the only difference between G and G' is in the weights of outgoing edges of vertex v_i: for all $v_i \to v_q \in E$, $w_{G'}(v_i \to v_q)$ is equal to the opposite weight of $w_G(v_q \to v_i)$. The next lemma reports properties similar to those of $R_{i\ *}(G)$.

Lemma 2. *Let G be a nonnegative graph and $i \in [0, n]$. Consider the graph $G' = m(R_{*\ i}(G))$. Then,*

1. *$R_{*\ i}(G)$ is a nonnegative graph.*
2. *If G is minimal then, for all edges $v_p \to v_q \in E$:*

$$w_{G'}(v_p \to v_q) = \begin{cases} w_G(v_p \to v_i) & if\ q = i \\ -w_G(v_q \to v_i) & if\ p = i \\ w_G(v_p \to v_i) - w_G(v_q \to v_i) & otherwise \end{cases} \qquad \square$$

Proof of Theorem 1

Proof. To prove the theorem, it is equivalent to show that, for all $k \in [0, n]$, the polyhedra:

$$Z_k^M = \bigwedge_{v_i V_0,\, v_i = v_k} (v_i - v_k \leq l_{ik} \wedge v_k - v_i \leq -l_{ik}) \wedge \bigwedge_{v_i, v_j \in V_0,\, v_i = v_j = v_k} (v_i - v_j \leq l_{ij})$$

and

$$Z_k^m = \bigwedge_{v_i V_0,\, v_i = v_k} (v_k - v_i \leq l_{ki} \wedge v_i - v_k \leq -l_{ki}) \wedge \bigwedge_{v_i, v_j \in V_0,\, v_i = v_j = v_k} (v_i - v_j \leq l_{ij})$$

are not empty sets ($Z_k^M \neq \emptyset$ and $Z_k^m \neq \emptyset$). In fact, let G_ρ be the constraint graph of $cf(Z_\rho)$ and $k \in [0, n]$. $cf(Z_\rho)$ is canonical then G_ρ is minimal. $Z_\rho \neq \emptyset$ implies that G_ρ is a nonnegative graph (proposition 2). Now, one can notice that the constraint graph $G(Z_k^M)$ (resp. $G(Z_k^m)$) associated to Z_k^M (resp. Z_k^m) is nothing else than the graph obtained from G_ρ by the transformation $R_{k\ *}()$ (resp. $R_{*\ k}()$) defined above: $G(Z_k^M) = R_{k\ *}(G_\rho)$ et $G(Z_k^m) = R_{*\ k}(G_\rho)$. So, according to the first point of the lemma 1 (resp. lemma 2), we deduce that $G(Z_k^M)$ (resp. $G(Z_k^m)$) is a nonnegative graph and by consequence, $Z_k^M \neq \emptyset$ (resp. $Z_k^m \neq \emptyset$). Furthermore, the second point of lemma 1 (resp, lemma 2) gives a method to compute the canonical form of Z_k^M (resp. Z_k^m). $\square$

Annex C: Proof of Corollary 1

Proof. The proof is a consequence of $TTrace(\rho) \subseteq TTrace(\rho')$ iff $cf(Z_\rho) \subseteq cf(Z_{\rho'})$. As $(\nu_k^M(Z_\rho))_{k \in [0,n]}$ and $(\nu_k^m(Z_\rho))_{k \in [0,n]}$ reach all bounds of $cf(Z_\rho)$, then if $\nu_k^M(Z_\rho) \in Z_{\rho'}$ and $\nu_k^m(Z_\rho) \in cf(Z_{\rho'})$, we can deduce that bounds of $cf(Z_\rho)$ are less than the bounds of $cf(Z_{\rho'})$. The density and convexity properties of sets $cf(Z_\rho)$ and $cf(Z_{\rho'})$ imply that all $\nu \in cf(Z_\rho)$ is also in $cf(Z_{\rho'})$. $\square$

Symbolic and on the Fly Testing
with Real-Time Observers

Rachid Bouaziz and Ousmane Koné

University of Toulouse - CNRS IRIT, 31062 Toulouse Cedex - France
{bouaziz, kone}@irit.fr

Abstract. Analyzing real-time specifications involves new difficulties in
the test generation process. In addition to usual combinatory explosion,
issues like tests executability and controllability become more problematic. To deal with such issues, the new method proposed in this paper combines both on the fly computation (not on line) and optimized
symbolic analysis with the underlying concept of real-time observers. A
symbolic forward analysis is used for test executability and a backward
analysis is performed to refine the tests controllability in view of avoiding
inconclusive verdicts. The featured observers and the backward computation are the basis for a more targeted test selection. To illustrate the
method, the work example is a process control communication system.
Finally, we introduce Real-time Ethernet and the related tests produced
with our method.

1 Introduction

Real time applications are those are applications for which real-time (i.e. *physical
time*) is the main execution constraint. They are concerned with business lines
such as aeronautics, aerospace, automotive or telecommunications. These applications often manage the systems and even the people security and therefore
must be designed with very rigorous techniques. For testing real-time systems,
one must carefully define when to submit an input to the Implementation Under
Test (IUT) and when to observe an output. A major issue of automatic test selection from a formal specification is the combinatory explosion of the analyzed
behaviour. In the presence of real time constraints, the test selection problem
is worsened as a huge number of time instances are relevant to test. Moreover
controllability and test executability becomes non trivial [10]. To deal with such
problems, the method proposed in this paper combines different strategies to
improve the efficiency of automatic test selection.

Symbolic techniques were firstly introduced in the field of formal methods, as
they produced reachability graph of reduced size [13, 4, 17, 14]. These techniques
have recently been used for the purpose of "real-time testing", with some idea
of on line testing [9]. But in these testing approaches, the meaning of "*on line*"
is that the test sequences are not computed in advance, before test execution.
There is no test suite available before, but the tests are calculated "on line", event
after event, during test execution. Since these approaches are based on the reuse

M.Ü. Uyar, A.Y. Duale, and M.A. Fecko (Eds.): TestCom 2006, LNCS 3964, pp. 306–323, 2006.
© IFIP International Federation for Information Processing 2006

of model-checking tools, the resulting tests can not be directly reproduced as they often correspond to diagnostic traces or run time executions. This also disables a reasonable *prediction* of test suite coverage. Meanwhile, an explicit test suite selection strategy can be performed when using test purposes to characterize the expected tests [8, 6, 7] and, above all, "to guide" the test selection process, "*off line*", before test execution. This is a different and now classical basis for test selection against combinatory explosion. Indeed, the "*on the fly traversal*" consists of searching the test pattern from the system model, while avoiding the whole exploration of the model. For timed systems, the principles of on the fly test selection from test purposes were presented a few years ago [6, 15]. The new features and improvements here are the combination of on-the-fly traversal and modern symbolic abstractions together with enhanced real-time observers, for improving the efficiency of test case selection. Here, real-time observers are used to target the expected tests. Then a symbolic forward analysis is used for test executability and a backward analysis is performed to refine the tests controllability in view of avoiding inconclusive verdicts. For illustration, an example of process control system is presented and finally the case study started with Real-time Ethernet protocol is introduced. The current work is being implemented in a prototype tool named OOTEST, and which architecture will be introduced. The papers is organized as follows. Section 2 presents the formalism and notations related timed automata. The kernel of the test methodology is presented in section 3. Section 4 and 6 are complementary comments and concluding remarks.

2 Timed Input Output Automata

Let $\mathbf{R}^+$ be a set of non-negative real numbers and let X be a set of non negative real-time valued variables called *clocks*. The set of *Guards* $G(X)$ is defined by the grammar $g := x \sim c \mid x - y \sim c \mid g \wedge g \mid true$, where x and $y \in X$, $c \in \mathbf{R}^+$ and $\sim \in \{<, \leq, >, \geq\}$. We denote by $\mathbf{T}$ the finite sequences of elements in $\mathbf{R}^+$ called *time domain*, and by Σ the finite set of *actions*. A *time sequence* over $\mathbf{T}$ is a finite no decreasing sequence $\rho = t_1, t_2, \ldots, t_n$ and a *timed word* $w = (a_1, t_1,), (a_2, t_2) \ldots (a_n, t_n)$ is an element of $(\Sigma \times \mathbf{R}^+)^*$. A *clock valuation* is a function $\nu : X \to \mathbf{R}^+$, if $\delta \in \mathbf{T}$ the $\nu + \delta$ denotes the valuation such that for each clock $x \in X, (\nu + \delta)(x) = \nu(x) + \delta$. If $r \subseteq X$ then $\nu[r := 0]$ denotes the valuation such that for each clock $x \in X \setminus r, \nu[r := 0](x) = \nu(x)$ and for each clock $x \in r, \nu[r := 0](x) = 0$. $[r := \infty]\nu$ denotes the valuation such that for each clock $x \in X \setminus r, [r := \infty]\nu(x) = \nu(x)$ and for each clock $x \in r, [r := \infty]\nu(x) = \infty$.

Definition. A *Timed Automaton* (TA) is a tuple $A = (L, L_0, L_f, X, \Sigma, E, I)$, where L is a finite set of locations, $L_0(L_f) \subset L$ is a subset of initial (final) locations, X is a finite set of clocks. Σ is a finite set of events. If the set of events (actions) is partitioned in two disjoint subsets $\Sigma^?$ and $\Sigma^!$, where $\Sigma^?$ is the set of *input* actions and $\Sigma^!$ is the set of *output* actions, the TA A is called *Timed Input Output Automaton* (TIOA). $E \subseteq L \times G(X) \times \Sigma \times R(X) \times L$ is a set of edges. We write $l \xrightarrow{a,g,r} l'$ iff $(l, a, g, r, l') \in E$, where $l, l' \in L$ are the source and destination locations, $g \in G(X)$ is a conjunction of constraints in $G(X)$, $a \in \Sigma$ is the action (or event), $r \in R(X)$ is the set of clocks to be *reset*. $I : l \rightarrow G(X)$ assigns invariants to locations.

We use the notation such as $l \xrightarrow{a}$ (resp. $l \xrightarrow{a}\!\!\!\!/\,$) to denote that there exists l' such that $l \xrightarrow{a} l'$ (resp. there is no such l'). This notation naturally extends to time sequences. We write $l \xrightarrow{(a,t)}$ if from location l, a can be executed at time t. A TIOA A is said to be *complete*, if it accepts every action in Σ at every time. It is said to be input-complete if it accepts every input action in $\Sigma_!$ at every time. A TIOA is called deterministic if $\forall l, l', l'' \in L \cdot \forall a \in \Sigma \cdot \forall t \in \mathbf{R}^+ \cdot l \xrightarrow{(a,t)} l' \wedge l \xrightarrow{(a,t)} l'' \Rightarrow l' = l''$. It is called non-blocking if $\forall l \in L, \forall a \in \Sigma^! \cup \mathbf{R}^+ \cdot l \xrightarrow{a}$.

A *Path* P in TA A is a finite sequence of consecutive transitions $l_0 \xrightarrow{g_1,a_1,r_1} l_1 \xrightarrow{g_2,a_2,r_2} l_2 \ldots$. It is said to be *Accepting* if it starts in an initial location ($l_0 \in L_0$) and ends in a final location ($l_f \in L_f$). A *Run* of the automaton along the path P is a sequence of the form $(l_0, \nu_0) \xrightarrow[t_1]{g_1,a_1,r_1} (l_1, \nu_1) \xrightarrow[t_2]{g_2,a_2,r_2} (l_2, \nu_2) \ldots$, where $\sigma = t_1, t_2 \ldots$ is a time sequence in $\mathbf{T}$, and $\nu_i (i = 1, 2 \ldots)$ is a clock valuation such that: $\nu_0(x) = 0, \forall x \in X$; $\nu_{i-1} + (t_i - t_{i-1}) \models g_i$; $\nu_i = [r_i := 0](\nu_{i-1} + (t_i - t_{i-1}))$. The label of the run is the timed word $\omega = (a_1, t_1), (a_2, t_2), \ldots (a_n, tn)$. The set of all timed words in A is denoted $\mathsf{Traces}(A)$. If the path P is accepting the timed word ω then it is said to be accepted by the TA A.

Example. The following presents a process control communication system in an energy production center. The system operates under real-time constraints with a fault recovery mechanism. The overall system (figure 1) is composed of a **Sensor** module, an **Actuator** module and a **Main** module. The different modules communicate through synchronization ports. In the sequel, we focus on the **Sensor** module only, which model in represented with figure 2. The sensor tries to detect the temperature signal from the environment (**?t**). If no signal was received within 4 time units, the famine signal (**!f**) is sent to the main module. Otherwise the system proceeds with checking the pressure (**?p**) and then sends the (**!v**) signal to the actuator, for subsequent operation. In case of failure during this walk, an error procedure (**!e**) is started for an new tentative. Finally the sensor resets (**!r**) and returns to initial state, etc.

The automaton of figure 1 contains six locations $L = \{S_0, S_1, S_2, S_3, S4, S_5\}$, where the set of initial locations is $L_0 = \{S_0\}$, the set of final locations is $L_f = \{S_0\}$. Two clocks x and y and ten transitions are used. An *accepted path* of the automaton can be represented by:

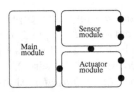

Fig. 1. The process control system - Communicating sensor and actuator

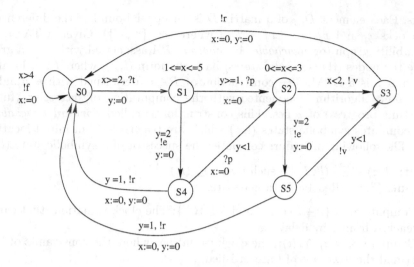

Fig. 2. A process control communication system

$$S_0 \xrightarrow{(x\ 2),?t,(y)} S_1 \xrightarrow{(y\ 1),?p,(x)} S_2 \xrightarrow{(y=2),!e,(y)} S_5 \xrightarrow{(y=1),!r,(x,y)} S_0.$$

An *accepting run* of this path is: $(S_0, (3, \infty)) \xrightarrow[3]{(x\ 2),?t,(y)} (S_1, (4.5, 1.5)) \xrightarrow[4.5]{(y\ 1),?p,(x)}$

$\cdots\cdots (S_2, (0.5, 2)) \xrightarrow[5]{(y=2),!e,(y)} (S_5, (1.5, 1)) \xrightarrow[6]{(y=1),!r,(x,y)} S_0(0,0).$

where $(4.5, 1.5)$ is the valuation ν associated to the clock x and y such that $\nu(x) = 4.5$ and $\nu(y) = 1.5$.

An *accepted timed word* of this run is : $(?t, 3), (?p, 4.5), (!e, 5), (!r, 6)$.

3   Test Design

3.1   From Symbolic Abstraction to Executability and Controllability

Since actual IUT runs correspond to test execution, the executability problem turns to standard reachability analysis that handles all the possible executions of the system. For timed automata, symbolic reachability is now a well established technique addressing the explosion problem of the reachability graph. Rather than enumerating all the states (e.g. like regions) the states are characterized and gathered by a Boolean formula. Different approaches exist for representing the nodes of the reachability graph (Zones represented by Difference Bound Matrix, Clock Difference Diagram, State classes etc). Currently, we have started to implement them as options in our OOTEST tool, in view of a further performance comparison and analysis. In the sequel, we illustrate the method with DBM (Difference Bound Matrix) which have an intuitive representation. For instance, the constraint $0 \leq x \leq 4 \wedge y = 0$ defines a zone that can be represented using a DBM. The DBM is $(n + 1) \times (n + 1)$ matrix where n is the number of clocks. Each element $D_{i,j}$ of a matrix D is an upper bound of the difference of two clocks x_i and x_j ($x_i - x_j \prec D_{i,j}$ where $\prec \in \{\leq, <\}$). Given a TA A, the reachability graph (or *Reachable Automaton*) RA associated with A is a graph where the nodes (the *symbolic states*) have the form (l, z), where l is a location of A and z is a DBM. The construction of RA is performed with a standard reachability algorithm, augmented with the computation of *successors* (future after time progress) of zones. This construction is called *Forward Reachability* as it computes symbolic states (l, z) which are reachable from initial locations of A. The following procedure computes the successor of a symbolic state (l, z).

Input: $(l, z) \xrightarrow{g,a,r} (l', I(l'))$ such that $z \subseteq I(l)$.
Output: (l', z'): Reachable target state.

1. Compute $z_1 = \{\nu + \delta \cdot \nu \in z \text{ and } \delta \in \mathbf{R}^+\}$; The clock valuations that can be reached from z by delaying.
2. Compute $z_2 = z_1 \cap I(l)$; The clock valuations where the constraints of both z_1 and the invariant of l are satisfied.
3. Compute $z_3 = z_2 \cap g$; The clock valuations where the constraints of both z_2 and the guard condition are satisfied.
4. Compute $z_4 = z_3[r := 0] = \{\nu[r := 0] \cdot \nu \in z_3\}$; The clock valuations obtained by resetting clocks in r in the zone z_3.
5. Compute $z' = z_4 \cap I(l')$; The initial reachable clock valuations in l'.

The *Forward Reachability* will guarantee the tests *executability*. But as explained in the following section, we are also interested in improving the *controllability* of the tests execution towards some specific (Pass) states. So we also need to compute the sufficient and necessary constraints that can lead from an initial configuration to such particular states. For that, the *backward* propagation of constraints from the target states to the initial states should be computed. Such computation is called *Backward Reachability* as it calculates symbolic states (l, z) with the *actual* necessary constraints for reaching a given state (l', z'). The following procedure computes the predecessor of a symbolic state (l', z').

Input: $(l', z') \xleftarrow{g,a,r} (l, z)$.
Output (l, z''): Actual predecessor state.

1. Compute $z_1' = \{\nu \cdot \exists \delta \in \mathbf{R}^+ \cdot \nu + \delta \in z'\}$; The set of clock valuations that can reach z' by delaying.
2. Compute $z_2' = z_1'[r := 0]$; The clock valuations just after performing the transition between l and l' and thus after resetting clocks in r.
3. Compute $z_3' = [r := \infty]z_2'$; The clock valuations just before executing the transition; All clocks in r are allowed to have any values.
4. Compute $z_4' = z_3' \cap g$; The set of clock valuations where the constraints of both z_2' and the condition that allowed to fire the transition are satisfied.
5. Compute $z'' = z_4' \cap z$; Concrete predecessor constraints in l obtained by the intersection between the clock valuations in z_4' and the invariant of l (included in z).

A full example will be presented later with figure 8.

3.2   Real-Time Observers

The current developments are aimed to be used with the observers currently designed in the framework of the Open source TOPCASED project[11, 12]. These observers have been initially defined for capturing several extra features like fault tolerance, diagnosis, etc, with an underlying structure of extended timed automata. In this paper, for the time being, we consider the aspects related to functional conformance requirements. In addition to standard timing requirements, the observers define not only the expected behaviour expressed in terms of *test purpose*, but also some aspects of the behaviour that we are not currently interested in. The latter are excluded during the search of the target test sequence and therefore enable restricting the behaviour to be computed. It is up to the test engineer to define a given test observer. We assume that he has the required knowledge for deciding which features are to be included or excluded. In the following, we present an approach to model and construct a real-time observer.

Modelling the Observer. The observer is used for checking the interactions within the test environment. Obviously, its design will be related to the one of the reference specification considered.

Let $A_S = (L_S, L_{0S}, L_{fS}, X_S, \Sigma_S^?, \Sigma_S^!, E_S, I_S)$ be a TIOA model of the Specification. An observer of specification is a TIOA $A_O = (L_O, L_{0O}, L_{fO}, X_O, \Sigma_O^?, \Sigma_O^!, E_O, I_O)$ with the following characteristics:

- The set of locations L_O is equipped with two new disjoint locations $Accept \in L_{fO}$ and $Reject \in L_{fO}$. If the location $Accept$ is reached, we conclude that the functionality modeled by the observer has been satisfied. An efficient set of test cases can be extracted from all paths reaching and *no traversing* a location $Accept$. All path reaching or traversing a $Reject$ location should be ignored (cf below).
- $\Sigma_O = \Sigma_S$ such that $\Sigma_O^! = \Sigma_S^!$ and $\Sigma_O^? = \Sigma_S^?$.
- A_O is *non blocking, deterministic,* and *complete* TIOA.

In this paper, we use passive observers. They basically model a test purpose which characterizes some expected functionality involving a *Pass verdict*, but they also capture undesired (even if correct) behaviour that lead to inconclusive verdicts. An example is presented in figure 3. The construction of an observer can be done by completing the test purpose basic functionality. From every location l in the test purpose, outgoing transitions are added according to the following cases :

1. For every $a \in \Sigma_S$ such that $l \xrightarrow{g, a}$ is an outgoing transition in the test purpose
   - If we'd like to test a under (g) only, then a transition $l \xrightarrow{g, a} Reject$ is added to (E_O).
     ($\neg g$ is the negation of g)
   - If we are interested in the occurrence of a under another constraint (g'), then

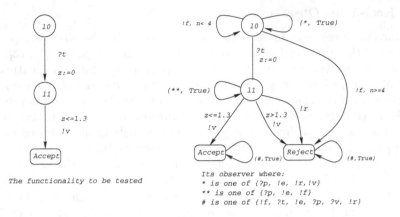

Fig. 3. An example of real-time observer

   (a) A loop $l \xrightarrow{a,g'} l$ is added;

   (b) A transition $l \xrightarrow{a,\ g\ \ g'} Reject$ is added.

2. If the action $a \in \Sigma_S$ is not specified in the outgoing transitions of l
   - If we are interested in the occurrence of a under the constraints (g) then a loop in $l \xrightarrow{a,g} l$ and a transition $l \xrightarrow{g,\ a} Reject$ are added to (E_O).
   - If we do not take care of the occurrence of a, a loop $l \xrightarrow{a,true} l$ is added to the (E_O).

Example. Let us consider the process control example of figure 2. A main characteristic of that real-time system is that only "fresh signals" must be transmitted. We may be interested in the following properties (test purposes).

1. The system sends signal $(!v)$ to the actuator in less than 1.3 time units after the reception of the temperature signal $(?t)$.
2. If an error occurs, it can be corrected within 1 time unit.
   etc ...

The figure 3 (right part) represents the observer modelled from the first functionality described above. In this observer all traces starting with $(!f, n \geq 4)$ are rejected. Such traces increase the behaviour to be analyzed while they are useless for the expected tests. On the other hand, $(!f, n < 4)$ should be preserved as it can detect conformance violation.

3.3   On the Fly Traversal

The main characteristic of our test selection approach is *to perform an on-the-fly traversal of the Reachable (symbolic) Automaton of specification A_S* until the current observer A_O is exhausted (Accept state reached). As usual, this can be formulated on the basis of a synchronous product of A_S and A_O. This is done in the following, in terms of full synchronization as the observer is assumed to be complete, by its construction.

Let $A_S = (L_S, L_{0S}, L_{fS}, X_S, \Sigma_S^?, \Sigma_S^!, E_S, I_S)$ and $A_O = (L_O, L_{0O}, L_{fO}, X_O, \Sigma_O^?, \Sigma_O^!, E_O, I_O)$. The synchronous product of A_S and A_O is a TIOA $A_{SP} = (L_{SP}, L_{0SP}, L_{fSP}, X_{SP}, \Sigma_{SP}, E_{SP}, I_{SP})$ where:

- $L_{SP} = L_S \times L_O$ is the set of states equipped by two distinguished sets of Accepting and Rejecting states which are defined as follows :
  - $Accept_{SP} = L_S \times \{Accept\} = L_{fSP}$;
  - $Reject_{SP} = L_S \times \{Reject\}$.

  An accepting state is an element from $Accept_{SP}$, and it has the form $(l_S, Accept)$, and a rejecting state is an element of $Reject_{SP}$ and it has the form $(l_S, Reject)$.
- $L_{0SP} = L_{0S} \times L_{0O}$ is the set of initial locations;
- $\Sigma_{SP} = \Sigma_{SP}^? \cup \Sigma_{SP}^!$ such that : $\Sigma_{SP}^? = \Sigma_S^? = \Sigma_O^?$, and $\Sigma_{SP}^! = \Sigma_S^! = \Sigma_O^!$ is the set of actions;
- $X_{SP} = X_S \cup X_O$ is the set of clocks ;
- E_{SP} is the set of transitions defined by the following minimal rule :

$$\frac{(l_S \xrightarrow{a, gs, rs} l_S') \wedge (l_O \xrightarrow{a, go, ro} l_O')}{(l_S, l_O) \xrightarrow{a, gs\ go, rs\ ro} (l_S', l_O')}$$

- I_{SP} is such that $I((l_S, l_O)) = I(l_S) \wedge I(l_O)$ is the invariants to locations in A_{SP}.

The synchronous product is illustrated with the specification of figure 2 and the observer of figure 3. For space and readability reasons, the picture has been decomposed into figure 4 and figure 5, and is partially drawn.

Optimizing the on the Fly Traversal. The intuition behind the use of the *Accept* and *Reject* states is to eliminate some behaviours and therefore improve (or optimize) the computation during the search of target test case. The search must be stopped if *Accept* state is reached while transitions towards *Reject* state are not considered. This improves the reachability of the resulting test graph since much of the behaviour is eliminated. The test graph (the tester structure) can be defined by the following operations on the synchronous product.

1. Each element of Accepting states in synchronous product TIOA must be transformed in *PASS* state.
2. Each element of Rejecting states in synchronous product TIOA must be transformed in *INCONCLUSIVE* state.
3. Each transition which initial location is a *PASS* must be removed.
4. Each transition which initial location is an *INCONCLUSIVE* state must be removed.

Starting from specification of figure 2 and observer of figure 3, and their product in figure 4 and figure 5, the operations below produce the test graph of

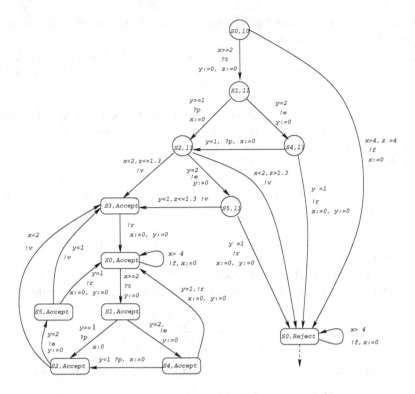

Fig. 4. Synchronous product of Specification and Observer

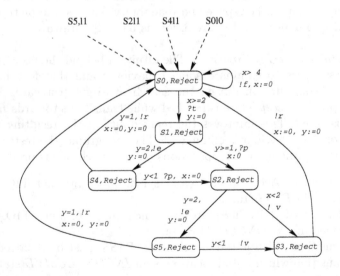

Fig. 5. Synchronous product of Specification and Observer (cont.)

figure 6 which is transformed into the final tester structure in figure 7. In the latter, the inputs of specification are transformed into outputs and vice versa, and a *Fail* state is added to capture all the unexpected outputs.

The test graph defines the **subset** of the product automaton that is considered for the reachability. Moreover, during the symbolic reachability analysis, the target is the (Accept/PASS) state and all the transitions that lead to (Reject/INCONCLUSIVE) are not to be traversed. This obviously improves the performance of the forward analysis. An on the fly traversal of the synchronous product (restricted to the test graph) produces the test path depicted in figure 8-(A). This accepting path is submitted to symbolic analysis for generating test cases, which can be performed during the on the fly traversal (next subsection).

3.4   Test Paths Executability and Controllability Improved

Forward Symbolic for Executability. Let us consider the accepting test path depicted in figure 8-(A). A forward symbolic reachability detects that a state such as $(S_1; (x = 3, y = 1.7))$ is not actually reachable in this path because whenever the automaton occupies location S1, the difference between x and y is at last 2 time units $(x - y >= 2)$. Therefore, a test run containing such state is unsound and may fail with a correct implementation. The following presents such unsound test case.

$$(S_0, (2, 2)) \xrightarrow[t_1]{x>=2,!t,(y)} (S_1, (3, 1.7)) \xrightarrow[t_2]{y>=1,!p,(x)} (S_2, (0.1, 1.8)) \xrightarrow[t_3]{x<2,?v,()} (S_3, (0.2, 1.8)).$$

To guarantee the executability and the test soundness, the "forward" procedure previously presented must be performed, throughout the sequence of transitions in the test path, so that to insure the correctness of the propagated constraints (in the related zones). Figure 8-(B) presents the computed (correct) symbolic states that enable test executability and soundness.

Backward Symbolic for Controllability. The future of an IUT run is potentially a tree structure and one can not always control its evolution towards a specific expected state. This often turns to inconclusive verdict (There was no error but the expected test could not be completed). To try to avoid such situation, one must compute the minimal constraints necessary to lead the IUT towards the expected state. Such constraint is computed in a backward manner with the target symbolic state as input. The "backward" procedure previously presented must be performed, throughout the sequence of transitions in the **executable** test path so that to refine the propagated constraints (in the related zones) until the beginning of the path.

Figure 8-(C) indicates that the PASS state is reachable only if the input !p is sent when the clock x takes a value in the interval $[3, 5]$ rather than in $[2, 5]$, and the clock y takes a value in $[1, 1.3]$ rather than $[1, 2]$. In state the S1 of figure 2, such interval $[1.3, 2]$ where $y > 1.3$ can not allow the interaction !p to

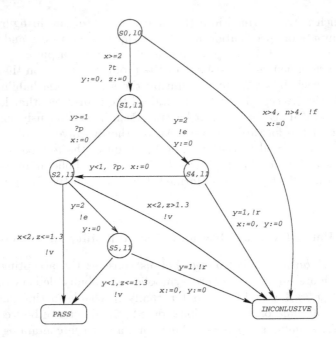

Fig. 6. Reduced/optimised graph of synchronous product

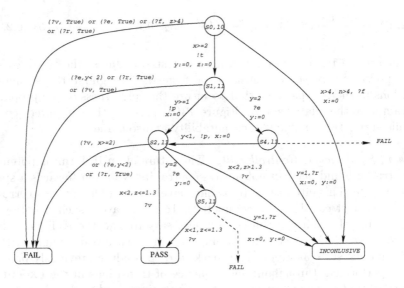

Fig. 7. Complete test graph

be executed in a timely manner, which implies an inconclusive outcome. Figure 9 shows the *controllable* zone in S1 in which we must submit the action $!p$, where Z is the reachable state space and Z_f is the controllable zone. If the action $!p$ is

A: ONE PATH LEADING TO PASS

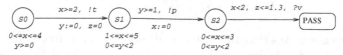

B: FORWARD REACHABILITY

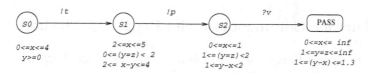

C: BACKWARD REACHABILITY

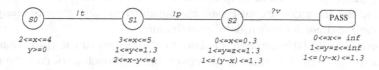

Fig. 8. Forward and backward reachability analysis for one accepting path

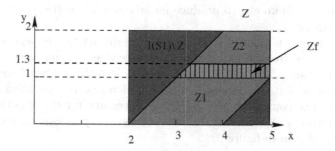

Fig. 9. Controllable zone of S1

sent in the Zone $Z1 \cup I(S1) \setminus Z$, an unsound test is generated, and if it is sent in the zone $Z2$, an INCONCLUSIVE test is generated. To avoid such situations, $!p$ should be submitted only in the controllable zone Z_f.

Finally, to instantiate concrete tests, particular valuations (ν) of clocks can be chosen and propagated in the backward reachability. One could for instance consider extreme values like "minimal/maximal clock valuation" etc.

Example of Successful Test Run. Let us consider the path shown in figure 8-(A). After computing forward and backward reachability graphs (figures 8-(B) and 8-(C)), we obtain the actual symbolic test path below.

$$R = S_0 \xrightarrow{(2 \leq x \leq 4), !t} S_1 \xrightarrow{(3 \leq x \leq 5, 1 \leq y \leq 1.3, 2 \leq (y-x) \leq 4), !p} S_2 \xrightarrow{(0 \leq x \leq 0.3, 1 \leq y \leq 1.3, 1 \leq (y-x) \leq 1.3), ?v} S_3$$

The symbolic test path above can be instantiated with the following test case:

$$S0 \xrightarrow[2]{x=2,!t} S1 \xrightarrow[3]{y=1,!p} S2 \xrightarrow[3.3]{x\leq0.3,?v} \text{PASS}$$

This test case means that the tester sends signal $!t$ at 2 time units, waits at last 1 time unit and submits signal $!p$. Then it should receive $?v$ no later than 0.3 time unit after $!p$ was performed.

4   Further Comments on the Proposed Method

Algorithms and Complexity - Efficiency, Savings of the Method. In this paper, the test selection algorithm proceeds in two steps: First, it uses an on the fly traversal of the product (specification, observer) with a standard DFS (Depth First Search) performed in conjunction with the symbolic forward computation. Second, the intermediate test pattern obtained at this step is analysed in a backward manner for controllability refinement.

The on the fly DFS algorithm is linear with respect to the transitions relation and state space of the product. Moreover, our approach performs the on the fly selection with an optimized graph, with a reduced transition relation and state space: Check the savings by comparing the reduced/optimized graph (figure 6) against the synchronous product (figure 4 continued in figure 5). Finally, symbolic computation is known to produce fewer reachable vertices, which leads to better performances.

The other aspect is related to controllability during test experiment. As the implementation is "free", it can happen that one does not manage to carry out the desired scenario. The test experiment thus is to be replayed several times in the hope of exhibit the desired behavior, which is very expensive in times of development. The controllability analysis is necessary for the selection of more targeted scenarios and for saving the coast of tests (cf previous comments on the refinements and gain in figure 9).

Fault Detection and Conformance Relation. For timed systems, the principal models of errors identified in the literature are : *output error* (when an unexpected output action arrived), *transfer error* (when an unexpected state is reached), and *time constraints errors* (when an output action arrived too early or too late). The method presented in this paper detects *output* and *time constraints errors*. Moreover, it can be related to the implementation relation, referred to as tioco below, since the detection of such errors imply a violation of this implementation relation. The tioco relation is a time extension of the ioco implementation relation used for input-output systems: Let A_I and A_S be two TIOAs modelling the IUT and the specification. The IUT *conforms* to its specification if for each behavior of specification, the possible outputs of the IUT after this behavior is a subset of possible outputs of the specification behaviors. To formally define this, we use the following notations: Given a run σ, Aafterσ is the set of all states of A that can be reached after the execution of σ. Formally,

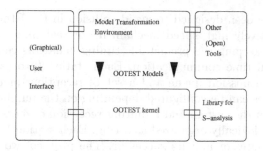

Fig. 10. Architecture of OOTEST environment

A after $\sigma = \{l \in L_A \cdot l_0 \xrightarrow{\sigma} l\}$. The set out refers to the set of all output actions or *delay* that can occur when the automaton reach l : $\mathsf{out}(l) = \{a \in \Sigma^! \cup \mathbf{R}^+ \cdot l \xrightarrow{a}\}$. The relation tioco is defined as follows:

$$A_I \text{ tioco } A_S \text{ iff } \forall \sigma \in \mathsf{Traces}(A_S) \cdot \mathsf{out}(A_I \text{ after } \sigma) \subseteq \mathsf{out}(A_S \text{ after } \sigma).$$

Coverage. The test purpose is a natural basis to coverage analysis. It defines/ specifies the tests to be computed from the system model. The method proposed in the paper computes at least one test case for a given test purpose, if it exists. One could try to compute all the test paths corresponding to a given test purpose. This option is easily implemented, but we incur combinatory explosion, and incur loosing the benefits of the on the fly search. Moreover, without additional hypotheses on the time domains, it is impossible to compute all the possible instances of timed tests because of dense time.

Architecture and Status of OOTEST. The prototype tool OOTEST is under a very "Beta version". The test paths generated before backward analysis are not always the shortest one (there is no search optimization implemented yet). The architecture is presented in figure 10. The tool is designed to be flexible, evolutive and must be connected to other platforms and tools. Currently the tool inputs are timed automata generated from a Graphical User Interface, developed in the french Averroes project [2]. For symbolic analysis, the tool reuses existing libraries for the manipulation of polyhedra. Many such libraries are available as open source and we have currently used some extensions of the Polylib library [5]. We can manipulate structures equivalent to DBM, and few modifications enable the manipulation of Clock Difference Diagrams, or state classes. The Averroes platform partially implements some model transformation and generates automata in XML format that can be parsed towards over model-checking tools.

5 Real-Time Ethernet Protocol

In this section we briefly comment the case study that we are currently carrying in our laboratory, for testing RT-EP (*Real Time Ethernet Protocol*). More details can be found in the report [3].

RT-EP [16] has been designed to avoid collision in the Ethernet media, and to achieve a relatively high speed mechanism for real time communication at a low cost, while keeping the predictable timing behaviors required in the distributed hard real time communication. Each station (processing or CPU) in RT-EP has a transmission queue and a set of reception queues. The number of reception queues can be configured depending on the number of applications threads running in the system and requiring reception of messages.

The network is logically organized as a ring. Each station knows which other station is its predecessor and its successor. The protocol works by rotating a token in this ring. The token holds information about the station having a highest priority packet to be transmitted and its priority value. The network operates in two phases. The first phase corresponds to the priority arbitration, and the second phase to the transmission of an application message.

The following operations show the functionality of RT-EP.

- Firstly, each station in RT-EP reads a configuration file describing the token ring and gets configured as one of its station. The station configured as initial *token-master* sends the *Initial Token (In-Token)* to the successor station.
- Each station listens for the arrival of any packet. When a packet is received, a check is done to determine its type:
  - If it is an information packet (*infos*) the information is written into the appropriate reception queue and the station becomes the *token-master*.
  - If it is a token packet, the station checks its type. (1) If it is a *Regular Token (Rg-Token)* the station compares the priority carried by the token with the highest priority element on its transmission queue, changes the regular token if its own priority is higher, and sends the *Update Token (Up-Token)* to the next station. (2) If the token is the *Initial Token* the station sends information (infos) if it has the highest priority, or sends the permission (action *Tr-Token*) to the highest priority station. (3) If the token is the *Transmit Token* the state has the highest priority on the ring and it is allowed to transmit it.

To recover faults due to the loss of packets, each station, after sending a packet (information or token) listens to the media for an acknowledge (action *ack*), which is the transmission of the next frame by the receiving station. If no acknowledge is received after some specified *timeout*, the station assumes that the packet is lost and retransmits it. The station repeats this process until an acknowledge is received or a specified number of retrials is produced. In the latter case the receiving station is considered as a failing station and will be excluded from the ring (action *dk*).

Each station in RT-EP can be modeled by the two concurrent automata: a sender module and a receiver module. The sender is shown in Figure 12, it uses 3 clocks (x, y, ω).

More details on RT-EP can be found in [16, 3].

For testing RT-EP, if we want to test the ability of the protocol to handle faults due to the loss of packets, we study the following examples:

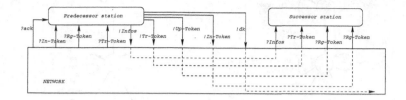

Fig. 11. Operations in RT-EP

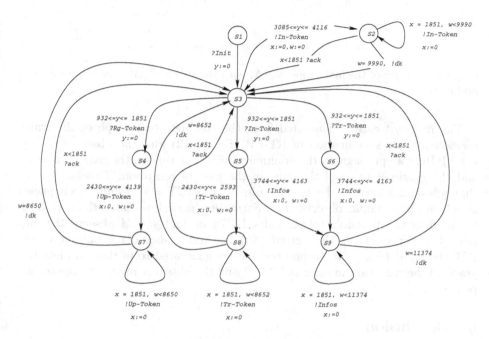

Fig. 12. Transmission module

1. A station should not be excluded from the ring only if it cannot response after 4 retransmission from the predecessor station.
2. When an Information packet is received, the station can submit an acknowledge no later than 1851 ns.

The observer related to the first property is shown in figure 13-(B).
As examples of test sequences leading to the PASS we have:

$$Tc_1 = (S1, L1) \xrightarrow[y:=0]{!Init} (S3, L1) \xrightarrow[y:=0]{y=932,!Tr-Token} (S6, L1) \xrightarrow[x:=0,w:=0,m:=0]{y=4163,?Infos} (S9, L2) \ldots$$

$$\ldots (S9, L2) \xrightarrow{w:=11374,m:=11374,?dk} PASS.$$

$$Tc_2 = (N1, F1) \xrightarrow[m:=0]{!Init} (N2, F1) \xrightarrow[m:=0]{y=2119,!Infos} (N2, F2) \xrightarrow{k<1851,m<1851,?ack} PASS.$$

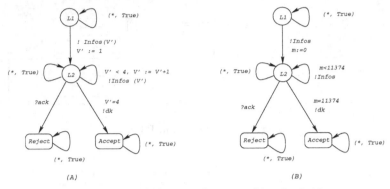

Fig. 13. Observer of the first property; (A) with variable and (B) with temporal interpretation

The first test case is generated from the complete test graph of the transmission module specification of RT-EP (Figure 12) and the observer of Figure 13. Here we propagated the minimal value of the time in the case of emission and the maximal value of the time in the case of reception. This test means that after initialization the tester emits to the IUT a transmit token packet at 930 *ns* and should observe the information packet no later than 4163 after sending the transmit token, waits 11374 *ns* and *should* observe the output (dk). For receiving the output (dk) the tester should not submit to the IUT the input (ack). The second test case is generated from the complete test graph of the reception module of RT-EP and the observer related to the second property.

6   Conclusion

We have presented a method to test selection for real-time systems. Some ideas in our test design process are inspired by techniques used, in other respects, in different research fields. The forward symbolic analysis was used for model-checking of timed systems. The backward analysis was a technique used in fault tolerance analysis to track the cause of a failure in the past of system execution. Combining such techniques for the purpose of testing is - to our knowledge - a new contribution for test selection improvement.

The current work is concerned with functional conformance requirements, and it does not address the full features of the real-time observers actually to be used in the current project. These observers are expected to model extra features like failures of the run-time environment, etc. For future work, our method and its implementation within the OOTEST tool will upgrade in a short term, to take some aspects of robustness testing into account.

References

1. R. Alur and D. Dill, *A Theory of Timed Automata*. Theoretical Computer Science 126:183-235, 1994.
2. *Analysis and VERification for the Reliability Of Embedded Systems*.(www.education.gouv.fr/rntl)
3. R.Bouaziz, O.Koné. *Design principles and applications of the* OOTEST *tool*. Technical Report, CNRS University of Toulouse 2006.
4. Lori Clarke and Debra Richardson. *Symbolic evaluation methods for program analysis. In Program Flow Analysis: Theory and Applications*, S. Muchnick and N. Jones, Eds. Prentice-Hall, Englewood Cliffs, NJ, 79–101. 1981
5. Ph. Clauss and V. Loechner *PolyLib: A Library for Manipulating Parameterized Polyhedra*. Technical Report, University of Strasbourg, 1999.
6. R. Castanet, O. Koné and P. Laurencot, *On-the-Fly Test Generation for Real-Time Protocols*. IEEE International Conference on Computer Communication and Networks. Lafayatte, 1998.
7. A. En-Nouaary and G. Liu : Timed Test Cases Generation Based on MSC-2000 Test Purposes, in Workshop on Integrated-reliability with Telecommunications and UML Languages (WITUL'04), part of the 15th IEEE International Symposium on Software Reliability Engineering (ISSRE), Rennes, France, November 2004.
8. Grabowski J.,Hogrefe D.,Nahm R. *Test case generation with test purpose specifications by MSCs*. 6th SDL Forum. Elsevier Science, North Holland, 1993. Pages 253-266.
9. K. Larsen, M. Mikucionis, and B. Nielsen, *On line Testing of Real-Time Systems*. Formal Approaches To Testing of Software, Link2, Austria. September 2004.
10. T. Higashino, A. Nakata, K. Taniguchi, and A. Cavalli, *Generating Test Cases for a Timed I/O Automaton Model*. IFIP (IWTCS'99) Budapest, 1999.
11. P.Gaufillet. *The TOPCASED project: a Toolkit in OPen source for Critical Aeronautic SystEm Design* ERTS2006 - 3rd Embedded Real Time Software Conference - Toulouse January 2006. http://www.topcased.org
12. Ph.Dhaussy, JC.Roger, H.Bonin, E.Saves and J.Honnoré. *Experimentation of Timed Observers for Validation of an Avionics Software*. Toulouse, January 2006.
13. William E. Howden. *Methodology for the Generation of Program Test Data*. IEEE Trans. Computers, 24(5): 554-560, 1975
14. T. Hinzinger, X. Nicollin, J. Sifakis, and S. Yovine. *Symbolic Model Checking for Real-Time Systems. Information and Computation.* 111(2): 193-244, June 1994.
15. O.Koné. A local approach to the testing of real-time systems. *The Computer Journal*, British Computer Society, Oxford Press. Vol. 44 N.5, 2001.
16. J. M. Martinez, M. G. Harbour, and J. J Gutierrez, *RT-EP : Real-Time Ethernet for analyzable distributed application an a minimum real-time POXIS-kernel*. 2nd International Workshop on Real-Time LANs in the Internet Age. RTLIA 2003.
17. K.L.McMillan. *Symbolic model-checking: An approach to the state explosion problem*. Kluwer Academic, 1993.

Using *TIMED*TTCN-3 in Interoperability Testing for Real-Time Communication Systems

Zhiliang Wang[1], Jianping Wu[1], Xia Yin[1], Xingang Shi[2], and Beihang Tian[1]

[1] Department of Computer Science and Technology, Tsinghua University,
Beijing, P.R. China, 100084
{wzl, yxia, tbh}@csnet1.cs.tsinghua.edu.cn,
jianping@cernet.edu.cn
[2] Network Research Center, Tsinghua University,
Beijing, P.R. China, 100084
shixg@cernet.edu.cn

Abstract. Interoperability testing is an important technique to ensure the quality of implementations of network communication software, and real-time protocol interoperability testing is an important issue in this area. *TIMED*TTCN-3 is a real-time extension of test specification language TTCN-3. In this paper, test notations for real-time interoperability testing are studied. Test behavior trees are constructed from specifications of system under test and then transformed to *TIMED*TTCN-3 test cases. We also investigate real-time TTCN and analyze the insufficiency of its capabilities in specifying time constraints. Possible extensions for real-time TTCN are given to specify real-time interoperability test cases. From the comparisons between the two real-time test notations, it can be concluded that *TIMED*TTCN-3 is more powerful and flexible than real-time TTCN and can be suitable for real-time interoperability testing.

1 Introduction

In order to ensure the quality of communication software, protocol test techniques are widely used. Conformance testing is the basic method of protocol testing, which can be used to test whether an implementation conforms to its protocol specification. As the complement of conformance testing, interoperability testing is often used to test whether two or more protocol implementations can communicate with each other correctly and inter-operate as a whole system to perform functions specified in protocol specifications. Interoperability testing is necessary because (1) It is difficult to perform exhaustive conformance testing, that is, a conformance test suite can hardly ensure 100% test coverage; (2) Many optional features may be contained in network protocols, and moreover vendors perhaps have their own extensions, so if two implementations implement different options, problems on interoperability will happen. Interoperability testing is also being performed by IETF and ETSI in the process of protocol design ([KD03]). Interoperability testing events have been organized by these organizations.

In the area of interoperability testing, [Hao97] proposed a TTCN-2 based framework for interoperability testing, [VBT01] presented a formal framework for

M.Ü. Uyar, A.Y. Duale, and M.A. Fecko (Eds.): TestCom 2006, LNCS 3964, pp. 324–340, 2006.
© IFIP International Federation for Information Processing 2006

interoperability testing and several interoperability relations were defined to guide test generation. Interoperability test generation is an important issue in this field. In most works of test generation, the basic idea is to model the interoperability system under test as a system of communicating finite state machines and generate test sequences for the composition of these machines ([RC90]). Based on this idea, a series of test generation techniques have been proposed ([KSK00, SKKJ03, TKS03, SKCK04, ETSY04]). Different from the above literatures, [HLSG04] proposed an efficient method only considering the specification of one protocol entity. By this method, there is no need to generate the composition machine and the state space explosion can be alleviated.

But in most real-life network protocols, not only the behaviors of input and output, but also their time of occurrence should be considered, that is, such protocols can be modeled by real-time systems. In order to test real-time systems, we should check if the I/O behaviors act under the specified time constraints. In the field of real-time testing, many methods of conformance testing have been proposed. In most of these works, Timed Automaton [AD94] or its variants have been used to specify real-time system. [SVD01, EDK02] converted timed automaton to grid automaton, and applied existing test generation methods for finite state machine (FSM) to it. But this method suffers from the state space explosion problem. [HNTC99] presented a test generation method of executability decision. [KJM03, KT04, LMN04, BB04] defined timed conformance relations, and proposed associated test generation methods. As far as we know, [WWY04] is the first work to study interoperability testing of real-time systems. An interoperability test generation method of time dependent protocols was presented in [WWY04].

In this paper, we focus on the problems of test notations suitable for interoperability testing of real-time communication systems, i.e., how to specify test cases. The *testing and test control notation version 3*(TTCN-3) ([TTCN3, GHRS03]) is a new test specification language standardized by ETSI (*European Telecommunications Standards Institute*), which is a new version and redesign of TTCN (*tree and tabular combined notation*) ([TTCN]). *TIMED*TTCN-3 ([DGN02]) is a real-time extension of TTCN-3. [DGN03] presented a method of generating *TIMED*TTCN-3 code from MSC test specifications; and [NDG04] used *TIMED*TTCN-3 in specifying real-time communication patterns. In this paper, we intend to use *TIMED*TTCN-3 in real-time interoperability testing. Following the test generation method presented in [WWY06], a parameterized test behavior tree will be generated from the formal model of system under test (SUT). Parameters in the test behavior tree are relative time intervals between IO events. In this paper, firstly the test behavior tree will be converted from the view of SUT to the view of test system, which is an intermediate notation of test cases. Then we give transformation rules to transform such a test behavior tree to a *TIMED*TTCN-3 test case.

We also investigate the real-time extension of TTCN – real-time TTCN ([WG99]) and intend to use real-time TTCN to specify test cases. But unfortunately, we find that real-time TTCN has no enough capabilities to specify timed interoperability test cases. We also give possible extensions of real-time TTCN on the syntactical and semantic levels to specify test cases. Based on the transformation results to *TIMED*TTCN-3 and extended real-time TTCN test cases, we compare the two test notations mainly on the capabilities of specifying *hard* real-time requirements.

The rest of the paper is structured as follows. Section 2 gives the formal model Communicating Multi-port TIOA (CMpTIOA) to specify interoperability system under test; and as a working example, a simple real-time communication protocol system is specified by using this model. In Section 3, test architecture is given and test behavior trees will be generated. In Section 4, we give transformation rules from test behavior trees to *TIMED*TTCN-3 test cases. In Section 5, we investigate real-time TTCN and draw a comparisons between *TIMED*TTCN-3 and real-time TTCN. Conclusion and future work are given in Section 6.

2 Preliminaries

2.1 Multi-port TIOA

Timed Automaton ([AD94]) is a widely-used model of real-time system. TIOA (*Timed Input Output Automata* [EDK02]) is a variant of Timed Automaton, which distinguishes whether an action is an input or output. To specify an entity interacting with more than one other entities, we extend TIOA to Multi-port TIOA as follows.

Definition 1. *Multi-port Timed Input Output Automaton (MpTIOA)*
A Timed Input Output Automaton with n ports (for short, np-TIOA) is a 6-tuple (L, I, O, l^0, C, T), where,

- L is a finite set of locations;
- It has n ports communicating with environment, which are denoted as $P_1, P_2, ..., P_n$ respectively;
- I is an n-tuple: $I=(I_1, I_2, ..., I_n)$, where $I_k(k=1,2,...,n)$ is the set of input action symbols of port P_k; $\overline{I}=I_1 \cup I_2 \cup \cdots \cup I_n$ is the set of input action symbols; An input action symbol occurring in port P_k can be denoted as $P_k?a$ $(a \in I_k)$;
- O is an n-tuple: $O=(O_1, O_2, ...O_n)$, where $O_k(k=1,2,...,n)$ is the set of output action symbols of port P_k; $\overline{O}=O_1 \cup O_2 \cup \cdots \cup O_n$ is the set of output action symbols. An output action symbol occurring in port P_k can be denoted as $P_k!b$ $(b \in O_k)$;
- $l^0 \in L$ is the initial location;
- C is a finite set of clocks $\{t_1, t_2, ..., t_{|C|}\}$, where, $|C|$ is the number of clocks;. $v_i \in R^+$(non-negative real numbers) is the clock value of t_i; $\vec{v} = (v_1, v_2, \cdots, v_{|C|})$ denotes a clock valuation;
- T is a set of transitions: $(l,a,P,R,l') \in T$, where, $l,l' \in L$ are the source and destination locations; $a \in \overline{I} \cup \overline{O}$ is an input or output action symbol; P is the time constraint, which is a Boolean conjunction over linear inequalities $P(\vec{v})$; The subset $R \subseteq C$ specifies the clocks to be reset to 0. The transition $(l,a,P,R,l') \in T$ can be also denoted as $l \xrightarrow{a[P]/R} l'$; □

In the model, we assume that time constraints of transitions are all the format of $\wedge(v_i \sim d)$, where $\sim \in \{<, >, \leq, \geq, =\}$, and $d \in R^+$. We distinguish two urgency types ([BST98]) of transitions implicitly: (1) *Lazy*, for transitions with input actions, means that input actions may be not taken because they are controlled by environment (such a property is also called "*Unforced Inputs*"); and (2) *Delayable*, for transitions with output actions, means that the corresponding output action must be taken during such transitions' enabling time.

The semantics of MpTIOA can be defined as a TIOTS (*Timed Input Output Transition System*) $(S, s_0, A_{in}, A_{out}, \rightarrow)$, where $A_{in} = \overline{I}$ and $A_{out} = \overline{O}$. We denote $Act = A_{in} \cup A_{out}$ as the set of all IO symbols. Its states are the pairs $s = (l, \vec{v})$, where $l \in L$ is a location, $\vec{v} = (v_1, v_2, \cdots, v_{|C|})$ is a clock valuation. S is the set of all possible states. $\rightarrow \subseteq S \times (Act \cup R^+) \times S$ is the set of transitions. There are two types of transitions: *Timed transitions* and *Discrete transitions*. *Timed transitions* model time progress, which are the form $(l, \vec{v}) \xrightarrow{d} (l, \vec{v}')$, where $d \in R^+$ is the delaying time, $\vec{v}' = \vec{v} + \vec{d} = \vec{v} + (d, d, \cdots, d)$, and in this period, no discrete transitions occur. *Discrete transitions* $(l, \vec{v}) \xrightarrow{a} (l', \vec{v}')$ correspond to execution of the transition (l, a, P, R, l') in MpTIOA, where P is satisfied by $\vec{v}$ ($P(\vec{v}) = \textbf{\textit{true}}$) and $\vec{v}'$ is obtained by updating $\vec{v}$ according to R.

2.2 Communicating Multi-port TIOA

To specify an interoperability system under test including two or more entities, we introduce a formal model Communicating MpTIOA (CMpTIOA). In the model, MpTIOA can model each single entity, and all these entities in the system can communicate with each others via channels between different MpTIOAs.

Definition 2. *Communicating Multi-port Timed Input Output Automata (CMpTIOA)*
A Communicating MpTIOA is composed of a set of MpTIOAs M and a set of channels Ch, where,

(1) $M = \{M_1, M_2, ..., M_m\}$ is a finite set of m MpTIOAs;
(2) $Ch = \{C_{ij} \mid i, j = 1, 2, ..., m \wedge i \neq j\}$ is a finite set of channels between MpTIOAs: $C_{ij} \in Ch$ represents the communicating channel from MpTIOA M_i to M_j. □

In the definition of CMpTIOA, channels behave like FIFO queues. Intuitively, the semantic of channels is that outputs of MpTIOA M_i can be transferred via channel C_{ij} to be inputs of M_j. In this paper, we assume that transfer time of actions in communicating channels can be neglected, that is, the channels are **lossless** and **non-delayed**.

Definition 3. *Port Mapping Relations of CMpTIOA*
Port Mapping Relations R of CMpTIOA M is an m-tuple: $R = (R_1, R_2, ..., R_m)$, where $R_k (k=1, 2, ..., m)$ is the Port Mapping Relations of MpTIOA M_k; R_k is a set of Port Mapping Relations for all ports of M_k: $R_k = \{r_1, r_2, ..., r_n\}$, where n is the port number of M_k, and $r_i (i=1, 2, ..., n)$ can be the format of 1) $P_i \rightarrow M_j:P_h (j \neq k)$, which means that the port

P_i of M_k is connected to the port P_h of M_j via the channel C_{kj}; 2) $P_i \rightarrow env$, which means that the port P_i of M_k is connected to the external environment of the system. □

According to the above definitions, we can get the abstract topology of the system under test. We furthermore denote the ports communicating with the external environment as "external ports"; and others as "internal ports". Inputs/outputs on external/internal ports are "external/internal inputs/outputs".

2.3 A Simple Real-Time Communication Protocol

We specify a simple real-time communication protocol by using MpTIOA. Fig. 1 (a) shows the specification of such a protocol, which is a 2p-TIOA with two ports (U and l) and two clocks $\{t_1, t_2\}$. $I_U=\{A\}$, $O_U=\{B,C\}$, $I_l=O_l=\{a,b,c\}$. The initial location is '0'. The protocol can be specified informally as follows:

(1) Initiate a connection to a remote entity actively
If an input 'A' is received from port U in the initial location 0, the protocol entity should initiate a connection to a remote entity actively; In this transition $(0,U?A,\boldsymbol{true},\{t_1,t_2\},1)$, the two local clocks t_1 and t_2 should be reset to 0. Within 2 time units, an output 'a' should be sent from port l to remote entity, and the clock t_1 should be reset to 0 (transition $(1,l!a,[t_1<2],\{t_1\},2)$). After that, three cases should be considered:

  (a) Receiving an input 'b' from port l in time, i.e., transition $(2,l?b,[t_1\geq1,t_2<2],\{\},3)$, indicates that the connection can be established;
  (b) Receiving an input 'b' from port l too late, i.e., transition $(2,l?b,[t_2\geq2],\{t_1\},4)$, indicates that the connection cannot be established;
  (c) Receiving an input 'c' from port l, i.e., transition $(2,l?c,\boldsymbol{true},\{t_1\},4)$, indicates that the connection cannot be established.

If the connection can be established, an output 'B' should be sent to port U, i.e., transition $(3,U!B,[2<t_2<3],\{\},0)$; else, an output 'C' should be sent to port U, i.e., transition $(4,U!C,[t_1<1],\{\},0)$.

(2) Respond a connection request from a remote entity passively

  (a) Sending an output 'b' to port l, i.e., transition $(5,l!b,[1<t_1<3],\{\},0)$;
  (b) Sending an output 'c' to port l, i.e., transition $(5,l!c,[t_1<3],\{\},0)$.

To test interoperability, we make an assumption that both specifications and implementations are input-complete, that is, they can accept any inputs at any locations. To make a specification input-complete, some self-loop transitions can be added to it, which indicates that a specification ignores the unspecified input actions. Fig. 1(b) shows an input-complete specification after adding self-loops to Fig. 1(a).

Fig. 2 shows an example of a system under test specified by CMpTIOA, which is a real-time communication system containing two real-time protocol entities. $M=\{M_1, M_2\}$, $Ch=\{C_{12}, C_{21}\}$. The specifications of M_1 and M_2 are both the MpTIOA of Fig. 1(b). We use subscript 1, 2 on ports and actions to distinguish them. Port Mapping Relations are $(\{U_1 \rightarrow env, l_1 \rightarrow M_2:l_2\}, \{U_2 \rightarrow env, l_2 \rightarrow M_1:l_1\})$.

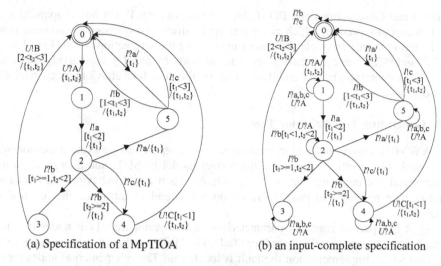

(a) Specification of a MpTIOA   (b) an input-complete specification

Fig. 1. A simple real-time protocol specified by using MpTIOA

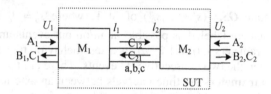

Fig. 2. An example of CMpTIOA

3 Test Behavior Tree

3.1 Test architecture

To test interoperability of protocol system, test architecture should be defined firstly. Fig. 3 shows test architecture that can be used to test SUT in Fig. 2. In the test architecture, there are two types of access points to SUT in the test system: PCO (Point of

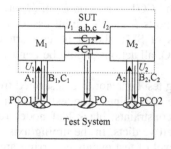

Fig. 3. Test architecture used to test SUT in Fig.2

Control and Observation) and PO (Point of Observation). PCOs have capabilities of control and observation, which can either apply stimuli to or receive responses from SUT; and POs only have capabilities of monitoring the interactions of SUT. In Fig. 3, PCO1 and PCO2 are connected to the external ports U_1 and U_2 respectively, and only one PO is contained in test architecture to monitor the IO behaviors in channel C_{12} and C_{21}.

3.2 Generating Test Behavior Tree

In [WWY06], based on timed interoperability relations, a test generation method was presented. This method starts from the formal model of SUT, and as a result, a parameterized test behavior tree can be generated. Such a test behavior tree is just an intermediate notation. In this paper, we do not intend to introduce this method in detail.

Fig. 4 is a part of resulting parameterized test behavior tree. Leaf nodes of a test behavior tree are the verdict "**pass**" or "**fail**". For "**fail**" verdict, it is also necessary to indicate which implementation the fault is located in. The other internal nodes represent the tester's knowledge of the SUT's current global states, denoted as $(s^1, s^2, ..., s^m)$, where $s^i = (l^i, \vec{v}^i)$ $(i = 1, 2, ..., m)$, representing local states of M_i. The root node is the initial global state $GS_0 = (s_0^1, s_0^2, ..., s_0^m)$ of SUT, where $s_0^i = (l_0^i, \vec{v}_0^i)$ $(i = 1, 2, ..., m)$. Edges between nodes are labeled as possible input/output events and their time constraints in SUT. Parameters $d_i (i=0,1,2,...)$ in a test behavior tree represent relative time intervals between the two consecutive IO events. There are two types of parameters: **controllable parameters** are time intervals between an external input event and its last IO event in the tree, and their values should be set in test cases in advance, so such parameters are controllable for test system, e.g., d_0 in Fig. 4; **uncontrollable parameters** are time intervals between an internal or external output event and its last IO event, and their values are dependent on SUT and only can be retrieved on the process of test execution and cannot be set in advance, so such parameters are uncontrollable for test system, e.g., d_1, d_2, d_3 in Fig. 4.

Parameterized test behavior tree in Fig. 4 is described from the view of SUT. To generate executable test case, at first, it should be converted to the test behavior tree which is described from the view of test system. The basic idea is to convert edges of the original tree to nodes of the resulting tree, and associate each IO event with one access point of test system; e.g., for test architecture of Fig. 3, IO events on the port U_1 of M_1 are associated with PCO1, events on U_2 of M_2 are associated with PCO2 and events on the two internal ports l_1 and l_2 are all associated with PO. On PCO1 and PCO2, input/output actions of SUT should be converted to sending/receiving test events of test system. On PO, all actions should be converted to receiving test events of test system.

Fig. 5 shows the resulting test behavior tree described from the view of test system: the root node represents the start point of the test; other internal nodes are labeled as test events and their time constraints; black leaf nodes represent pass verdicts, and gray leaf nodes represent fail verdicts. In the timing axis on the right of the tree, the global time for the same level of test events occurring are denoted as $T_i (i=0,1,2,...)$, so relative time intervals between two consecutive events are $d_i = T_{i+1} - T_i$ $(i=0,1,2,...)$.

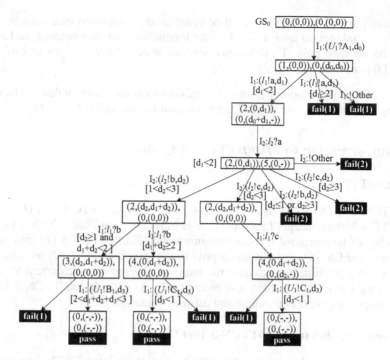

Fig. 4. A part of parameterized test behavior tree (from the view of SUT)

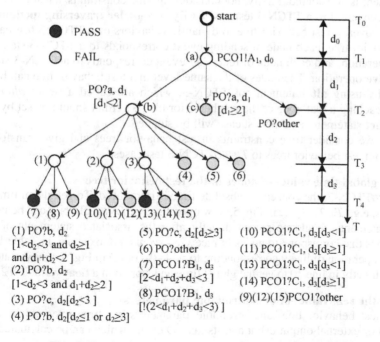

(1) PO?b, d_2
[$1<d_2<3$ and $d_2 \geq 1$
and $d_1+d_2<2$]
(2) PO?b, d_2
[$1<d_2<3$ and $d_1+d_2 \geq 2$]
(3) PO?c, $d_2[d_2<3$]
(4) PO?b, $d_2[d_2 \leq 1$ or $d_2 \geq 3$]

(5) PO?c, $d_2[d_2 \geq 3]$
(6) PO?other
(7) PCO1?B_1, d_3
[$2<d_1+d_2+d_3<3$]
(8) PCO1?B_1, d_3
[!($2<d_1+d_2+d_3<3$)]
(9)(12)(15)PCO1?other

(10) PCO1?C_1, $d_3[d_3<1]$
(11) PCO1?C_1, $d_3[d_3 \geq 1]$
(13) PCO1?C_1, $d_3[d_3<1]$
(14) PCO1?C_1, $d_3[d_3 \geq 1]$

Fig. 5. Test behavior tree (from the view of test system)

We have proved in [WWY06], all time constraints on each test event can be represented by a conjunction over a set of linear inequalities on parameters (see **Lemma 1**), e.g., on the time point T_3, time constraints of node (1) are [$1<d_2<3$ and $d_2\geq1$ and $d_1+d_2<2$] ([$1<d_2<3$ and $d_1+d_2<2$]).

Lemma 1. On the time point T_k, time constraints of the test event nodes can be represented by a conjunction over a set of linear inequalities on $d_i(i=0,1,...,k-1)$.   □

4   Transformation to *TIMED*TTCN-3 Codes

4.1 *TIMED*TTCN-3

*TIMED*TTCN-3 ([DGN02]) is a real-time extension of TTCN-3([TTCN3]). In *TIMED*TTCN-3, the concept of absolute time is introduced, so *TIMED*TTCN-3 provides a capability of testing *hard* real-time requirements. *TIMED*TTCN-3 (1) introduces a new verdict **conf** to indicate *functional pass* but *no-functional fail*; (2) introduces the concept of absolute time and provides mechanisms of retrieving the current local time and delaying the execution of a test component; (3) extends the TTCN-3 logging mechanism; (4) supports both online and offline evaluation.

4.2 Transformation to *TIMED*TTCN-3 Test Cases

Now we consider how to convert a test behavior tree to a *TIMED*TTCN-3 test case. In this paper, for the sake of simplicity, only test architecture with one single main test component is considered. Firstly, not considering time constraints, a test behavior tree can be converted to a TTCN-3 test case easily: **Pre-order traversing method** can be used to covert a test behavior tree to dynamic behaviors of a TTCN-3 test case. In a test behavior tree, each node of sending event corresponds to a TTCN-3 statement of **send** operation, and each node of receiving event corresponds to a TTCN-3 statement of **receive** operation. The nodes of the same level in a test behavior tree can be represented by using **alt** statements of TTCN-3, which are called a set of *alternatives*. When reaching a leaf node of the test behavior tree, a verdict should be set by using a **setverdict** statement, and the test case will be stopped.

Now we consider time constraints in test behavior trees and give transformation rules from test behavior trees to *TIMED*TTCN-3 test cases.

(1) Get global time values of nodes in the test behavior tree
In *TIMED*TTCN-3, the concept of absolute time is introduced. To get global time values of nodes, e.g., T_0, T_1, T_2 ... in Fig. 5, **now** operations in *TIMED*TTCN-3 can be used. The **now** statements should be placed directly after the associated statements of **receive** operations that correspond to nodes of receiving events. For example, in Table 1, for the **receive** operation of line 9 corresponding to node (b) or (c) in Fig. 5, a **now** statement is placed directly in line 10 to get the global time and store it in a **float** variable T2.

(2) Get the real values of uncontrollable parameters
In the test behavior tree, uncontrollable parameters are time intervals between an internal or external output event and its last IO event, which can be calculated only in

Table 1. The *TIMED*TTCN-3 test case for the test behavior tree in Fig. 5

```
1   testcase testcase1() runs on simple_rtp {        42          }
2     var float T0,T1,T2,T3,T4; //global clock       43          []PCO1.receive{ //node(12)
3     var float d1,d2,d3; //time interval            44            setverdict(fail);
4     T0 := self.now;                                45          }
5     T1 := T0 + 1;                                  46        } //end of alt 4
6     resume(T1) ; //wait until T1 point             47      } else { //node(4)
7     PCO1.send(A1); //node (a)                      48        setverdict(conf);
8     alt { //alt 1                                  49      }
9     []PO.receive(a) {                              50    }
10      T2 := self.now;                              51    []PO.receive(c) {
11      d1 := T2 – T1;                               52      T3 := self.now;
12      if(d1 < 2) {    //node (b)                   53      d2 := T3 - T2;
13        alt { //alt 2                              54      if(d2 < 3) { //node (3)
14          []PO.receive(b) {                        55        alt { //alt 5
15            T3 := self.now;                        56          [] PCO1.receive(C1) {
16            d2 := T3 – T2;                         57            T4 := self.now;
17            if((d2>1) and (d2<3)                   58            d3 := T4 - T3;
              and ((d1+d2)<2)){ //node(1)            59            if(d3 < 1) { //node(13)
18              alt { //alt 3                        60              setverdict(pass);
19                []PCO1.receive(B1) {               61            } else { //node(14)
20                  T4 := self.now;                  62              setverdict(conf);
21                  d3 := T4 – T3;                   63            }
22                  if( ((d1+d2+d3)>2) and           64          }
                    ((d1+d2+d3)<3)){//node(7)        65          [] PCO1.receive {//node (15)
23                    setverdict(pass);              66            setverdict(fail);
24                  } else {   //node (8)            67          }
25                    setverdict(conf);              68        } //end of alt 5
26                  }                                69      } else { //node (5)
27                }                                  70        setverdict(conf);
28                []PCO1.receive { //node(9)         71      }
29                  setverdict(fail);                72    }
30                }                                  73    []PO.receive { //node (6)
31              } //end of alt 3                     74      setverdict(fail);
32            } else if((d2>1) and (d2<3)            75    }
              and ((d1+d2)>=2)){ //node(2)           76        } //end of alt 2
33              alt { //alt 4                        77      } else {
34                []PCO1.receive(C1) {               78        setverdict(conf);
35                  T4 := self.now;                  79      }
36                  d3 := T4 – T3;                   80    }
37                  if(d3 < 1) { //node (10)         81    []PO.receive {
38                    setverdict(pass);              82      setverdict(fail);
39                  } else { //node (11)             83    }
40                    setverdict(conf);              84  } //end of alt 1
41                  }                                85 } //end of test case
```

the process of test execution. In *TIMED*TTCN-3, **assignment** statements can be used to get the real values of such parameters. These values can be calculated by expressions on global time values, in fact, $d_i=T_{i+1}-T_i$ $(i=0,1,2,\ldots)$. For example, an uncontrollable parameter d_1 can be calculated by an **assignment** statement in line 11.

(3) Implementation of controllable parameters

In the test behavior tree, controllable parameters are time intervals between an external input event and its last IO event, which should be set in advance. To implement such parameters, **resume** statements can be used. For example, we set the controllable parameter d_0 in Fig. 5 to 1 time unit, so this parameter can be implemented by line 4~7 in Table 1, which means that after waiting 1 time unit from the time point T_0, PCO1 will send A1 to SUT.

(4) Online evaluations and verdicts setting

Only online evaluation can be used to test *hard* real-time requirements in real-time interoperability testing. According to **Lemma 1**, Time constraints of the test event nodes on the time point T_k can be represented by a conjunction over a set of linear inequalities on parameters $d_i(i=0,1,\ldots,k-1)$. Until the time point T_k, values of all uncontrollable parameters have been calculated by transformation rule (2) and values of all controllable parameters have been set in advance, so during the testrun, Mathematical formulae on the values of $d_i(i=0,1,\ldots,k-1)$ can be used in online evaluations on the time point T_k. In the example of Table 1, the **if** statement of line 17 checks if the test event node (1) is reached by using the condition 1<d2<3 and d1+d2<2.

When verdict nodes of the test behavior tree are reached, verdicts should be set by using **setverdict** statements. Besides **pass** or **fail** verdicts, if functional requirements are satisfied, i.e., received messages are correct, but non-functional requirements are violated, i.e., time constraints are not satisfied, a **conf** verdict should be set. In the example of Table 1, **setverdict** statement in line 23 set a **pass** verdict for node (7) in Fig. 5; and **setverdict** statement in line 25 set a **conf** verdict for node (8) in Fig. 5.

According to above transformation rules, the test behavior tree in Fig. 5 can be converted to the *TIMED*TTCN-3 test case with about 85 lines codes shown in Table 1.

5   Comparisons with Real-Time TTCN

5.1   Real-Time TTCN

Real-time TTCN ([WG99]) is a real-time extension of TTCN-2, which is a previous version of TTCN-3. Table 2 is an example of real-time TTCN behavior description. Real-time TTCN extends TTCN-2 both on the level of syntax and semantics. In real-time TTCN, an assumption is made that execution of each statement is instantaneous. On the syntactical level, real-time TTCN adds two columns in dynamic behavior description table: Time and Time options column. Time columns are used to define the earliest and latest execution times (EET and LET) to constrain relative time interval between the associated test event statement and a previous or earlier test event. In real-time TTCN, two types of methods for specifying EET and LET are defined: (1) Define the two values by using time expressions directly, which indicate relative time interval between the execution time of the associated statement and its previous statement (just the parent node in the test behavior tree). In the example of Table 2, line 1 is defined by two constants directly, and line 3 is defined by using a time name, which should be defined in Time Declaration table and be evaluated by an **assignment** statement before use (line 2). (2) Define the two values by using Labels, which indicate relative time interval between the execution time of the associated statement and its earlier statement that labeled as this Label. For example, in Table 2, line 5 defines time constraints (L1+WFN, L1+LET), which means that the time interval between the execution time of line 5 and line 1 are from WFN to LET. The two types of specifications are different from the starting time points for relative time interval. In real-time TTCN, entries in Time Options columns are combinations of M and N. On the semantics level, [WG99] defines its operational semantics and formal semantics based on timed transition system.

Table 2. An example of real-time TTCN behavior description ([WG99])

Test Case Dynamic Behaviour							
Nr	La	Time	Time Options	Behaviour Description	C	V	Comments
1	L1	2, 4	M	A ? DATA_ind			Time Label
2				(NoDur := 3)			Time Assignment
3		2, NoDur		A ! DATA_ack			
4				A ? DATA_ind			
5		L1+WFN, L1+LET	M, N	B ? Alarm			

5.2 Problems in Transformation from Test Behavior Tree to Real-Time TTCN

We consider how to transform a test behavior tree to a real-time TTCN test case. Not considering time constraints, the method of transformation is similar to the one for TTCN-3. Now consider time constraints of each node in the test behavior tree of Fig. 5. For controllable parameters, three cases should be considered:

Case 1: Node (3) PO?c, $d_2[d_2<3]$
In this case, d_2 is just the time interval between the execution time of this statement and its previous statement corresponding to its parent node in the test behavior tree, so the first method of specifying EET and LET in real-time TTCN can be used: $EET_{(3)} = 0$, $LET_{(3)} = 3$.

Case 2: Node (7) PCO1?B_1, $d_3[2<d_1+d_2+d_3<3]$
In this case, $d_1+d_2+d_3 = T_4 - T_1$, is just the time interval between the execution time of this statement and the statement corresponding to node (a), so the second method of specifying EET and LET in real-time TTCN can be used: Label the statement corresponding to node (a) as L1, then $EET_{(7)} = L1+2$, $LET_{(7)} = L1+3$. See Table 3.

Case 3: Node (2) PO?b, $d_2[1<d_2<3$ and $d_1+d_2\geq2]$
This case is most complicated. In this case, time constraints of the node are $1<d_2<3$ and $d_1+d_2\geq2$: on the one hand, $1<d_2<3$ indicates the time interval between the execution time of this statement and its parent statement; on the other hand, $d_1+d_2=T_3 - T_1$, so $d_1+d_2\geq2$ indicates the time interval between the execution time of this statement

and the statement corresponding to node (a) must be not less than 2 time units. Thus neither the two methods of specifying EET and LET can satisfy the two real-time requirements at the same time. If real-time TTCN should be used in real-time interoperability testing, real-time TTCN must be extended to solve this problem.

For uncontrollable parameters, their values should be set in advance, so the relative time intervals corresponding to such parameters must be fixed values. In this case, EET and LET are the same. For example, in Table 4, line 1 represents the node (a) in the test behavior tree, so its EET and LET are both 1 time unit.

Table 3. Real-time TTCN representation of Node (7) in Fig. 5

Nr	La	Time	TOpt	Behaviour Description	C	V	Comments
				Test Case Dynamic Behaviour			
1	L1			PCO1!A1			Node (a)
2				......			
3		L1+2,L1+3		PCO1?B1		P	Node (7)
4				......			

Table 4. the real-time TTCN test case for the test behavior tree in Fig. 5

Nr	La	Time	TOpt	Behaviour Description	C	V	Comments
				Test Case Dynamic Behaviour			
1	L1	1		PCO1!A1			Node (a)
2	L2	0,2		PO?a			Node (b)
3		1, 2-T(L2)+T(L1)		PO?b			Node (1)
4		L1+2,L1+3	M	PCO1?B1		P	Node (7)
5				PCO1?otherwise		F	Node (9)
6		max(2-T(L2)+T(L1),1), 3		PO?b			Node (2)
7		0,1		PCO1?C1		P	Node (10)
8				PCO1?otherwise		F	Node (12)
9		0,3		PO?c			Node (3)
10		0,1		PCO1?C1		P	Node (13)
11				PCO1?otherwise		F	Node (15)
12		0,1		PO?b		F	Node (4)
13				PO?otherwise		F	Node (6)
14				PO?otherwise		F	

5.3 Possible Extensions of Real-Time TTCN

In this section, we give a suggestion of possible extensions of real-time TTCN for interoperability testing of real-time communication system.

In the **Case** 3 of Section 5.2, time constraints can also be represented as $max(2-d_1,1)<d_2<3$, where the return value of the function **max()** is the maximal value of parameters. With different values of d_1, the values of $max(2-d_1,1)$ are possibly different: in fact, when $d_1<1$, $max(2-d_1,1)=2-d_1$, so time constraints can be represented as $2-d_1<d_2<3$; when $d_1\geq1$, $max(2-d_1,1)=1$, so time constraints are $1<d_2<3$. If the above various cases are distinguished in dynamic behavior descriptions, test cases will become very fussy. To avoid such a problem, a uniform syntax can be used. Here, the concept of absolute time must be introduced in real-time TTCN just like in *TIMED*TTCN-3. Possible extensions for real-time TTCN on the syntactical level can be as follows.

(1) Introduce a timestamp recording function **T**( ): for a lable **L**, **T**(**L**) returns the absolute time value of the execution time for the statement labeled as **L**;
(2) Introduce the third type of method for specifying EET and LET: the meanings of EET and LET are the same with the first type of specifying method, i.e., time constraints of the relative time interval between the execution time of the associated statement and the previous statement corresponding to its parent node; in the expressions of EET and LET, function **T**( ), **max**( ) and **min**( ) can be used.

Proposition 1. In real-time interoperability testing, all time constraints of each node in test behavior trees can be represented as EET and LET by the above syntactical extensions.

Proof. According to **Lemma 1**, on the time point T_k, time constraints of the test event nodes can be represented by a conjunction over a set of linear inequalities on $d_i (i=0,1,\ldots,k-1)$. So on the time point T_{k+1} ($k=0,1,2\ldots$), time constraints of d_k can be reduced to the two following formats: (1) $d_k \le f(d_0,d_1,\cdots,d_{k-1})$ or (2) $d_k \ge g(d_0,d_1,\cdots,d_{k-1})$, here, $f(d_0,d_1,\cdots,d_{k-1})$ and $g(d_0,d_1,\cdots,d_{k-1})$ are both linear expressions on $d_0, d_1,\ldots, d_{k-1}$; thus EET and LET of the statement corresponding to this node can be represented as **max**{ $g(d_0,d_1,\cdots,d_{k-1})$ } and **min**{ $f(d_0,d_1,\cdots,d_{k-1})$ } respectively. Because of $d_i = T_{i+1} - T_i$ ($i=0,1,2,\ldots$), so EET and LET also can be represented as **max**{ $g'(T_0,T_1,\cdots,T_k)$ } and **min**{ $f'(T_0,T_1,\cdots,T_k)$ }, here, $g'(T_0,T_1,\cdots,T_k)$ and $f'(T_0,T_1,\cdots,T_k)$ are expressions by using $T_{i+1} - T_i$ in instead of d_i in $g(d_0,d_1,\cdots,d_{k-1})$ and $f(d_0,d_1,\cdots,d_{k-1})$ respectively, and they are all linear expressions on $T_0, T_1,\ldots, T_k$. If a label is attached to the corresponding statement, T_i ($i=0,1,2,\ldots$) can be get by function **T**( ). Thus EET and LET of the statement can be specified by using function **T**( ), **max**( ) and **min**( ).   □

On the semantics level, we can also refine operational semantics for the syntactical extensions. Before evaluating a set of alternatives, the values of EET and LET for each alternative should be evaluated at first. If the value of EET for one statement is greater than LET, the corresponding statement should be ignored in the process of evaluation, and test execution should not be stopped.

By using these extensions of real-time TTCN, the test behavior tree shown in Fig. 5 can be converted to a real-time TTCN test case in Table 4.

5.4 Comparisons Between *TIMED*TTCN-3 and Real-Time TTCN

Since *TIMED*TTCN-3 is a real-time extension of TTCN-3, it has also the characteristics of TTCN-3. In this section, we compare *TIMED*TTCN-3 with real-time TTCN only from the aspect of the capability of real-time testing. From the discussions in Section 4 and 5, we can see that

(1) *TIMED*TTCN-3 is powerful enough to specify time constraints in real-time in-
    teroperability testing; even more complicated time constraints can be evalu-
    ated easily by retrieving current absolute time, storing its value in a variable
    and passing it to **expression** statements. In real-time TTCN, no concept of ab-
    solute time is introduced; only Time and Time Options columns are added to
    Dynamic Behavior table to specify the earliest and latest execution times,
    which are constraints of time intervals relative to a fixed time point. So
    real-time TTCN cannot specify some complicated real-time requirements; one
    example of such situations has been analyzed in **Case 3** of Section 5.2. To
    remedy this gap, real-time TTCN should be extended both on the syntactical
    and semantic levels.

(2) *TIMED*TTCN-3 is more flexible than real-time TTCN for its style like common
    programming languages in specifying real-time requirements. However, the
    style of real-time TTCN is more compact and formal.

(3) The semantics of *TIMED*TTCN-3 is straightforward and simple, just like a com-
    mon programming language. However, the semantics of real-time TTCN is a lit-
    tle more complicated, especially the two time options are fussy and impenetrable.

(4) *TIMED*TTCN-3 supports both online and offline evaluations, so it has the ca-
    pabilities of evaluating both *hard* and *soft* real-time requirements and it can be
    used not only real-time testing but performance testing. However, real-time
    TTCN has only capabilities of evaluating *hard* real-time requirements.

6   Conclusion

*TIMED*TTCN-3 is a real-time extension of TTCN-3. In this paper, we use
*TIMED*TTCN-3 in real-time interoperability testing. From system specifications, test
behavior trees can be generated. Then transformation rules from such intermediate
notations to *TIMED*TTCN-3 test cases are given. We also investigate a real-time extension of TTCN – real-time TTCN. Since this notation has not enough capabilities of
specifying time constraints in real-time interoperability testing, we extend real-time
TTCN to fill in such a gap and transform test behavior trees to extended real-time
TTCN test cases. From the comparisons between the two real-time test notations, it
can be concluded that *TIMED*TTCN-3 is more powerful and flexible than real-time
TTCN and can be more suitable for real-time interoperability testing.

We have implemented initial prototypes of test execution for both real-time TTCN
and *TIMED*TTCN-3. In our future work, we plan to study real-time interoperability
testing under distributed test architecture and use the *TIMED*TTCN-3 based test system
in real-life timed interoperability testing.

Acknowledgments

This work is partially supported by the National Natural Science Foundation of China
under Grant No. 90104002 and No. 60572082/F010110, and 973 Program of China
under Grant No. 2003CB314801.

References

[AD94]   R. Alur and D. Dill. A theory of timed automata. Theoretical Computer Science, 1994, 126(2): 183–235.

[BB04]   L. B. Briones, E. Brinksma. A Test Generation Framework for quiescent Real-Time Systems. Workshop on Formal Approaches to Testing of Software (FATES) 2004: 64-78.

[BST98]   S.Bornot, J.Sifakis, S. Tripakis. Modeling Urgency in Timed Systems. COMPOS'97, LNCS 1536, Springer Verlag, 1998.

[DGN02]   Z. Dai, J. Grabowski, and H. Neukirchen. Timed TTCN-3 -- A Real-Time Extension for TTCN-3. Testcom2002: 407-424.

[DGN03]   Z. Dai, J. Grabowski, H. Neukirchen. Timed TTCN-3 Based Graphical Real-Time Test Specification. TestCom 2003: 110-127.

[EDK02]   A. En-Nouaary, R. Dssouli, F. Khendek. Timed Wp-method: testing real-time systems. IEEE Transactions on Software Engineering, 2002, 28(11): 1023 -1038.

[ETSY04]   K. El-Fakih, V. Trenkaev, N. Spitsyna and N. Yevtushenko. FSM Based Interoperability Testing Methods for Multi Stimuli Model. TestCom 2004: 60-75.

[GHRS03]   J. Grabowski, D. Hogrefe, G. Réthy, I. Schieferdecker, et al. An introduction to the testing and test control notation (TTCN-3). Computer Networks, 2003, 42(3): 375-403.

[Hao97]   R. Hao. Research on Protocol Conformance and Interoperability Testing based on Formal Methods (In Chinese). PhD thesis, Tsinghua University, P. R. China, 1997.

[HLSG04]   R. Hao, D. Lee, R.K. Sinha and N. Griffeth. Integrated System Interoperability Testing With Applications to VoIP. IEEE/ACM Transactions on Networking, 2004, 12(5): 823-836.

[HNTC99]   T. Higashino, A. Nakata, K. Taniguchi, and A. R. Cavalli. Generating test cases for a timed I/O automaton model. IFIP TC6 12th International Workshop on Testing Communicating Systems, 1999: 197-214.

[KD03]   P. Krémer and S. Dibuz. Framework and Model for Automated Interoperability Test and Its Application to ROHC. Testcom2003: 243 - 257.

[KJM03]   A. Khoumsi, T. Jéron and H. Marchand. Test cases generation for nondeterministic real-time systems. Workshop on Formal Approaches to Testing of Software (FATES) 2003, LNCS 2931: 131-146.

[KSK00]   S. Kang, J. Shin, and M. Kim. Interoperability Test Suite Derivation for Communication Protocols. Computer Networks, 2000, 32(3): 347-364.

[KT04]   M. Krichen and S. Tripakis. Black-Box Conformance Testing for Real-Time Systems. SPIN 2004: 109-126.

[LMN04]   K. Larsen, M. Mikucionis, B. Nielsen. Online Testing of Real-time Systems Using Uppaal. Workshop on Formal Approaches to Testing of Software (FATES) 2004: 79-94.

[NDG04]   H. Neukirchen, Z. Dai, J. Grabowski. Communication Patterns for Expressing Real-Time Requirements Using MSC and Their Application to Testing. TestCom 2004: 144-159.

[RC90]   O. Rafiq and R. Castanet. From conformance testing to interoperability testing. The 3rd Int. Workshop on Protocol Test Systems, 1990.

[SKCK04]   S. Seol, M. Kim, S. T. Chanson, and S. Kang. Interoperability Test Generation and Minimization for Communication Protocols Based on the Multiple Stimuli Principle. IEEE Journal on Selected Areas in Communications (JSAC), 2004, 22(10): 2062-2074.

[SKKJ03]   S. Seol, M. Kim, S. Kang and J. Ryu. Fully Automated Interoperability Test Suite Derivation for Communication Protocols. Computer Networks, 2003, 43(6): 735-759.

[SVD01]   J. Springintveld, F. Vaandrager, and P.R. D'Argenio. Testing Timed Automata. Theoretical Computer Science, 2001, 254(1-2): 225-257.

[TKS03]   V. Trenkaev, M Kim, and S. Seol. Interoperability Testing Based on a Fault Model for a System of Communicating FSMs. TestCom 2003, LNCS 2644: 226–242.

[TTCN]   ITU-T Recommendation X.292 (1998): OSI Conformance Testing Methodology and Framework for Protocol Recommendations for ITU-T Applications—The Tree and Tabular Combined Notation (TTCN). ITU-T, Geneva (Switzerland).

[TTCN3]   ETSI European Standard (ES) 201 873-1 V2.2.1 (2002-08): The Testing and Test Control Notation version 3; Part 1: TTCN-3 Core Language. European Telecommunications Standards Institute (ETSI), Sophia-Antipolis (France), 2002.

[VBT01]   C. Viho, S.Barbin and L. Tanguy. Towards a formal framework for interoperability testing. FORTE 2001: 51-68.

[WG99]   T. Walter, J. Grabowski. A framework for the specification of test cases for real-time distributed systems. Information & Software Technology, 1999, 41(11-12): 781-798.

[WWY04]   Zhiliang Wang, Jianping Wu, Xia Yin. Towards Interoperability Test Generation of Time Dependent Protocols: a Case Study. IEEE Globecom2004, Vol. 2: 589-594.

[WWY06]   Zhiliang Wang, Jianping Wu and Xia Yin. A Formal Framework to Interoperability Testing for Real-time Systems. Submitted.

Test Generation for Network Security Rules

Vianney Darmaillacq[1], Jean-Claude Fernandez[2], Roland Groz[1],
Laurent Mounier[2], and Jean-Luc Richier[1,*]

[1] Laboratoire LSR-IMAG,
BP 72, 38402 St Martin d'Hères, France
[2] Laboratoire Vérimag,
Centre Equation - 2, avenue de Vignate, 38610 Gières, France
{Vianney.Darmaillacq, Jean-Claude.Fernandez, Roland.Groz,
Laurent.Mounier, Jean-Luc.Richier}@imag.fr

Abstract. Checking that a security policy has been correctly deployed over a network is a key issue for system administrators. Since policies are usually expressed by rules, we propose a method to derive tests from a set of rules with a single modality. For each element of our language and each type of rule, we propose a pattern of test, which we call a tile. We then combine those tiles into a test for the whole rule.

1 Introduction

Network and system administrators are in charge of implementing and controlling the security policies of their organisations. Enforcing a policy typically relies on the adequate configuration of Policy Enforcement Points (PEP) in dedicated equipment (such as firewalls, ciphering chips) or specific software (e.g. account managers, mail scanners). Since most networks and systems would undergo daily changes, maintaining the consistency of the rules actually implemented by the PEPs and their conformance to the specified security objectives and the rules expressed in the policy is not an easy task.

One way to guarantee a correct policy enforcement is to derive configurations from a description of the policy (top-down approach). In order to do that systematically, the description must be formal enough to provide unambiguous rules and translations of them into configurations of security devices. Since most policies consist of combination of rules with various scopes and potentially conflicting modalities (e.g. restricting access for generic subjects but authorising it for specific categories), the descriptions usually include constructs or semantic rules to solve conflicts between policy statements. Ponder [5] and OrBAC [1] are typical such description languages with associated methods to allow deployment on networks.

In this paper, we investigate another approach. We consider that testing a given network configuration for compliance with a stated policy is a kind of conformance testing. Therefore, we aim at deriving tests from a formal specification

---

* This work has been partly funded by the POTESTAT project of the national research programme on security (ACI Sécurité) and by the IMAG project Modeste.

M.Ü. Uyar, A.Y. Duale, and M.A. Fecko (Eds.): TestCom 2006, LNCS 3964, pp. 341–356, 2006.
© IFIP International Federation for Information Processing 2006

of the security policy to check whether the implementation is correct. These tests could be used either after some initial deployment of a policy to check whether it can be and actually is well implemented, or typically on a more regular basis to see whether any update on the configurations might have breached security rules. Some sort of testing is actually performed by system administrators to check for known vulnerabilities: portscans and password crackers fall into that category. However, such tests are usually limited to just a single security mechanism. With existing techniques, it is difficult to address the issue of consistency of configurations on distributed devices. Although some work has been published on the analysis of the consistency of firewall rules (typically to minimise them or to detect conflicts), this is still limited to specific points of security policies. [12] is an application of the idea of generating conformance tests for a single firewall.

In our work, we address conformance with respect to a global specification of a security policy for a network of interconnected systems. Although at first view this might appear as just another instance of conformance testing of an implementation w.r.t. a given formal specification, there are a number of significant differences with the framework of such standards as IS9646 and its formal counterpart FMCT [8]. We investigated a few of these differences in [6]: typically, protocol conformance testing is done on a well defined protocol level, whereas security policies are often implemented with a mixture of mechanisms at various levels of communication and O/S interfaces. One major issue for test generation is that we cannot expect to start from an updated comprehensive model of the system: this is why we adopt a method based on (security) requirements.

In this paper, we focus on a method to derive tests from a policy expressed as a set of rules. More precisely, we consider an LTL-like specification language that can be used to formalise a reasonable subset of the rules usually found in network security policies. This restricted language allows us to design a "tile-based" generation method. For each element of our language and each type of rule, we propose a pattern of test, which we call a tile. The simple form of rules makes it possible to compose a global test by simply combining the tiles associated to the elements.

This paper is organised as follows. In section 2, we present our approach, in particular the types of rules that we cover and their relation to proposed formal methods for the description of security policies. Our notation for rules is presented in section 3. In section 4, we define the test generation tiles and the algorithm to derive tests. The whole approach is illustrated in section 5 on a typical example taken from an e-mail security architecture. Finally, in section 6 we present our perspectives for development from this basis.

2   Approach

Our aim is to automatically generate test cases from network policy rules. We intend to adapt the approach used in protocol conformance testing. We first started by looking for a formal description of security policies that would make it possible to generate tests. In order to identify typical requirements and the corresponding tests, we worked from a case study.

2.1   A Case Study

We carried out during the summer of 2004 a case study to identify the security policies used in the IMAG network, which connects our laboratories. This study included the analysis of documents provided to the administrators and the users, and interviews. We collected information at different levels of management, from the laboratory to the access supplier. The analysis resulted in a rather broad and detailed description of a typical network security policy in a university environment. In this paper, we shall present a small subset of rules, centred on electronic mail. This set was selected to be rich enough to show the majority of the concepts to be studied: several levels of policies and inter-connected organisations, variety of the services and of the access methods.

No	Requirement
1	Mail relays accepting messages from the exterior should be located at the entry of the network, in the DMZ if possible.
2	There should be no user account on hosts located in the DMZ.
3	Mailbox servers containing user accounts should be in the private zone. There can be as many of these servers as necessary.
4	Relays in the DMZ are the only machines allowed to communicate with the exterior world using the SMTP protocol. Relay of inbound mails (to mailboxes) and of outbound mails (to exterior) is done using these relays.
5	At the entry of the site, a filtering default policy is applied, which forbid all traffic not explicitly authorised.
6	It is forbidden to hosts of the network to relay mails from an external host to an external host.
7	All messages coming from the exterior are redirected to mail relays located at the entry of the site (using DNS MX records), probably in the DMZ.
8	It is forbidden to communicate with a host belonging to the blacklist updated daily provided by the MAPS (Mail Abuse Prevention System) partner.
9	Antivirus and spam filters shall be installed on hosts acting as relays or mailboxes.
10	A mail shall be checked by antivirus software before being opened.
11	All mails entering the network infected by a virus shall be disinfected.
12	All mail shall be modified if it contains a potentially dangerous attached file. The original file is kept somewhere in the network. Recipients are notified.
13	All mail analysed as a possible spam shall be flagged.

Fig. 1. Security rules for e-mail

In a network, security distinguishes inside and outside. The inside of the network is the set of machines under the responsibility of the organisation. The outside consists of uncontrolled machines considered dangerous for security purposes. This can be refined by considering other zones, differing by degrees of administration, reliability and trust. The sub-zones of the internal zone correspond in general to architectural criteria, for example separations between physical sub-networks, whereas the outside sub-zones correspond to different levels of trust. One distinguishes moreover among the internal hosts those which provide services accessible from the outside. Due to their visibility, these hosts are more often subject to attacks. Therefore the standard approach is to define for these hosts a strongly controlled buffer zone, often called DMZ (demilitarized zone). Only the hosts in this zone can communicate with the outside world, and all the traffic between the DMZ and the rest of the internal network is controlled.

We give in figure 1 a sample set of rules for the electronic mail drawn from our case study. This sample illustrates: flows of information, separation in zones, possibility, obligation or prohibition of certain actions, at different levels of details. To simplify the problem, we suppose that certain global hypotheses are true and do not have to be clarified: correct routing, systems up to date w.r.t vulnerability patches ...

2.2   Description Techniques for Network Security

Most of the rules in the case study (which actually covers much more than e-mail) express some constraints about the possible behaviour of the system. More specifically, they are of the form "*Mod P*", where *Mod* is a modality among obligation, permission or interdiction, and P is either a predicate on the system or a behaviour.

The deontic logic of von Wright [14] is a modal logic whose modal operator semantic is that of obligation. With only this operator authorisation of a formula is defined as the negation of the obligation of the negation of the formula, while interdiction is defined as the negation of the obligation of the formula. However deontic logic raises a number of paradoxes that have hindered its wider use [10]. Nevertheless modalities are a key issues in formal models of security policies, in particular in formal description techniques such as PDL, Ponder and Or-BAC, which have been proposed to address network security policies.

PDL is a language created to model network management policies, including security requirements. It is based on the idea that a policy is a specification of the behaviour the network should have according to what happened in the network [9]. A rule states that an action is triggered by an event, provided a condition holds. PDL was used to monitor switches in a network, to guarantee quality of service [13].

Ponder is a language created to specify network security policies [5] and proposes a full choice of different modalities: conditional obligation, authorisation, interdiction, delegation and refrain policies could be specified.

Or-BAC [1] is another access control model, loosely based on RBAC. While RBAC abstracts subjects into roles, Or-BAC abstracts moreover objects into views and actions into activities, and introduces the notion of organisation. Or-BAC has been used to model a network security policy, and to generate firewall rules from it [4, 2].

All these formalisms include structuring and typing constructs, resolution of conflict mechanisms and various aspects which are important. In this paper, we go for a simpler course, since we want to derive tests from the rules. These rules could be extracted from description in the above formalisms. All we need is a description with just enough expressive power to represent the modalities used in PDL, Ponder or Or-BAC, at least as they can be tested through network events.

2.3 Approach for Test Generation

Compared to classical conformance testing for communication protocols, test generation in our case exhibits two major differences.

- The policy is not described by a comprehensive model such as a global LTS or state machine, but by a collection of rules. This is similar to deriving tests from requirements. Much work has been done in test generation from test purposes which are confronted to a formal model of the protocol. Here, we do not assume any formal model of the network, we derive tests from the rules. Our approach is rule based: it generates separate tests for each rule.
- The policy is described at a much higher level than the actual events that can be observed or controlled in the network; whereas the specification of a communication protocol would refer to PDUs or SDUs even though some of them might occur at non-observable interfaces. We need to establish a correspondence between the basic predicates appearing in a rule and sequences of events that can represent a test or instances of such predicates.

To each high-level predicate (such as $externRelay(h)$ meaning that machine h can relay mails sent from outside the domain considered) we associate a test pattern which we call a *tile*. For instance, in the case of external relaying of mails, a tester would have to establish a SMTP connection to the machine from an external machine and try to send a mail. However, there are different types of predicates. Some may have to be tested dynamically through interaction with the system. Others might be checked without PDU exchanges, for instance if we have access to the configuration files of the system when the test is set up. In the case of $externRelay(h)$, this could be checked in the configuration of the mail system on h. The choice of one method or another may depend on accessibility to the system, but on trust as well: typically, information on configurations might not be reliable.

In [7], we investigated a refinement approach to derive control at PDU level from security rules. In this paper, we will consider that the tiles are provided. A policy would be described by combining elementary predicates which would be well-known elements for security policies, so that the corresponding tiles would be readily available.

In this paper, we concentrate on the combination of tiles, based on the structure of the formula in the rule. We address formulas of a restricted form with just one modality, as this corresponds to the usual style of security policy rules; Ponder, PDL and OrBAC also propose a single modality on each rule. We first propose a formal description of the rules in the next section, then we describe the combination of tiles in the following one.

3 Rules Formalisation

We first give the syntax and semantics of a rather general formalism and then we restrict it to a smaller subset used in this work to generate test cases.

Preliminaries. A *labelled transition system* (LTS, for short) is a quadruplet (Q, A, T, q^0) where Q is a set of states, A a set of labels, T the transition relation $(T \subseteq Q \times A \times Q)$ and q^0 the initial state $(q^0 \in Q)$. We will use the following definitions and notations: $(p, a, q) \in T$ is noted $p \xrightarrow{a}_T q$ (or simply $p \xrightarrow{a} q$). An *execution sequence* ρ is a composition of transitions: $q^0 \xrightarrow{a_1} q_1 \xrightarrow{a_2} q_2 \cdots \xrightarrow{a_n} q_n$. We denote by σ^ρ (resp. α^ρ) the sequence of states (resp. observables actions) associated with ρ. The sequence of actions α^ρ is called a *trace*. We note by Σ_S, the set of finite execution sequences starting from the initial state q^0 of S. For any sequence λ of length n, λ_i or $\lambda(i)$ denotes the i-th element and $\lambda_{[i \ n]}$ denotes the suffix $\lambda_i \cdots \lambda_n$.

We also consider in the sequel Boolean Labelled Transition Systems in order to obtain a more compact representation of test cases. A Boolean Labelled Transition System (BLTS for short) is a tuple (X_b, Q, A, T, q^0) where X_b is a set of Boolean variables, Q is a set of states, A is a set of actions, q^0 is the initial state and $T \subseteq Q \times (Bexp \times A \times Bcmd) \times Q$ is the set of transitions, where:

- *Bexp* is a *guard*, i.e. a Boolean expression constructed using the following grammar b ::= True | x | False | b ∧ b | ¬ b (where x∈ X_b);
- *Bcmd* is either an assignment x := b (where x∈ X_b, b ∈ *Bexp*) or the null command skip.

As usual, we note $p \xrightarrow{(b,a,c)} q$ for $(p, (b, a, c), q) \in T$. We can omit b (*resp.* c) when b is True (*resp.* c is skip).

The semantics of a BLTS is given by a LTS. We define a notion of configuration (p, γ), where p is a state of the BLTS and $\gamma : X_b \mapsto Bool$ a valuation, where $Bool = \{True, False\}$, is the set of Boolean values. Valuation γ are extended to *Bexp* in the usual way (i.e., $\gamma : Bexp \mapsto Bool$). Given a BLTS $B = (X_b, Q, A, T, q^0)$ and the initial valuation γ_0, where $\gamma_0(x) = False$ for all $x \in X_b$, the underlying LTS $S_B = (Q_1, A, T_1, q_1^0)$ is defined as follows:

$$Q_1 \subseteq Q \times Bool^{X_b},$$
$$(p, \gamma) \xrightarrow{a} (p_1, \gamma_1) \text{ iff } (p, (b, a, c), q) \in T \text{ and } \gamma(b) = True,$$
$$\gamma_1 = \begin{cases} \gamma[v/x] & \text{if } c \text{ is } x:=e, \gamma(e) = v \text{ and } \gamma(b) = true \\ \gamma & \text{if } c \text{ is skip} \end{cases}$$
$$q_1^0 = (q^0, \gamma_0)$$

3.1   Syntax of Security Rules

The formalism we adopt to express security policy rules is based on a variant of the LTL temporal logic [11], with only the $\mathcal{F}$ and $\mathcal{G}$ modalities, and mixing state-based and event-based atomic predicates. Each rule is expressed by a logical *formula* (φ), built upon *literals*. Each literal can be either a *condition literal* $(p_c \in P_c)$, or an *event literal* $(p_e \in P_e)$. A conjunction of condition literals is simply called a *condition* (C), whereas a conjunction of a single event literal and a condition is called a *(guarded) event* (E). Finally, we also use a modal operator $\mathcal{F}$, the dual one $\mathcal{G}$, and the usual Boolean connectors ¬ and ⇒.

The abstract syntax of a formula is then given by the following grammar:

$$\varphi ::= C \mid E \mid \neg\varphi \mid \varphi \Rightarrow \varphi \mid \mathcal{F}\varphi \mid \mathcal{G}\varphi$$
$$C ::= p_c^1 \wedge \cdots \wedge p_c^n \quad \text{where } p_c^i \in P_c$$
$$E ::= p_e[C] \qquad \text{where } p_e \in P_e$$

3.2   Semantics

Formulas are interpreted over LTS. Intuitively, an LTS S satisfies a formula φ iff *all* its execution sequences ρ do, where condition literals are interpreted over *states*, event literals are interpreted over *labels* and the modal operator $\mathcal{F}\varphi$ means that *there exists* a suffix $\rho_{[i..|\rho|]}$ of ρ such that φ holds on $\rho_{[i..|\rho|]}$, where $|\rho|$ is the number of elements of ρ. We first introduce two interpretation functions for condition and event literals:

$f_c : P_c \rightarrow 2^Q$, associates to p_c the set of states on which p_c holds;
$f_e : P_e \rightarrow 2^A$, associates to p_e the set of labels on which p_e holds.

The satisfaction relation of a formula φ on an execution sequence ρ ($\rho \models \varphi$) is then (inductively) defined as follows:

- $\rho \models C$ for $C = p_c^1 \wedge \cdots \wedge p_c^n$ iff $\forall i.\ \sigma^\rho(1) \in f(p_c^i)$
- $\rho \models E$ for $E = p_e[C]$ iff $\alpha^\rho(1) \in g(p_e) \wedge \sigma^\rho(2) \models C$
- $\rho \models \neg\varphi$ iff $\rho \not\models \varphi$
- $\rho \models \varphi_1 \Rightarrow \varphi_2$ iff $((\rho \models \varphi_1) \Rightarrow (\rho \models \varphi_2))$
- $\rho \models \mathcal{F}\varphi$ iff $\exists i \in [1, |\rho|].\ \rho_{[i\ \ |\rho|]} \models \varphi$
- $\rho \models \mathcal{G}\varphi$ iff $\forall i \in [1, |\rho|].\ \rho_{[i\ \ |\rho|]} \models \varphi$

Finally, $S \models \varphi$ iff $\forall \rho \in \Sigma_S.\ \rho \models \varphi$.

3.3   Expression of Security Rules

Our purpose is to use the specification language defined in the previous paragraph to express *security rules* to be satisfied by a network. In this particular context its semantics should be interpreted as follows:

- The network behaviour is expressed by the LTS S: each state of S represents the global state of the network at a given time (network configuration and topology, transiting PDUs, etc.), and each label of S represents an observable action performed at the network level (modification of the configuration/topology, PDU reception, PDU emission, etc.).
- A condition literal p_c expresses a (*static*) predicate on a network state, at the security policy level. For instance *externRelay*($h1$) holds on a state iff machine $h1$ is configured as an external mail relay in this state, or *infected*($m1$) holds on a state iff message $m1$ contains a virus.
- An event literal p_e expresses a (*dynamic*) predicate on a network transition, from a given state, at the security policy level. For instance *enterNetwork*(m) holds iff the current transition corresponds to reception of mail m by the network, and *chkVirus*(m) holds iff the current transition corresponds to a virus check on mail m.
- The $\mathcal{F}$ operator is used here to express an *obligation*, meaning that a given formula *should eventually* hold later, in a bounded future. For instance

enterNetwork(*m*) $\Rightarrow$ $\mathcal{F}$*chkVirus*(*m*) means that whenever mail *m* enters the network then it should be checked.

As a matter of fact, it happens that all the formulas found in the case study could be expressed using only a restricted subset of this language. In particular formulas can be classified into three types, according to the following grammar:

$$\varphi ::= \mathcal{G}\, \mathcal{C}-\mathbf{Rule} \mid \mathcal{G}\, \mathcal{F}-\mathbf{Rule} \mid \mathcal{G}\, \mathcal{G}-\mathbf{Rule}$$
$$\mathcal{C}-\mathbf{Rule} ::= C \Rightarrow C \mid E \Rightarrow C$$
$$\mathcal{F}-\mathbf{Rule} ::= E \Rightarrow \mathcal{F}E$$
$$\mathcal{G}-\mathbf{Rule} ::= C \Rightarrow \mathcal{G}\neg E \mid E \Rightarrow \mathcal{G}\neg E$$

A $\mathcal{C}-\mathbf{Rule}$ expresses a static conditional implication, an $\mathcal{F}-\mathbf{Rule}$ expresses a (triggered) obligation and a $\mathcal{G}-\mathbf{Rule}$ expresses that, when a given condition holds or when a given event happens, then a particular event is always *prohibited*.

4   Test Generation

In this section, we propose a "tile-based" approach to generate *abstract test cases* from a formula expressing a security rule. The test generation principle is the following: assuming that elementary test cases (i.e., *tiles*) t_i are provided for each (condition and event) literals appearing in a formula φ, the test case t associated to φ is obtained by combining test cases t_i with "test operators" (defined below), corresponding to the logical operators appearing in φ. This allows defining a structural correspondence between formulas and test cases.

4.1   Test Cases and Test Execution

We can define the notion of test case as a BLTS extended with two special actions (to deal with *timers*), and three special states (called *verdicts*): action *timerset* means a timer initialisation to a given value, and action *timeout* means the timer expiration; verdict states are "sink states", indicating the end of a successful (*pass*), unsuccessful (*fail*) or inconclusive (*inconc*) test execution. We denote by A^t the set $A \cup \{timeout, timerset\}$, and by V the set $\{pass, fail, inconc\}$. We denote by Σ_S^{pass} (resp. Σ_S^{fail}, $\Sigma_S^{\mathrm{inconc}}$) the set of execution sequences, starting from the initial state and ending in the state *pass* (resp. *fail*, *inconc*). A *test case t* is then a BLTS $t = (X_p, Q, A^t, T, q^0, V)$, with $V \subseteq Q$.

A test case is supposed to be executed by a tester against a network whose behaviour can be modelled by an LTS $I = (Q^I, A^I, T^I)$ (we ignore initial states for the network). Usually, in "black-box" conformance testing, this behaviour is observed/controlled by the tester only through a restricted interface. For the sake of simplicity we assume here that *any* output action (*resp.* input action) performed by the network can be observed (*resp.* controlled) by the tester. Thus, execution of a test t on an IUT I, noted $\mathrm{Exec}(t, I)$, is expressed as a set of common execution sequences of S_t and I, defined by a composition operator $\otimes$:
Let $\rho_I = q_0^I \xrightarrow{a_1} q_1^I \xrightarrow{a_2} q_2^I \cdots \xrightarrow{a_n} q_n^I \cdots \in \Sigma_I$ and $\rho_{S_t} = q^{0,t} \xrightarrow{a_1} q_1^t \xrightarrow{a_2} q_2^t \cdots \xrightarrow{a_n} q_n^t \in \Sigma_{S_t}$, then

$$\rho_{S_t} \otimes \rho_I = (q^{0,t}, q_0^I) \xrightarrow{a_1} (q_1^t, q_1^I) \cdots \xrightarrow{a_n} (q_n^t, q_n^I) \in \mathrm{Exec}(t, I).$$

For $\rho \in \mathrm{Exec}(t, I)$, we define the verdict function: $\mathrm{VExec}(\rho) = pass$ (resp. $fail, inconc$) iff there is $\rho_{S_t} \in \Sigma_{S_t}^{\mathrm{pass}}$ (resp $\Sigma_{S_t}^{\mathrm{fail}}, \Sigma_{S_t}^{\mathrm{inconc}}$) and $\rho_I \in \Sigma_I$ such that $\rho_{S_t} \otimes \rho_I = \rho$.

4.2 Test Generation Functions

Let φ be a formula, let p_e, p_c^i its literals, and t_{p_e}, $t_{p_c^i}$ their corresponding elementary test cases. Note that an elementary test case is reduced to a simple verdict state when it corresponds to a literal that can be immediately checked on the network behaviour (without requiring any interaction sequence with a tester). The test generation function gentest(φ) is inductively defined on the syntax of the formula, where X_1 denotes either a condition C or an event E, and test operators $\triangleright_{\mathrm{rrF}}$, $\triangleright_{\mathrm{lrF}}$, $\triangleright_{\mathrm{rrC}}$, $\triangleright_{\mathrm{lrC}}$, $\triangleright_{\mathrm{llX}}$ and Inv are defined below:

$$\mathrm{gentest}(X_1 \Rightarrow C_2) = \mathrm{gentest\_lX}(X_1) \triangleright_{\mathrm{lrC}} \mathrm{gentest\_rC}(C_2)$$

$$\mathrm{gentest}(E_1 \Rightarrow \mathcal{F}E_2) = \mathrm{gentest\_lX}(E_1) \triangleright_{\mathrm{lrF}} \mathrm{gentest\_rF}(E_2)$$

$$\mathrm{gentest}(X_1 \Rightarrow \mathcal{G}\neg E_2) = Inv(\mathrm{gentest\_lX}(X_1) \triangleright_{\mathrm{lrF}} \mathrm{gentest\_rF}(E_2))$$

$$\mathrm{gentest\_lX}(p_c^1 \wedge \cdots \wedge p_c^n) = (((t_{p_c^1} \triangleright_{\mathrm{llX}} t_{p_c^2}) \triangleright_{\mathrm{llX}} \dots) \triangleright_{\mathrm{llX}} t_{p_c^n})$$

$$\mathrm{gentest\_lX}(p_e[C]) = t_e \triangleright_{\mathrm{llX}} \mathrm{gentest\_lX}(C)$$

$$\mathrm{gentest\_rC}(p_c^1 \wedge \cdots \wedge p_c^n) = (((t_{p_c^1} \triangleright_{\mathrm{rrC}} t_{p_c^2}) \triangleright_{\mathrm{rrC}} \dots) \triangleright_{\mathrm{rrC}} t_{p_c^n})$$

$$\mathrm{gentest\_rC}(p_e[C]) = t_e \triangleright_{\mathrm{rrC}} \mathrm{gentest\_rC}(C)$$

$$\mathrm{gentest\_rF}(p_e[C]) = t_e \triangleright_{\mathrm{rrF}} \mathrm{gentest\_rC}(C)$$

4.3 Test Operators

In the following we assume that $t_1 = (X_{b_1}, Q_1, A_1, T_1, q_1^0, \{pass_1, fail_1, inconc_1\})$ and $t_2 = (X_{b_2}, Q_2, A_2, T_2, q_2^0, \{pass_2, fail_2, inconc_2\})$ are two test cases. For each binary test operator $\triangleright$ we define the test case $t = (X_b, Q, A, T, q^0, V)$ such that $t = t_1 \triangleright t_2$ and $V = \{pass, fail, inconc\}$. For each operator, a graphical presentation is proposed on figure 2 and we give hereafter the formal definition of only three operators, the three others being similar.

Operator $\triangleright_{\mathrm{llX}}$ ($t = t_1 \triangleright_{\mathrm{llX}} t_2$). This operator is used to combine two test cases appearing on the left-hand side of an implication. Therefore, t *pass* iff t_1 and t_2 does, and t is *inconclusive* otherwise (the entire formula cannot be tested). More formally:

$$X_b = X_{b_1} \cup X_{b_2},$$
$$Q = (Q_1 \setminus V_1) \cup (Q_2 \setminus V_2) \cup V,$$
$$A = A_1 \cup A_2,$$
$$q^0 = q_1^0,$$
$$\begin{aligned}
T = \; & T_1 \setminus \{(p, a, q) \mid q \in V_1\} \cup T_2 \setminus \{(p, a, q) \mid q \in V_2\} \\
& \cup \{(p, a, q_2^0) \mid (p, a, pass_1) \in T_1\} \;\cup\; \{(p, a, pass) \mid (p, a, pass_2) \in T_2\} \\
& \cup \{(p, a, inconc) \mid (p, a, q) \in T_1 \wedge q \in \{inconc_1, fail_1\}\} \\
& \cup \{(p, a, inconc) \mid (p, a, q) \in T_2 \wedge q \in \{inconc_2, fail_2\}\}
\end{aligned}$$
$$\quad (\alpha)$$

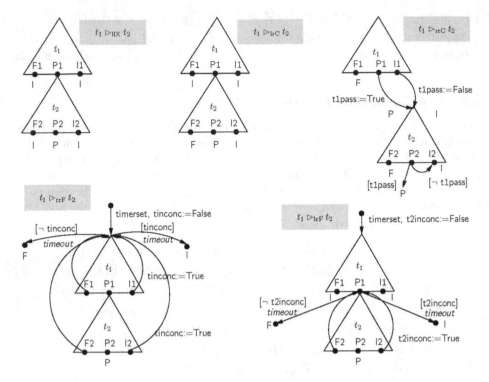

Fig. 2. Test operators

Operator $\triangleright_{\mathrm{lrC}}$ $(t = t_1 \triangleright_{\mathrm{lrC}} t_2)$. This operator is used to combine the left-hand side (t_1) and the right-hand side (t_2) of an implication for a $\mathcal{C}-$**Rule**. Therefore t can be expressed by a sequential execution of t_1 and t_2, and it *pass* iff t_1 *and* t_2 does, it *fails* when t_1 pass and t_2 fails (implication), and it is *inconclusive* otherwise. Formal definition of test t is then simply obtained by replacing the line (α) of the previous definition (operator $\triangleright_{\mathrm{llX}}$) by the following one:

$$\{(p, a, inconc) \mid (p, a, inconc_2) \in T_2\} \cup \{(p, a, fail) \mid (p, a, fail_2) \in T_2\}$$

Operator $\triangleright_{\mathrm{rrC}}$ $(t = t_1 \triangleright_{\mathrm{rrC}} t_2)$. This operator is used to combine two test cases appearing on the right-hand side of an implication, for a $\mathcal{C}-$**Rule**. Therefore t *pass* when t_1 *and* t_2 does, t *fails* when t_1 *or* t_2 fails, and it is *inconclusive* otherwise. Thus, t can be obtained by executing t_1 first, followed by t_2 (when t_1 does not fail). A Boolean variable $\texttt{t1pass}$ is used to store the verdict of t_1.

Operator $\triangleright_{\mathrm{lrF}}$ $(t = t_1 \triangleright_{\mathrm{lrF}} t_2)$. This operator is used to combine the left-hand side (t_1) and the right-hand side (t_2) of an implication for an $\mathcal{F}-$**Rule**. It is therefore similar to the $\triangleright_{\mathrm{lrC}}$ operator, excepted that, due to the $\mathcal{F}$ temporal modality, t *pass* iff t_1 pass and t_2 pass *at some point later in the future* (remember that the right-hand side of an $\mathcal{F}-$**Rule** is necessarily an *event*). t is then obtained by executing t_1 first, and then repeatedly executing t_2 until either it passes or a given timeout is reached (to ensure that the test execution remains finite). A Boolean variable `t2inconc` is used to keep track of an occurrence of an inconclusive verdict of t_2 (during its repeated execution), hence leading to an inconclusive verdict of t. More formally:

$$
\begin{aligned}
X_b &= X_{b_1} \cup X_{b_2} \cup \{\texttt{t2inconc}\}, \\
Q &= (Q_1 \setminus V_1) \cup (Q_2 \setminus V_2) \cup V \cup \{q_0\}, \\
A &= A_1 \cup A_2, \\
T &= T_1 \setminus \{(p, a, q) \mid q \in V_1\} \cup T_2 \setminus \{(p, a, q) \mid q \in V_2\} \\
&\quad \cup \{(q_0, (timerset, \texttt{tinconc} := \texttt{False}), q_1^0)\} \\
&\quad \cup \{(p, a, inconc) \mid (p, a, q) \in T_1 \wedge q \in \{inconc_1, fail_1\}\} \\
&\quad \cup \{(p, a, q_2^0) \mid (p, a, pass_1) \in T_1\} \\
&\quad \cup \{(p, a, q_2^0) \mid (p, a, fail_2) \in T_2\}\} \ \cup \ \{(p, a, pass) \mid (p, a, pass_2) \in T_2\} \\
&\quad \cup \{(p, (a, \texttt{t2inconc} := \texttt{True}), q_2^0) \mid (p, a, inconc_2) \in T_2\} \\
&\quad \cup \{(q_2^0, (\neg\texttt{t2inconc}, timeout), fail)\} \cup \{(q_2^0, (\texttt{t2inconc}, timeout), inconc)\}
\end{aligned}
$$

Operator $\triangleright_{\mathrm{rrF}}$ ($t = t_1 \triangleright_{\mathrm{lrF}} t_2$). This operator is used to combine two test cases appearing on the right-hand side of an implication for an $\mathcal{F}-$**Rule**, where t_1 tests the occurrence of an *event literal* p_e and t_2 a (static) *condition* C (possibly restricting p_e). Therefore, t *pass* iff both the expected event occurs at some point (t_1 pass) *and* condition C holds on the same time. t is then obtained by repeating t_1 followed by t_2 until both pass. Here again, a timeout ensures that execution of t always remains finite, and a Boolean variable `tinconc` is used to keep track of an inconclusive verdict of t_1 or t_2.

Operator *Inv* ($t = Inv(t_1)$). This operator simply "reverts" the *pass* and *fail* verdicts produced by a test case.

4.4 Soundness of the Test Generation Function

It now remains to establish that an abstract test case produced by function $\mathrm{gentest}(\varphi)$ is always *sound*, i.e., it delivers a *fail* verdict when executed on a network behaviour I only if formula φ does not hold on I.

Two hypotheses are required in order to prove this soundness property:

H1. First, for any formula φ, we assume that each elementary test case t_i provided for the (event or condition) literals p_i appearing in φ is *strongly sound* in the following sense:

$$
\forall \rho \in \mathrm{Exec}(t_i, I) \cdot (\mathrm{VExec}(\rho) = Pass \Rightarrow \rho \models p_i) \wedge (\mathrm{VExec}(\rho) = Fail \Rightarrow \rho \not\models p_i)
$$

H2. Second, we assume that the whole execution of a (provided or generated) test case t associated to a condition C is *stable* with respect to condition literals: the valuation of these literal does not change during the test execution. This simply means that the network configuration is supposed to remain stable when a condition is tested. Formally:

$$
\forall p_i \in P_c. \ \forall \rho \in \Sigma_I \cdot \rho_{S_t} \otimes \rho \in \mathrm{Exec}(t, I) \Rightarrow (\sigma^\rho \subseteq f_c(p_i) \vee \sigma^\rho \cap f_c(p_i) = \emptyset)
$$

where σ^ρ denotes here tacitly a set of states instead of a sequence.

We now formulate the soundness property:

Proposition: Let φ a formula, I an LTS and $t = \mathrm{gentest}(\varphi)$. Then:

$$\rho \in \mathrm{Exec}(t, I) \wedge \mathrm{VExec}(\rho) = fail \Longrightarrow I \not\models \varphi.$$

The proof of this proposition relies on the following lemma:

Lemma 1. Test cases generated by auxiliary function gentest_lX are *strongly sound*, and test cases generated by auxiliary functions gentest_rC and gentest_rF are *sound*.

Proof sketch of Lemma 1. Let t a test case generated by function gentest_lX. The proof that t is *strongly sound* is performed by recurrence on the number of elementary tests cases t_i appearing in t (assuming that each t_i itself is *strongly sound* according to hypothesis **H1**). A similar proof can be done for functions gentest_rC and gentest_rF.

Proof of Proposition. By structural induction on the formulas φ (we only detail here some representative induction steps).

- $\varphi = C_1 \Rightarrow C_2$. By definition of function gentest there exists test cases t_1 and t_2 such that $t_1 = \mathrm{gentest\_lX}(C_1)$, $t_2 = \mathrm{gentest\_rC}(C_2)$, and $t = t_1 \rhd_{\mathrm{lrC}} t_2$. Let ρ be an execution sequence of $\mathrm{Exec}(t, I)$ such that $\mathrm{VExec}(\rho) = fail$. Then, by definition of operator $\rhd_{\mathrm{lrC}}$, there exist ρ_1 and ρ_2 such that: $\rho = \rho_1.\rho_2$, $\rho_1 \in \mathrm{Exec}(t_1, I)$, $\rho_2 \in \mathrm{Exec}(t_2, I)$, $\mathrm{VExec}(\rho_1) = pass_1$ and $\mathrm{VExec}(\rho_2) = fail_2$. Therefore, by Lemma 1, $\rho_1 \models C_1$, hence $\sigma^\rho(1) \models C_1$ and similarly $\rho_2 \not\models C_2$ and $\sigma^\rho(|\rho_1|) \not\models C_2$. By hypothesis **H2** we obtain $\sigma^\rho(1) \not\models C2$ and then $\sigma^\rho(1) \not\models \varphi$.
- $\varphi = E_1 \Rightarrow \mathcal{F}E_2$. By definition of function gentest there exist test cases t_1 and t_2 such that $t_1 = \mathrm{gentest\_lX}(E_1)$, $t_2 = \mathrm{gentest\_rF}(E_2)$, and $t = t_1 \rhd_{\mathrm{lrF}} t_2$. Let ρ be an execution sequence of $\mathrm{Exec}(t, I)$ such that $\mathrm{VExec}(\rho) = fail$. Then, by definition of operator $\rhd_{\mathrm{lrF}}$, there exist ρ_1, ρ_2 such that: $\rho = q_0 \xrightarrow{a_0} q_0^1 \xrightarrow{\rho_1} pass_1(\xrightarrow{\rho_2} fail_2)^* \xrightarrow{a_1} q_0^2 \xrightarrow{a_2} fail$, with $a_0 = (timerset, \mathtt{t2inconc} := \mathtt{False})$, $a_1 = \mathtt{t2inconc} := \mathtt{False}$, and $a_2 = (timeout, \neg\mathtt{t2inconc})$. Moreover, $\rho_1 \in \mathrm{Exec}(t_1, I)$ and $\rho_2 \in \mathrm{Exec}(t_2, I)$. Therefore (by Lemma 1) $\sigma^\rho(1) \models E_1$ and, for all i in $[|\rho_1|, |\rho|]$, $\sigma^\rho(i) \not\models E_2$. We conclude that $\sigma^\rho(1) \not\models (E_1 \Rightarrow \mathcal{F}E_2)$.

5   Case Study Application

This section shows how the approach presented above can be applied to generate concrete tests for some examples from the case study of section 2.1.

5.1 $\mathcal{C}-$Rule

Consider the requirement "*External relays shall be in the DMZ*", which can be modelled by the $\mathcal{C}-$**Rule**:

$$externRelay(h) \Rightarrow inDMZ(h)$$

The goal of the test is to verify that each external relay is in the DMZ. As noted in section 4.2, an elementary test case is reduced to a simple verdict state when it corresponds to a literal that can be checked without requiring an interaction sequence. Such a case arises when the value can be checked by an analysis of the configuration of devices in the network and/or administrators' databases. For example, if the value of $externRelay(h)$ is true, that means that h is defined as an external relay in the administrators' database and/or by the configuration of the network. This is known and trusted, not to be tested.

On the other side, the value is unsure if one has no knowledge about the fact that h is an external relay from the analysis of configurations, or if these data are untrusted. In this case the behaviour of the network should be tested to decide whether h acts as an external relay.

The following table shows the different formulas that may be built depending on which literals can be immediately asserted:

	externRelay(h)=true	... = false	... unsure
inDMZ(h)=true	t_{pass}	t_{inconc}	$t_{externRelay(h)} \triangleright_{\mathrm{lrC}} t_{pass}$
inDMZ(h)=false	t_{fail}	t_{inconc}	$t_{externRelay(h)} \triangleright_{\mathrm{lrC}} t_{fail}$
inDMZ(h) unsure	$t_{pass} \triangleright_{\mathrm{lrC}} t_{inDMZ(h)}$	t_{inconc}	$t_{externRelay(h)} \triangleright_{\mathrm{lrC}} t_{inDMZ(h)}$

If both values of $externRelay(h)$ and $inDMZ(h)$ are known and trusted, there is nothing to test. Also, no test is needed if $externRelay(h)$ is *false*, as we cannot put the system in the desired state, and the verdict is *inconc*. If the value of $inDMZ(h)$ (resp. $externRelay(h)$) is unsure, then it should be tested whether h behave like a host in the DMZ (resp. an external relay). These tests are then composed as described in section 4 into the formula $t_{externRelay(h)} \triangleright_{\mathrm{lrC}} t_{inDMZ(h)}$, as illustrated in figure 3.

5.2 $\mathcal{F}$−Rule

Consider the requirement *"If an electronic mail is infected by a virus, the virus shall be deleted from the mail"*. It can be modelled by the $\mathcal{F}$−**Rule** rule:

$$enterNetwork(m)[infected(m)] \Rightarrow$$

$$\mathcal{F}\ transfer(h_1, h_2, m)[interior(h_2) \wedge \neg infected(m)]$$

The goal of this test is to verify that if a mail infected by a virus is sent to a user in the network, eventually one of the hosts the mail is passing through will suppress the virus, before a certain time elapses.

As always in our approach, a choice is made concerning which predicates are sure or unsure. The formula $t = (t_{enterNetwork(m)} \triangleright_{\mathrm{llX}} t_{pass}) \triangleright_{\mathrm{lrF}} (t_{transfer(h_1, h_2, m)} \triangleright_{\mathrm{rrF}} (t_{pass} \triangleright_{\mathrm{rrF}} t_{\ infected(m)}))$ corresponds to the case when we build a test tile with a parameter m made of an infected message. This is the case because we choose to actively test the conformity of this particular rule against infected messages. One could also use a "passive" mode, checking the $infected(m)$ literal in the left part of the formula on incoming messages.

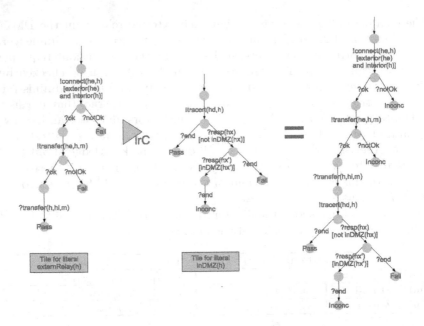

Fig. 3. Composition for example of $\mathcal{C}-$**Rule**

On the other hand the literal $\neg infected(m)$ in the right part shall be tested. The event predicate $enterNetwork(m)$ and the static predicate $\neg infected(m)$ are tested by the tiles shown in figure 4. Using these tiles, the formula $t_{enterNetwork(m)}$ $\triangleright_{\mathrm{lrF}} (t_{transfer(m)} \triangleright_{\mathrm{rrF}} t_{\ infected(m)})$ gives the test on figure 4. The `tinconc` and `t2inconc` variables have been suppressed since they cannot be true, and also the corresponding transitions.

6 Perspectives

In this paper, we have proposed a "tile-based" approach to derive test cases from rules expressed using a restricted set of logical operators that can be applied to network security policies. Complete test cases (dedicated to a whole formula) are obtained by combinations of more elementary ones (the tiles), following a syntax driven approach (a test combinator is associated to each logical operator of the formula). Elementary test cases, allowing to test basic events or predicates appearing in the security policy, have to be provided by the user (the way of testing such predicate depends on the network architecture and protocols involved). Our test generation method is based on the fact that security policies are most of the time expressed by rules which can be captured by a restricted logic as the one we described in section 3.

This approach is not limited to security, the idea of combining test tiles into a new one can be applied to other domains, for example testing software architecture, or testing systems built from existing components. However the com-

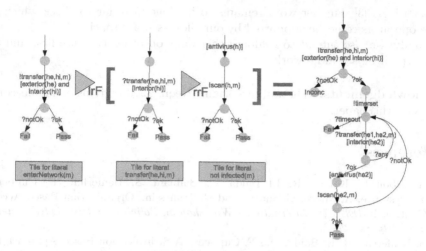

Fig. 4. Composition for example of $\mathcal{F}-$**Rule**

bination operators are based on the form of the rule being tested, and thus may be dependent on the domain. Up to now our examples use general purpose modalities, but we may have to consider more specialised cases.

At this point this work should be viewed essentially as a first step towards a formal approach to (automatically) test the compliance of a network with a given security policy. Therefore it should be extended into several directions.

First of all, the test cases produced are still very *abstract*. Turning them into executable test cases needs to take into consideration a concrete test architecture. Assuming that each elementary test case complies with this architecture, it would remain to ensure that it is also the case for the complete test case (or alternatively to take this architecture into account during the combination process). Moreover, these abstract test cases also need to be *instantiated* with concrete data (e.g. by selecting particular machines of the network). Suitable selection strategies should therefore be investigated (for instance a test could focus on the more recent changes in a network configuration, as in regression testing).

Furthermore, the proposed generation technique itself could be improved. In particular, the test case currently produced to test a condition (i.e., a disjunction of static predicates) consists in executing each corresponding elementary test case in sequence (according to the definition of our test combinators). An alternative way would have been to consider the *parallel* execution of such test cases (when it is compatible with the test architecture).

Another improvement could be to extend the formalism we considered to specify the security rules. This initial choice was motivated by our case study, and it was sufficient to demonstrate the effectiveness of the approach. However, it is clear that this formalism may be not sufficient to deal with arbitrary security rules, and that more specific operator/modalities need to be considered. One can think for instance of a triggered obligation bounded by an event (and not by an arbitrary timeout), or of some of the general operators proposed in the

NOMAD logic [3]. Further work remains to be done in order to check which of these operators could be supported by our tile-based approach.

Finally, we also intend to evaluate this work on other case studies, and to prototype it on a real network.

Acknowledgements. The authors thank Keqin Li for giving valuable comments on this paper.

References

1. A. Abou El Kalam, R. El Baida, P. Balbiani, S. Benferhat, F. Cuppens, Y. Deswarte, A. Miège, C. Saurel, and G. Trouessin. Organization Based Access Control. In *IEEE 4th International Workshop on Policies for Distributed Systems and Networks*, 2003.
2. S. Benferhat, R. E. Baida, and F. Cuppens. A Stratification-Based Approach for Handling Conflicts in Access Control. In *8th ACM Symposium on Access Control Models and Technologies*, 2003.
3. F. Cuppens, N. Cuppens-Boulahia, and T. Sans. Nomad: A security model with non atomic actions and deadlines. In *18th IEEE Computer Security Foundations Workshop, (CSFW-18 2005)*, pages 186–196, Aix-en-Provence, France, 2005.
4. F. Cuppens, N. Cuppens-Boulahia, T. Sans, and A. Miège. A formal approach to specify and deploy a network security policy. In *Second Workshop on Formal Aspects in Security and Trust (FAST)*, 2004.
5. N. Damianou, N. Dulay, E. Lupu, and M. Sloman. The Ponder Policy Specification Language. In *International Workshop on Policies for Distributed Systems and Networks*, 2001.
6. V. Darmaillacq, J.-C. Fernandez, R. Groz, L. Mounier, and J.-L. Richier. Éléments de modélisation pour le test de politiques de sécurité. In *Colloque sur les RIsques et la Sécurité d'Internet et des Systèmes, CRiSIS*, Bourges, France, 2005.
7. V. Darmaillacq and N. Stouls. Développement formel d'un moniteur détectant les violations de politiques de sécurité de réseaux. In *AFADL2006 - Approches Formelles dans l'Assistance au Développement de Logiciels*, Paris, March 2006.
8. ITU. Framework on formal methods in conformance testing. ITU-T Recommendation Z.500, ITU, 1997.
9. J. Lobo, R. Bhatia, and S. Naqvi. A Policy Description Language. In *AAAI'99*, 1999.
10. J.-C. Meyer, F. Dignum, and R. Wieringa. The Paradoxes of Deontic Logic Revisited: A Computer Science Perspective. Technical Report UU-CS-1994-38, Utrecht University, 1994.
11. A. Pnueli. The Temporal Logic of Programs. In I. C. S. Press, editor, *18th Annual Symposium on Foundations of Computer Science*, 1977.
12. D. Senn, D. Basin, and G. Caronni. Firewall Conformance Testing. In *TestCom 2005, 17th IFIP TC6/WG6.1 International Conference on Testing of Communicating Systems, Montréal, LNCS 3502*, June 2005.
13. A. Virmani, J. Lobo, and M. Kohli. Netmon: Network Management for the SARAS Softswitch. In *IEEE/IFIP Network Operations and Management Symposium*, 2000.
14. G. H. von Wright. Deontic Logic. *Mind*, 60:1–15, 1951.

Message Confidentiality Testing of Security Protocols – Passive Monitoring and Active Checking*

Guoqiang Shu and David Lee

Department of Computer Science and Engineering, The Ohio State University,
Columbus, OH 43210, USA
{shug, lee}@cse.ohio-state.edu

Abstract. Security protocols provide critical services for distributed communication infrastructures. However, it is a challenge to ensure the correct functioning of their implementations, particularly, in the presence of malicious parties. We study testing of message confidentiality – an essential security property. We formally model protocol systems with an intruder using Dolev-Yao model. We discuss both passive monitoring and active testing of message confidentiality. For adaptive testing, we apply a guided random walk that selects next input online based on transition coverage and intruder's knowledge acquisition. For mutation testing, we investigate a class of monotonic security flaws, for which only a small number of mutants need to be tested for a complete checking. The well-known Needham-Schroeder-Lowe protocol is used to illustrate our approaches.

1 Introduction

Security protocols have been playing an important role in the critical distributed systems such as E-Commerce and military infrastructure. Most security protocols use cryptography to achieve data transmission, authentication and key distribution [16, 17] in a hostile environment [1]. The existence of diverse intruders renders the resilience of those protocol systems more significant, and more challenging. Various formal modeling and analysis techniques, such as BAN logic, model-checking and strand spaces [14, 18, 19] have been developed in the recent years to ensure the correctness of security protocol system. These works are focused on validating the protocol specification. However, errors can also be introduced to the system in implementation phases, even if the specification is proven to be flawless. Furthermore, interconnected communication system interfaces may result in security problems, such as message content exposure. Systematic testing approaches for security protocols have been largely neglected by the research community, even though numerous reports show programming errors in security-critical systems are very common [22,23].

Testing for system security, often known as penetration testing [22] or red-team testing, refers to the activity of executing a predefined test script with the goal of finding a security exploit. Thomson in [23] classified four general penetration testing

---

* This work was supported in part by the U.S. National Science Foundation (NSF) under grant awards CNS-0403342, CNS-0548403, and by the U.S. Department of Defense under grant award N41756-06-C-5541.

M.Ü. Uyar, A.Y. Duale, and M.A. Fecko (Eds.): TestCom 2006, LNCS 3964, pp. 357–372, 2006.

methods: (1) Testing dependency; (2) Testing unanticipated user input; (3) Expose design vulnerabilities; and (4) Expose implementation vulnerabilities. Under these guidelines practical testing has been conducted in industry and proved to be very helpful. Nonetheless, most of the current penetration testing activities are ad-hoc and rely on expert knowledge of target systems or existing exploits [7]; the cost of a comprehensive testing is high and the response time is too long. On the other hand, current testing methods are largely at system level on system misconfiguration [20] or unexpected side effect of operations [2]. Protocol level penetration testing has not drawn adequate attention yet is crucial for discovering security protocol implementation errors. Particularly, automated test selection and execution techniques are desirable for complex protocols and for real-time response to security flaws.

In this paper we focus on automated testing of the key property of security protocols: message confidentiality. Several unique characteristics of security protocols make the traditional conformance testing approaches insufficient and pose new challenges for both modeling and test generation tasks. First, security protocols have a huge and special data portion. The I/O messages are from a language defined by cryptographic primitives such as public/private encryption and decryption. The formidable size of the alphabet makes generating a complete checking sequence infeasible. Therefore, tradeoff is usually made to focus only on a special type of nonconformance – security flaws. We use Extended Finite State Machine (EFSM) based formal model [5, 12] to specify the security protocol and augment the model to include security protocol message types as the parameter of I/O symbols. On the other hand, security properties can be tested only with a precise intruder model. We use EFSM to formally specify the intruder's behaviors based on the well-known Dolev-Yao model [4], which models most powerful and yet realistic intruder. Consequently, the whole protocol system is modeled as the communication system composed of the intruder and a set of legitimate principals. Furthermore, we define the notion of intruder's knowledge and message confidentiality requirement, and use it as the goal of testing.

Based on this formal model several testing approaches are proposed. We first give a simple passive monitoring procedure and then describe an active guided random walk algorithm. The algorithm is inspired by the earlier work [11] where heuristics are used to achieve high coverage of transitions in a CFSM model. Here our algorithm uses a new heuristic transition selection criterion that favors both new transition and new knowledge acquisition by the intruder. Both testing algorithms are unstructured in terms of the global system model, and the composite EFSM does not need to be computed. We also study mutation testing, since it is known to be efficient for a range of particular types of errors in software testing [3, 15]. Wimmel's et al's work based on their elegant validation tool AutoFocus [9, 24] is among the first attempts to apply the idea of mutation testing to security system. The greatest challenge (unaddressed by [9, 24]) of mutation testing is to control the number of mutants. This paper defines mutation functions with special property such that only mutants with single fault need to be considered for test generation. As a case study, we model the predicate (guard) absence fault type F_{PA} with this property, then present and analyze the test generation algorithm. We use the well-known Needham-Schroeder-Lowe (NSL) mutual authentication protocol [18] to illustrate our formal model and testing algorithms.

2 Modeling and Methodologies

After describing a formal model of security protocol systems we present our testing methods for both passive monitoring and active testing for message confidentiality.

2.1 Security Protocol Model

We define the security protocol message type as follows. First, there are three atom types: *Int, Key* and *Nonce*. A value of type *Int* is a non-negative bounded integer. A value of type *Key* ranges over a finite set K of keys. A value of type *Nonce* ranges over a finite set N of nonces. A protocol message is recursively defined as: (1) An atom; (2) Encryption of a message with a key; or (3) Concatenation of atoms and encryptions. A message can be represented by a string. For example $E(k_b,(k_a.n_a))$ is a messages that is formed by encrypting the concatenation of k_a and n_a with another key k_b. Given $A=<K, N>$, a set of keys and nonces, denote $L(A)$ as the message type and the set of messages formed using atoms in A. $L(A)$ is obviously infinite, and even if we restrict the number of atoms that a message contains, its size is exponential.

Among the atom types, *Key* and *Nonce* are treated as symbols in the sense that they can not be composed or calculated using other atom values. Also, *Key* type contains both symmetric keys and asymmetric key pairs. For the latter we use ku to represent a public key and kr for a private key. There are some basic operations defined on message type. Let *msg* be a message, function $Elem(msg,i)$ calculates the ith component in *msg*, and $D(k,msg)$ returns m' when $msg = E(k,m')$. Both functions are partial and they are undefined for messages with incompatible formats.

We define an extended finite state machine model that uses protocol message set $L(A)$ as input and output alphabet.

Definition 1. An Extended Finite State Machine (EFSM) with symbolic message type is a 7-tuple $M=<S, s_{init}, A, I, O, X, T>$ where

1. S is a finite set of states;
2. s_{init} is the initial state;
3. A is the set of atoms, and $L(A)$ is the set of messages formed using atoms in A;
4. $I = \{i_0, i_1,..., i_{P-1}\}$ is the input alphabet of size P; each input symbol i_k $(0 \le k < P)$ contains a parameter $\pi(i_k)$ of type $L(A)$;
5. $O = \{o_0, o_1,..., o_{Q-1}\}$ is the output alphabet of size Q ;each output symbol o_k $(0 \le k < Q)$ contains a parameter $\pi(o_k)$ of type $L(A)$;
6. X is a vector denoting a finite set of variables of type $L(A)$;
7. T is a finite set of transitions; for $t \in T$, $t = <s, s', i, o, p(x, \pi(i)),a(x, \pi(i), \pi(o))>$ is a transition where s and s' are the start and end state, respectively; $\pi(i)$ and $\pi(o)$ are the input/output symbol parameters; $p(x, \pi(i))$ is a predicate, and $a(x, \pi(i), \pi(o))$ is an action on the current variable values and parameters.

For practical protocol systems, the machine is often partially specified because a transition can only be triggered by a message with expected format. We use predicates to model the basic type checking capability. Upon receiving a message, the machine can reconstruct each element of it if the message format is correct. The special case is that some of the encrypted messages might be opaque to a machine because it does not possess the required key. For each combination of state and input symbol, there is one transition with a special predicate "else", meaning that it is enabled when all other transitions are not. We further assume that upon an input if none of the transitions are triggered, and then an implicit "else" transition make the machine stay at the current state and output nothing. For simplicity, we assume the machine contains a reliable reset symbol that takes the machine back to the initial state s_{init} and resets all state variables. In order to model the global uniqueness of nonce, a fresh new set of nonce $N' \subseteq N$ will be used whenever the machine is reset, so that different runs (sessions) of the protocol will use different nonces. Since N is finite we can only model finite number of sessions and each session only uses finite nonces. Finally, we only consider deterministic EFSM model.

A security protocol is specified by a set of communicating EFSM $\{M_1, M_2, \cdots, M_C\}$ that share the same message type L. Each component machine M_k represents a principal in the protocol system. It is possible that two machines are the same, meaning there are symmetric peers in the protocol. Moreover, for each transition in a component machine M_k, the input (output) symbol carries an extra parameter of the sender (receiver) identifier. We denote an input message received from M_a as $M_a?i$ and an output to M_a as $M_a!o$. The semantics of message sending/receiving follow the typical communicating FSM model [11]: the input/output is synchronized as a rendezvous and executed simultaneously.

2.2 Intruder Model

We model an intruder as an additional EFSM M_I in the protocol system that runs a special protocol. To model general behaviors of the intruder, we adapt Dolev-Yao's assumptions for two party message exchange protocols [4] that define a widely accepted powerful intruder model. It has been proved that one intruder poses the same security threat as multiple intruders and we model only one in our study.

An intruder is first a legitimate principal of the communication system; it can not only initiate a session with any other component machine M_a but also be the (passive) peer of any session. Furthermore, the intruder is capable of intercepting messages between any two legitimate principals. The important effect of this behavior is that the semantics of message sending and receiving in the original communicating EFSM model are altered. A transition in M_a with output message $M_b!msg$ now will be jointly executed with a transition in M_I that takes input $M_a \rightarrow M_b?msg$, instead of the transition in the intended receiver M_b. This should be clearly distinguished with the first case where the intruder M_I is the intended receiver (e.g. M_a outputs $M_I!msg$). Similarly, the intruder can inject any message, impersonating any other machines. That is, M_I can send output $M_a \rightarrow M_b!msg$ to M_b and this output matches the transition of M_b with input $M_a?msg$.

Besides the capability of catching and injecting normal protocol traffic, the intruder is also assumed to be able to generate any new message based on all and only the messages it possesses. Formally, we define the knowledge of the intruder as a set of messages $\Omega = Encl(\Omega_0 + MSG)$ where Ω_0 is the initial knowledge known to the intruder containing the public and intruder's own information, and MSG represents the set of messages the intruder has received. Function $Encl(L)$ is defined as the enclosure of L under the functions $Elem()$, $D()$ and $E()$. Therefore, Ω can be regarded as all the messages that the intruder is able to construct, using only the messages it obtains. Once the intruder gains a message it will not forget it and Ω never shrinks. As far as a realistic testing scenario is concerned, we have to assume the intruder has the capability of recognizing the message format, either by guessing the data field, or by reading the meta-info such as an XML schema.

M_I
State Variable
$\Omega: L$
Parameters
$a,b \in [1..C]$
$msg \in L$

Intercept $(M_a \rightarrow M_b?msg)$ / -
$[a \neq I]$ / $\Omega = Encl(\Omega + \{msg\})$

S_0

- / *Inject* $(M_a \rightarrow M_b!msg)$
$[msg\ in\ \Omega]$ & $[b \neq I]$ / $\{\}$

Fig. 1. EFSM model for the intruder

Fig. 1 shows the EFSM model of the intruder. M_I contains only one state and two transitions for message interception and injection respectively. *Intercept* transition takes any message *msg* sent from M_a to M_b as input. The guard ensures *msg* is not from M_I itself and the action updates the knowledge set. *Inject* transition outputs a message in the current knowledge set to another machine M_b under the disguise of M_a. The model of M_I is obviously independent of the other component machines. Note that messages meant to be delivered to M_I will also be caught by the *Intercept* transition, and in case the intruder does not want to intercept a message, the *Inject* transition is fired right after *Intercept* transition with the same message.

2.3 Testing Security Requirement: Message Confidentiality

Given the specification of protocol roles $\{M_1, M_2,..., M_C\}$ and the intruder M_I, the global behavior of the whole protocol system under investigation is described by the Cartesian product of all the machines: $M_1 \times M_2 \times \cdots \times M_C \times M_I$ with the input/output matching rule we define in the previous subsection. Since all the transitions involve one of the two intruder transitions, an I/O trace produced by the system can be described by an interleaving sequence of *Intercept* and *Inject* transitions each with a message in the parameters.

In this paper we focus on black-box penetration testing. The implementations of all protocol principals are treated as pure black boxes, i.e. $B = \{B_1, B_2,..., B_C\}$, each B_i can be a correct or faulty implementation of M_i. The tester plays the role of intruder and simulates the machine M_I. This is an active testing process because the tester can choose arbitrarily the parameters of *Inject* transition, namely the sender, receiver and message. A test sequence *seq* is defined as an I/O trace produced by the communicating system of M_I and B. Starting from the initial states, denote the value of Ω in M_I after a test sequence *seq* is applied as $\Omega(B, seq)$, which is the knowledge that an intruder gains by performing penetration test *seq*.

For a given security protocol system, there are many security requirements depending on the specific application needs. Typically, they include message confidentiality, message integrity, authentication, and non-repudiation [20]. In this paper we focus on the message confidentiality requirement that is the key property of a security protocol system. Other requirements can be handled, for instance, by appropriate hash functions, and we shall not digress here.

Definition 2. A protocol implementation $B = \{B_1, B_2, ..., B_C\}$ is insecure with regard to the confidentiality of messages $M^* \subset L$ if and only if there exists a test sequence seq and a message $m \in M^*$ such that $m \in \Omega(B, seq)$.

An implementation is flawed if and only if message content can be uncovered by the intruder after a test sequence is applied. We model the intruder following Dolev-Yao's approach, our definition 2 is consistent with their notion of security of two party cascade and name stamp protocols [4]. One natural question is that whether the protocol specification itself is secure. When the implementation of each component machine is equivalent to its specification, i.e. $B_i = M_i$, the intruder might still be able to obtain the secret if the protocol design itself is flawed [14]. Since our goal is testing rather than validation, we assume the protocol design and specification are secure.

As an example of modeling security protocols, we consider the well-known Needham-Schroeder-Lowe (NSL) mutual authentication protocol [18]. Among many of its variants, we use a simplest one with three message exchanges [14]. Two principles, the initiator and the responder, are involved and they are specified as M_A and M_B, respectively. The message sequence of a successful run is shown below.

$A \rightarrow B$ (Ask): $A.B.E(KU_B,(N_A.A))$
$B \rightarrow A$ (Rpl): $B.A.E(KU_A, (N_A.N_B.B))$
$A \rightarrow B$ (Cfm): $A.B.E(KU_B, (N_B))$

The protocol functions as follows. The initiator A encrypts a nonce with the responder B's public key and sends it to B. B then decrypts it and encrypts it together with another nonce using A's public key. Finally A gets the second nonce and sends it back. The purpose of NSL protocol is to allow both principles authenticate each other and exchange some secrets (two nonces), which later on can be used to construct shared keys. Fig. 2 shows the two complete EFSM specifications. We assign index 0, 1 and 2 to M_A, M_B and the intruder M_I. The intruder can participate legally as both the initiator and responder. The atom messages in this protocol include the public keys (KU_A, KU_B and KU_I) and the nonces. In order to express the security requirement conveniently, we distinguish the nonces used for different peers. For instance, the nonce M_B uses to challenge M_I is $N_B[I]$. Initially the intruder only knows its own key and nonces, i.e., $\Omega_0 = \{KU_I, N_I[A], N_I[B]\}$. The secret message set is $M^* = \{N_A[B], N_B[A]\}$; the intruder should not obtain the nonces that are only supposed to be shared only between A and B. Note that in Fig. 2 the parameter (message) of each I/O symbol is expanded by its structure, which is a short notation for format checking of the message. A special symbol Rst is used to reset the session when invalid message is processed.

To summarize this section, we essentially reduce the security testing problem to searching for special I/O sequences produced by a mixed communicating system, which contains M_I and one or more black boxes as principals. The characteristic of those sequences is that they lead the reachability graph of M_I to a state where the value of variable Ω contains message content/secret. The tester has full control over M_I but can only observe the I/O behaviors of the other protocol principals.

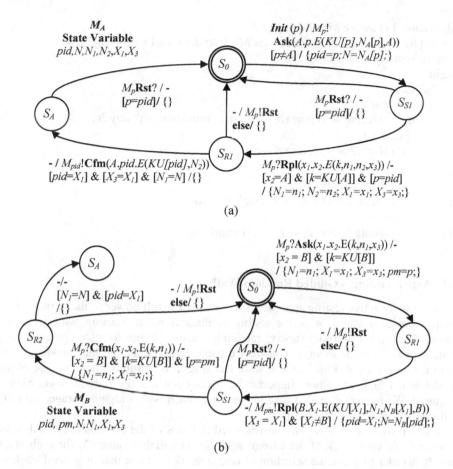

M_A
State Variable
pid,N,N_1,N_2,X_1,X_3

S_0

$Init (p) / M_p!$
$Ask(A.p.E(KU[p],N_A[p],A))$
$[p{\neq}A] / \{pid{=}p;N{=}N_A[p];\}$

$M_p\mathbf{Rst}? / -$
$[p{=}pid]/ \{\}$

S_A

$M_p\mathbf{Rst}? / -$
$[p{=}pid]/ \{\}$

S_{S1}

$- / M_p!\mathbf{Rst}$
else/ $\{\}$

S_{R1}

$- / M_{pid}!\mathbf{Cfm}(A.pid.E(KU[pid],N_2))$
$[pid{=}X_1]$ & $[X_3{=}X_1]$ & $[N_1{=}N] /\{\}$

$M_p?\mathbf{Rpl}(x_1.x_2.E(k,n_1,n_2,x_3)) / -$
$[x_2{=}A]$ & $[k{=}KU[A]]$ & $[p{=}pid]$
$/ \{N_1{=}n_1; N_2{=}n_2; X_1{=}x_1; X_3{=}x_3;\}$

(a)

S_A

$M_p?\mathbf{Ask}(x_1.x_2.E(k,n_1,x_3)) / -$
$[x_2 = B]$ & $[k{=}KU[B]]$
$/ \{N_1{=}n_1; X_1{=}x_1; X_3{=}x_3; pm{=}p;\}$

$-/-$
$[N_1{=}N]$ & $[pid{=}X_1]$
$/\{\}$

$- / M_p!\mathbf{Rst}$
else/ $\{\}$

S_0

$- / M_p!\mathbf{Rst}$
else/ $\{\}$

S_{R1}

S_{R2}

$M_p?\mathbf{Cfm}(x_1.x_2.E(k,n_1)) / -$
$[x_2 = B]$ & $[k{=}KU[B]]$ & $[p{=}pm]$
$/ \{N_1{=}n_1; X_1{=}x_1;\}$

$M_p\mathbf{Rst}? / -$
$[p{=}pid]/ \{\}$

M_B
State Variable
pid, pm,N,N_1,X_1,X_3

S_{S1}

$-/ M_{pm}!\mathbf{Rpl}(B.X_1.E(KU[X_1],N_1,N_B[X_1],B))$
$[X_3 = X_1]$ & $[X_1{\neq}B] / \{pid{=}X_1;N{=}N_B[pid];\}$

(b)

Fig. 2. Needham-Schroeder-Lowe protocol (a) Initiator M_A (b) Responder M_B

3 Message Confidentiality Testing

After presenting a simple passive monitoring algorithm, we describe an active testing procedure that is based on a guided random walk.

3.1 A Simple Passive Monitoring Algorithm

A passive tester or monitor of security protocol implementation is easy to devise. The intruder (tester) intercepts all messages among the component machines, updates its knowledge, replies if the message is directed to itself, and otherwise forwards it without any modification. The testing terminates when the intruder derives any secrets. The procedure is shown in Algorithm 1. As inherent to all passive testing approaches, this algorithm only utilizes part of the intruder's capability and it is suitable when the intruder could only conduct eavesdropping [1].

Algorithm 1 (Passive Monitoring)
Input: $\{B_1, B_2, \ldots, B_C\}$, message secrets M^*, Intruder initial knowledge Ω_0.
Output: security flaw if observed.
begin
1. $\Omega = \Omega_0$;
2.   while (true)
3.   try to execute *Intercept* $(B_i \rightarrow B_j?msg)$ transition with any B_i;
4.   if succeed
5.   if $(M^* \cap \Omega \neq \phi)$ return *flaw*;
6.   if $(j=I)$
7.   generate reply *msg'* ;
8.   execute *Inject* $(M_I \rightarrow B_i?msg')$ transition;
9.   else
10.   execute *Inject* $(B_i \rightarrow B_j?msg)$ transition;
End

3.2 Active Testing – Guided Random Walk

Now we study active testing approaches that utilize the full power of the intruder. One simple-minded method of active testing is random walk. Starting with an initial knowledge set, the intruder (tester) randomly chooses either to intercept a message from a pair of principals or to construct a message using its current knowledge and send it to a principal. Pure random walk has several limitations; the coverage of the model is not high and, more importantly, it does not use the intruder's knowledge acquired. We present a guided random walk approach with a high coverage and fully utilizing the intruder's knowledge acquired.

The approach is adaptive and unstructured in terms of the composite (global) state machine. We keep track of the current state S_i and variable values X_i for each black box B_i in order to guide the selection of next transitions. Note that in general tracking current state is not always possible even under the assumption that B_i contains no transition errors; it is due to the fact that part of the message is encrypted and intruder can not utilize the information to infer the current transition and state if he does not have the key. In this case, the algorithm makes a random guess.

At each step, the intruder always tries to intercept the messages coming from every machine B_i. Once a message is intercepted, the state of the sender as well as the intruder's knowledge is updated. Then the intruder constructs a message and injects it to a machine to fire a carefully selected transition. Our algorithm selects transition and message based on the following criteria. First the transitions of all component EFSM should be covered fairly. The algorithm keeps track of a counter $cnt[t]$ for each transition t, and at each step the one that has been executed least is favored. Moreover, we only select the transitions that could possibly be enabled by some input message and ignore those transitions that will definitely not be triggered (the current state variables themselves disable the predicate). We calculate T_{true} as the set of all possible transitions:

$$T_{true} = \{ \ t<S_i, S'_i, I, p, a> \ | \ t \in M_i \ \text{and} \ \exists \ msg \in \Omega: p(X_i, msg) = true\}$$

Once a transition t is determined, we construct an enabling input message for t using a greedy algorithm. Ideally we want an input message that will lead the machine to a state that can generate more new knowledge. That is, for all candidate messages we calculate the destination state S' of t, and select one that enables at least one output transition t' with parameter msg' not in Ω. We use subroutine $lookahead(\Omega, S, X, t)$ to calculate such messages. If such messages do not exist or there are ties, an enabling message is randomly picked:

$lookahead(\Omega, S, X, t<S_i, S'_i, I, p, a>) = \{msg \mid p(X_i,msg) =true$ and $(\exists\ t'<S_i',S_i'',O(msg'),p',a'>:p'(X_i')=true$ and $msg'\notin\Omega)\}$

Algorithm 2 (Active Testing - Guided Random Walk)
Input: $\{B_1, B_2,..., B_C\}$, secrets M^*, Intruder initial knowledge Ω_0.
Output: Adaptive test sequence.
begin
1.   initialize each M_i, for all transition t, $cnt[t] = 0$;
2. $X=<X_1,...,X_C>$, $S=<S_1,...,S_C>$;
3. $\Omega=\Omega_0$, $seq=\phi$;
4.   while (seq.len $< L$)
5.   foreach component B_i
6.   try to execute $Intercept(B_i \rightarrow B_j?msg)$ with B_i;
7.   if succeed
8.   deduce or guess the transition t;
9.   update X_i, S_i, $cnt[t] = cnt[t] +1$, $seq = seq + \{t\}$;
10.   calculate T_{true} ,select $t\in T_{true}$ with smallest $cnt[t]$;
11.   select msg from $lookahead(\Omega, S, X, t)$;
12.   try to execute $Inject$ transition with t using msg;
13.   if succeed
14.   update X_i, S_i, $cnt[t] = cnt[t] +1$, $seq = seq + \{t\}$;
15.   if ($M^*\cap\Omega \neq \Phi$) return seq;
16.   return seq;
end

To avoid infinite tests, the algorithm terminates when either the secret message content is obtained or the length of test sequence reaches a preset limit. This algorithm is more effective than random walk because the greedy heuristics take into account both coverage and intruder knowledge acquisition. However, it still has many inherent limitations. For example, calculation of T_{true} and $lookahead()$ is rather expensive. Also, the effectiveness of the heuristic relies on the estimation of current state and variable values, and if it fails the algorithm behaves the same as random walk. Advanced passive testing techniques [8, 10] that estimate data portion more accurately could be applied here to improve the performance.

3.3 Experiment

We conduct an experiment of Algorithm 2 on NSL protocol specified as Fig. 2. Two implementations are created with a common programming error in each. Then we treat them as black-boxes and run the algorithm to test for confidentiality violations.

Implementation X: The responder does not verify the encrypted identifier of the initiator after it receives Ask message, and proceeds as if it were correct.

Implementation Y: The initiator does not verify the encrypted identifier of the responder after it receives Rpl message, and proceeds as if it were correct. This error was first uncovered by Lowe [14] as a design flaw in the original Needham-Schroeder protocol.

For both Implementation X and Y errors have been detected. Table 1 (a) and (b) show the successful test sequences for them. In the first test sequence, at the beginning the intruder intercepts an Ask message from M_0 to M_1, and updates the state to $<S_{s1},S_0>$. Now three transitions are feasible and as the result M_1?Ask is selected. *Lookahead()* returns a random message that enables M_1?Ask because no message will further trigger an output transition. In the second round we intercept an Rpl message, and the intruder will obtain a secret ($N_0[1]$) and terminate the test. The sequence for Y is more complex. After injecting an Ask message to M_1 and intercepting the response, we have two transitions in T_{true}. M_1!Cfm is chosen and executed with a random message. At next step M_0 happens to initiate a session with M_1. This is a rare event yet critical for detecting errors in this implementation. The only transition that could be enabled is M_0?Rpl, and now the intruder happens to have a message to enable it. The last step is the interception of Cfm message from M_0 that exposes the nonce – secret $N_1[0]$.

Table 1. Detection of Errors in Implementation X (a) and Y (b)

States	Action	Note
$<S_0, S_0>$	*Intercept* $M_0 \rightarrow M_1$? Ask $(0.1.E(KU[1], N_0[1], 0))$	$\Omega^+ = \{E(KU[1],N_0[1],0)\}$
$<S_{s1}, S_0>$	*Inject* $M_2 \rightarrow M_1$! Ask $(2.1.E(KU[1], N_0[1], 0))$	$T_{true} = \{M_0?Rpl, M_0?Rst, M_1?Ask\}$ $t = M_1?Ask$
$\underline{<S_{s1}, S_{R1}>}$	*Intercept* $M_1 \rightarrow M_2$? Rpl $(1.2.E(KU[2], N_0[1], N_1[2],1))$	$\Omega^+ = \{N_0[1], N_1[2]\}$ $N_0[1] \in M^*$

(a)

States	Action	Note
$<S_0, S_0>$	*Inject* $M_0 \rightarrow M_1$! Ask $(0.1.E(KU[1], N_2[1], 0))$	$T_{true} = \{M_1?Ask\}$ $t = M_1?Ask$
$<S_0, S_{R1}>$	*Intercept* $M_1 \rightarrow M_0$? Rpl $(1.0.E(KU[0], N_2[1], N_1[0], 1))$	$\Omega^+ = \{ E(KU[0], N_2[1], N_1[0], 1)\}$
$<S_0, S_{S1}>$	*Inject* $M_0 \rightarrow M_1$! Cfm $(0.1.E(KU[1], N_2[1], 0))$	$T_{true} = \{M_1?Rst, M_1?Cfm \}$ $t = M_1?Cfm$
$<S_0, S_{R2}>$	*Intercept* $M_0 \rightarrow M_2$? Ask $(0.2.E(KU[2], N_0[2], 0))$	$\Omega^+ = \{N_0[2]\}$
$<S_{S1}, S_{R2}>$	*Inject* $M_2 \rightarrow M_0$! Rpl $(2.0. E(KU[0], N_2[1], N_1[0], 1))$	$T_{true} = \{M_0?Rpl\}$ $t = M_0?Rpl$
$\underline{<S_{R1}, S_{R2}>}$	*Intercept* $M_0 \rightarrow M_2$? Cfm $(0.2.E(KU[2], N_1[0], 0))$	$\Omega^+ = \{N_1[0]\}$ $N_1[0] \in M^*$

(b)

4 Mutation Testing

In this section we investigate mutation testing of security protocol, and design structured and preset test sequences. As introduced earlier mutation testing is a powerful technique for detecting specific types of security errors. Given the specification M_{spec} = $\{M_1, M_2,..., M_C\}$, we introduce some faults, resulting in a mutant $\{M_1{'}, M_2{'},..., M_C{'}\}$. Given a set of mutants P, a test suite is generated such that for each mutant p, there is at least one test sequence that distinguishes (detects) it with the specification (correct implementation). A main challenge of mutation testing, when applied to software in general, is that the number of mutants (therefore the number of tests required) is huge. The situation is not mitigated in our EFSM model given its equivalent computing power of Turing machine. We model a security flaw as a mutation function δ on a specification EFSM, and a type of fault F as a set of similar mutation functions. A mutant under F is the application of one or more such functions. If the type F contains k functions, then the number of mutants is $O(2^k)$.

One can take two hypotheses to reduce the number of mutants generated [3]. First, competent programmer hypothesis assumes that an implementation only contains a small number (C) of faults. This reduces the number of mutants to $O(k^C)$, which is still quite large. Second, coupling effect hypothesis states that the test sequences used to distinguish mutants with simple fault are sensitive enough to also uncover complex fault. Clearly this is not always true. Given an arbitrary mutation function, a test sequence that obtains the secret on $\delta_1(M_{spec})$ may not be effective for $\delta_2\delta_1(M_{spec})$. In fact, mutant $\delta_2\delta_1(M_{spec})$ could even be secure. On the other hand, if we could select test sequence that satisfies this property, then the number of mutants could be further reduced to k. For message confidentiality testing, we can reduce the number of mutants based on this observation.

4.1 A Fault Model: Predicate or Guard Absence

There are generally two categories of security sensitive fault in the protocol model. The first is message format fault. For example, one might use the private key to encrypt part of the message instead of the public key, or attach an unnecessary part, both giving the intruder more information. This type is easier to observe since it changes the alphabet of some component machines. The second category of fault is related to the predicate or action of the transitions, but has no effect on the message types. Based on the observation of security protocols, a commonly encountered implementation error is neglecting critical condition checking. Usually an action is taken place only if some condition – predicate - is satisfied by the current state and/or the input message. For example in the NSL protocol, the responder only replies to the message $Ask(x_1.x_2.E(k,n_1,x_3))$ when the x_2 is equal to its own index, and similarly the initiator only generates to the Cfm message when it verifies the responder's reply with the same nonce as the one it sends out. If the programmer neglects to check such condition such as in Implementation X and Y in section 3, it is likely that the resulting implementation is insecure. This type of fault is reflected in the EFSM model as the absence of part of the predicate in a transition - or often called a guard. Assuming the predicate is specified as a conjunctive normal form of Boolean expressions (i.e. $b_1\&b_2\&b_3$) , we formally define this fault type.

Definition 3. For all the transitions t_j, $j=0,1,...$, from a state s with a same input/output symbol y, a predicate absence (PA) mutation function $\delta_{PA}(s,y,t,b)$ with regard to a Boolean expression b in the predicate p_j of $t=t_i$, is obtained by removing b from p_j and adding $(!p_i)$ to p_j for all $i \neq j$.

Basically the mutation function removes one Boolean expression from a transition. In order to keep the resulting machine deterministic, we add its negation to all other transitions with the same start state and input/output symbol. Fig. 3 shows an example of a mutant of the function $\delta_{PA}(S_1, Y, t, [a=1])$.

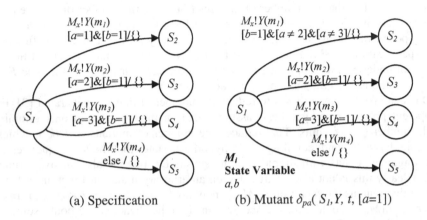

(a) Specification   (b) Mutant $\delta_{pa}(S_1, Y, t, [a=1])$

Fig. 3. Example of mutant δ_{PA}

Definition 4. For a protocol specification M_{spec}, a predicate absence (PA) fault type F_{PA} is obtained by applying one or more PA mutation functions $\delta_{PA}(s, y, t, b)$ on M_{spec}. A mutant under F_{PA} is defined as $\delta_S(M_{spec}) = \delta_1\delta_2... \delta_n(M)$, where $S = \{\delta_1, \delta_2,..., \delta_n\} \subseteq F_{PA}$, and for any $\delta_a(s, y, t_a, b_a)$, $\delta_b(s, y, t_b, b_b) \in S$, $t_a = t_b$.

A mutant under the PA fault type is the result of application of a set of PA mutation functions, each removing a Boolean expression from a predicate. Note that although this definition does not limit the number of faults in one mutant, it relies on the competent programmer hypothesis to assume that for each combination of component machine, state and I/O symbol, only a predicate from one transition could be removed. Consequently, if each transition contains a constant number of Boolean expressions, there are totally $O(T)$ mutation functions and $O(2^{(C \times N \times P)})$ mutants where T is the number of transitions, C is the number component machines, N is the maximum number of states and P is the number of I/O symbols.

Intuitively a mutant with more predicate missing should allow more transitions to be executed and therefore the security flaws are "monotonically" increasing with inclusion of more faults in F_{pa}. This is formulated in the following proposition.

Definition 5. A progressive I/O sequence of a communicating system is an I/O sequence that does not trigger any "else" transition of any component machine.

Proposition 1 (Monotonicity). For any two mutants $\delta_{S1}(M)$ and $\delta_{S2}(M)$ under F_{pa} with $S_1 \subseteq S_2$, if a progressive I/O sequence *seq* could be generated by M_I and $\delta_{S1}(M)$, then *seq* could also be generated by M_I and $\delta_{S2}(M)$.

Sketch of proof: The proof of this proposition is quite straightforward using an induction on the length of the sequence. Suppose a prefix of *seq* has already been executed by $\delta_{S2}(M)$ and the next message in *seq* will trigger transition t in M_i. if $\delta_{S2}(M)$ has the same t as $\delta_{S1}(M)$ then t will be executed. If $\delta_{S2}(M)$ further removes some expressions from t, then the current states and input message will satisfy the guard of the new transition, since t is not the "else" transition, and, therefore, t is executed.

An important implication of Proposition 1 is that if a progressive test sequence discovers a message secret for M, and we apply some other mutation functions to introduce more errors, the same test sequence can still expose the message content on the new mutant. In other words, faults do not cancel the evidence of each other with regard to a progressive test sequence. We remark that singularity about "else" transition does not decrease the applicability of this model because this special type of transition is usually used to model the behavior in abnormal conditions, and will not be included in an I/O sequence that achieves the functionalities of the protocol.

4.2 Mutation Test Generation Algorithm

Now we describe the procedure of generating test sequences for monotonic flaw type of F_{PA}. The goal is to generate a set of test sequence that distinguishes all mutants under F_{PA}. One valid concern would be that not all mutants are necessarily insecure according to the confidentiality requirement and it is reasonable to only focus on mutants, which lead to message confidentiality violations. This is a well-studied validation problem and we shall not digress here. For simplicity, we treat all mutants as potentially insecure and generate tests to detect each of them:

Algorithm 3 (Test Generation for Fault Type F_{PA})
Input: $M_{spec} = \{M_1, M_2, ..., M_C\}$, secrets M^*.
Output: test suite S, fault type F'_{PA}
begin
1. $S = \{\}; F'_{PA} = \{\};$
2.   remove all "else" transitions from M_{spec}
3.   calculate and minimize $M_I \times M_{spec}$;
4.   foreach mutation function δ_i
5.   calculate $\delta_i(M_{spec})$;
6.   calculate and minimize $M_I \times \delta_i(M_{spec})$;
7.   if $(M_I \times M_{spec} \mathrel{!=} M_I \times \delta_i(M_{spec}))$
8. t = separating sequence of $M_I \times M_{spec}$ and $M_I \times \delta_i(M_{spec})$
9. $S = S + \{t\};$
10. $F'_{PA} = F'_{PA} + \{\delta_i\};$
11.   return S
end

Algorithm 3 applies each mutation function alone to the specification and calculates a progressive separating sequence. This is done by removing all "else" transitions, minimizing the Cartesian product of the mutant and intruder machine, and calculate a separating sequence. The comparison in Line 7 refers to an equivalence test of two machines. The algorithm produces a new fault type F'_{PA} which only contains the mutation functions if the corresponding mutants are distinguishable. The number of test sequences generated by Algorithm 3 is no more than the number of mutants in F'_{PA}. The time needed for minimization is $O(N\log N)$ with online minimization algorithm [13], and the calculation of separating sequence requires $O(N^2)$ where N is the number of states in the reduced machine. We propose an optimization technique for generating separating sequence online in [21], which will reduce the cost of this algorithm for average case but the worst case complexity is the same.

As far as the fault detection capability is concerned, the test suite generated includes a test case to distinguish every mutant that is derived by applying one mutation function in F'_{PA}. Since all test sequences are progressive sequence, from Proposition 1, we have:

Proposition 2. Tests generated from Algorithm 3 detect all mutants under F'_{PA} in time $O(N^2)$ where N is the number of states in the reduced machine.

Algorithm 3 also applies to all other fault models that satisfy proposition 1. Note that the test suite does not discover all faulty mutants in F_{PA}; if a mutation function itself is not distinguishable, then Algorithm 3 simply discards it.

4.3 Experiment

We again conduct the experiment using NSL protocol. In the specification (Fig. 2) a total of 19 Boolean expressions are identified, as shown in Fig. 4. These expressions are used to construct the fault type F_{PA} and the mutants. Among them δ_{b12} and δ_{b7} correspond to the three implementations X and Y in Section 4, respectively. Algorithm 3 produces $F'_{PA} = F_{PA} - \{\delta_{b18}, \delta_{b19}\}$ and the set of 17 test sequences. The last two Boolean expressions are not associated with any I/O behaviors and are not observable. The lengths of those sequences are shown in Table 2 and the details are omitted. All the sequences are short (less than 4). This set of test sequences detect all implementations with one or more Boolean expressions missing.

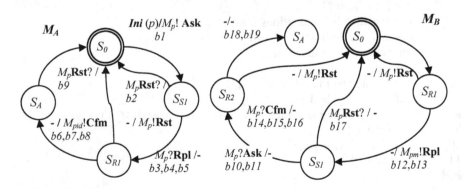

Fig. 4. Boolean Expressions in NSL Specification

Table 2. Test Sequence Lengths Generated by Algorithm 3

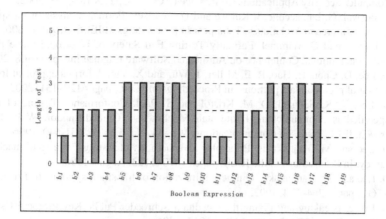

5 Conclusion

This paper studies the problem of testing message confidentiality of security protocols. EFSM with symbolic message type is used to model security protocol system with an omnipotent intruder. A formal definition of message confidentiality property and the black box testing model are provided. Passive monitoring, guided random walk and mutation testing approaches are presented with case studies.

A lot of issues remain to be explored, such as efficient modeling for intruder knowledge acquisition for more powerful testing results, thorough and structured active testing procedures, and more general mutation testing with more focus on message confidentiality violation yet with less computation costs. On the other hand, systematic experiments are to be conducted on the de-facto security protocols, such as Kerberos, electronic payment, and IPSec.

References

1. Achilles Proxy. http://www.mavensecurity.com/achilles
2. S. Chen, Z. Kalbarczyk, J. Xu and Ravishankar K. Iyer. A Data-Driven Finite State Machine Model for Analyzing Security Vulnerabilities. International Conference on Dependable Systems and Networks (DSN'03), page 605, 2003.
3. R. DeMillo, R. Lipton, and F. Sayward. Hints on Test. Data Selection : Help For The Practicing Programmer. IEEE Computer, vol. 1 l(4), pages 34-41, 1978.
4. D. Dolev and A. Yao. On the security of public-key protocols. IEEE Transaction on Information Theory 29, pages 198-208, 1983.
5. A. Duale and M. Ümit Uyar. A Method Enabling Feasible Conformance Test Sequence Generation for EFSM Models. IEEE Trans. Computers 53(5): pages 614-627, 2004.
6. S. Fabbri, J. Maldonado, T. Sugeta, and P. Masiero. Mutation testing applied to validate specifications based on statecharts. In International Symposium on Software Reliability Systems (ISSRE), pages 210-219, 1999.

7. D Geer and J. Harthorne. Penetration Testing: A Duet. In Proceedings of. the 18th Annual Computer Security Applications Conference (ACSAC), pages 185–198, 2002.
8. S. Jaiswal, G. Iannaccone, J. Kurose and D. Towlsey. Formal Analysis of Passive Measurement Inference Techniques. To appear in Proceedings of IEEE Infocom 2006.
9. J. Jurjens and G. Wimmel. Formally Testing Fail-Safety of Electronic Purse Protocols. IEEE International Conference on Automated Software Engineering, page 408, 2001.
10. D. Lee, D. Chen, R. Hao, R. E. Miller, J. Wu, and X. Yin. A formal approach for passive testing of protocol data portions. In Proceedings of ICNP, pages 122–131, 2002.
11. D. Lee, K. K. Sabnani, D. M. Kristol and S. Paul. Conformance Testing of Protocols Specified as Communicating Finite State Machines - a Guided Random Walk Based Approach. IEEE Trans. on Communications, Vol. 44, No. 5, pages 631-640, 1996.
12. D. Lee and M. Yannakakis. Principles and methods of testing finite state machines - A survey. In Proceedings of the IEEE, pages 1090–1123, August 1996.
13. D. Lee and M. Yannakakis. Online minimization of transition systems. In Proceedings of STOC, pages 264–274, 1992.
14. G. Lowe. Breaking and Fixing the Needham-Schroeder Public-Key Protocol Using FDR. In Proceedings of TACAS'96, LNCS 1055, 1996.
15. B. Marick. The Weak Mutation Hypothesis. Proceedings of The ACM SIGSOFT Symposium on. Testing, Analysis, and Verification, October, 1991.
16. C. Meadows. Applying formal methods to the analysis of a key management protocol, J. Comput. Security 1, pages 5-53, 1992.
17. C. Meadows. Formal methods for cryptographic protocol analysis: emerging issues and trends. IEEE Journal on Selected Areas in Communications, 21(1), pages 44-54, 2003.
18. R. Needham, M. Schroeder. Using encryption for authentication in large networks of computers, Communications of the ACM, 21(12), pages 993-999, 1978.
19. S. Schneider. Security Properties and CSP, Proceedings of the 1996 IEEE Symposium on Security and Privacy, page 174, 1996.
20. O. Sheyner, J. Haines, S. Jha, R. Lippmann and J. Wing. Automated Generation and Analysis of Attack Graphs. IEEE Symposium on Security and Privacy, 2002.
21. G. Shu and D. Lee. Network Protocol System Fingerprinting – A Formal Approach. To appear in Proceedings of IEEE Infocom 2006.
22. H. Thompson. Application Penetration Testing. IEEE Security & Privacy. 3(1), pages 66–69, 2005.
23. H. Thompson. Why Security Testing Is Hard. IEEE Security and Privacy. 1(4), pages 83-86, July-August, 2003.
24. G. Wimmel, J. Jürjens, Specification-Based Test Generation for Security-Critical Systems Using Mutations. Proceedings of ICFEM pages 471-482, 2002.

Author Index

Lecture Notes in Computer Science

For information about Vols. 1–3877

please contact your bookseller or Springer

Vol. 3925: A. Valmari (Ed.), Model Checking Software. X, 307 pages. 2006.

Vol. 3924: P. Sestoft (Ed.), Programming Languages and Systems. XII, 343 pages. 2006.

Vol. 3923: A. Mycroft, A. Zeller (Eds.), Compiler Construction. XIII, 277 pages. 2006.

Vol. 3922: L. Baresi, R. Heckel (Eds.), Fundamental Approaches to Software Engineering. XIII, 427 pages. 2006.

Vol. 3921: L. Aceto, A. Ingólfsdóttir (Eds.), Foundations of Software Science and Computation Structures. XV, 447 pages. 2006.

Vol. 3920: H. Hermanns, J. Palsberg (Eds.), Tools and Algorithms for the Construction and Analysis of Systems. XIV, 506 pages. 2006.

Vol. 3918: W.K. Ng, M. Kitsuregawa, J. Li, K. Chang (Eds.), Advances in Knowledge Discovery and Data Mining. XXIV, 879 pages. 2006. (Sublibrary LNAI).

Vol. 3917: H. Chen, F.Y. Wang, C.C. Yang, D. Zeng, M. Chau, K. Chang (Eds.), Intelligence and Security Informatics. XII, 186 pages. 2006.

Vol. 3916: J. Li, Q. Yang, A.-H. Tan (Eds.), Data Mining for Biomedical Applications. VIII, 155 pages. 2006. (Sublibrary LNBI).

Vol. 3915: R. Nayak, M.J. Zaki (Eds.), Knowledge Discovery from XML Documents. VIII, 105 pages. 2006.

Vol. 3914: A. Garcia, R. Choren, C. Lucena, P. Giorgini, T. Holvoet, A. Romanovsky (Eds.), Software Engineering for Multi-Agent Systems IV. XIV, 255 pages. 2006.

Vol. 3910: S.A. Brueckner, G.D.M. Serugendo, D. Hales, F. Zambonelli (Eds.), Engineering Self-Organising Systems. XII, 245 pages. 2006. (Sublibrary LNAI).

Vol. 3909: A. Apostolico, C. Guerra, S. Istrail, P. Pevzner, M. Waterman (Eds.), Research in Computational Molecular Biology. XVII, 612 pages. 2006. (Sublibrary LNBI).

Vol. 3908: A. Bui, M. Bui, T. Böhme, H. Unger (Eds.), Innovative Internet Community Systems. VIII, 207 pages. 2006.

Vol. 3907: F. Rothlauf, J. Branke, S. Cagnoni, E. Costa, C. Cotta, R. Drechsler, E. Lutton, P. Machado, J.H. Moore, J. Romero, G.D. Smith, G. Squillero, H. Takagi (Eds.), Applications of Evolutionary Computing. XXIV, 813 pages. 2006.

Vol. 3906: J. Gottlieb, G.R. Raidl (Eds.), Evolutionary Computation in Combinatorial Optimization. XI, 293 pages. 2006.

Vol. 3905: P. Collet, M. Tomassini, M. Ebner, S. Gustafson, A. Ekárt (Eds.), Genetic Programming. XI, 361 pages. 2006.

Vol. 3904: M. Baldoni, U. Endriss, A. Omicini, P. Torroni (Eds.), Declarative Agent Languages and Technologies III. XII, 245 pages. 2006. (Sublibrary LNAI).

Vol. 3903: K. Chen, R. Deng, X. Lai, J. Zhou (Eds.), Information Security Practice and Experience. XIV, 392 pages. 2006.

Vol. 3901: P.M. Hill (Ed.), Logic Based Program Synthesis and Transformation. X, 179 pages. 2006.

Vol. 3900: F. Toni, P. Torroni (Eds.), Computational Logic in Multi-Agent Systems. XVII, 427 pages. 2006. (Sublibrary LNAI).

Vol. 3899: S. Frintrop, VOCUS: A Visual Attention System for Object Detection and Goal-Directed Search. XIV, 216 pages. 2006. (Sublibrary LNAI).

Vol. 3898: K. Tuyls, P.J. 't Hoen, K. Verbeeck, S. Sen (Eds.), Learning and Adaption in Multi-Agent Systems. X, 217 pages. 2006. (Sublibrary LNAI).

Vol. 3897: B. Preneel, S. Tavares (Eds.), Selected Areas in Cryptography. XI, 371 pages. 2006.

Vol. 3896: Y. Ioannidis, M.H. Scholl, J.W. Schmidt, F. Matthes, M. Hatzopoulos, K. Boehm, A. Kemper, T. Grust, C. Boehm (Eds.), Advances in Database Technology - EDBT 2006. XIV, 1208 pages. 2006.

Vol. 3895: O. Goldreich, A.L. Rosenberg, A.L. Selman (Eds.), Theoretical Computer Science. XII, 399 pages. 2006.

Vol. 3894: W. Grass, B. Sick, K. Waldschmidt (Eds.), Architecture of Computing Systems - ARCS 2006. XII, 496 pages. 2006.

Vol. 3893: L. Atzori, D.D. Giusto, R. Leonardi, F. Pereira (Eds.), Visual Content Processing and Representation. IX, 224 pages. 2006.

Vol. 3892: A. Carbone, N.A. Pierce (Eds.), DNA Computing. XI, 440 pages. 2006.

Vol. 3891: J.S. Sichman, L. Antunes (Eds.), Multi-Agent-Based Simulation VI. X, 191 pages. 2006. (Sublibrary LNAI).

Vol. 3890: S.G. Thompson, R. Ghanea-Hercock (Eds.), Defence Applications of Multi-Agent Systems. XII, 141 pages. 2006. (Sublibrary LNAI).

Vol. 3889: J. Rosca, D. Erdogmus, J.C. Príncipe, S. Haykin (Eds.), Independent Component Analysis and Blind Signal Separation. XXI, 980 pages. 2006.

Vol. 3888: D. Draheim, G. Weber (Eds.), Trends in Enterprise Application Architecture. IX, 145 pages. 2006.

Vol. 3887: J.R. Correa, A. Hevia, M. Kiwi (Eds.), LATIN 2006: Theoretical Informatics. XVI, 814 pages. 2006.

Vol. 3886: E.G. Bremer, J. Hakenberg, E.-H.(S.) Han, D. Berrar, W. Dubitzky (Eds.), Knowledge Discovery in Life Science Literature. XIV, 147 pages. 2006. (Sublibrary LNBI).

Vol. 3885: V. Torra, Y. Narukawa, A. Valls, J. Domingo-Ferrer (Eds.), Modeling Decisions for Artificial Intelligence. XII, 374 pages. 2006. (Sublibrary LNAI).

Vol. 3884: B. Durand, W. Thomas (Eds.), STACS 2006. XIV, 714 pages. 2006.

Vol. 3882: M.L. Lee, K.-L. Tan, V. Wuwongse (Eds.), Database Systems for Advanced Applications. XIX, 923 pages. 2006.

Vol. 3881: S. Gibet, N. Courty, J.-F. Kamp (Eds.), Gesture in Human-Computer Interaction and Simulation. XIII, 344 pages. 2006. (Sublibrary LNAI).

Vol. 3880: A. Rashid, M. Aksit (Eds.), Transactions on Aspect-Oriented Software Development I. IX, 335 pages. 2006.

Vol. 3879: T. Erlebach, G. Persinao (Eds.), Approximation and Online Algorithms. X, 349 pages. 2006.

Vol. 3878: A. Gelbukh (Ed.), Computational Linguistics and Intelligent Text Processing. XVII, 589 pages. 2006.